THE CAVE PAINTING: A Parable of Science

THE CAVE PAINTING

A Parable of Science

Roddy M. Bullock, JD, BSME
Executive Director, Intelligent Design Network of Ohio

Foreword by

John H. Calvert, JD, BA (Geology)
Managing Director, Intelligent Design Network, Inc.

Third Edition
Published by Access Research Network
Colorado Springs, CO, USA
www.arn.org

Published by Access Research Network
Colorado Springs, CO, USA
www.arn.org

The Cave Painting: A Parable of Science

Third Edition

ISBN-10: 1-931796-27-0
ISBN-13: 978-1-931796-27-9

Printed in the United States

For Bruce, a very wise man.

Acknowledgements

I would like to thank the following people for their encouragement and helpful contributions to the final form of this book. My family was the first line of approval for early versions, followed by Bruce and the first of many proof readers and critics, Shaunna. Marcie provided additional proof reading and commentary as did Jeff, Laurie and Jared. Thanks to the "BOG" kids Kathryn, Joy, Erin, Alicia J., Alicia P., Megan, and Lindsay for their valuable input. Friends and colleagues Dave and Vlad offered additional encouragement.

Special appreciation goes to John Calvert for his insightful comments and critical commentary that resulted in changes to the final version.

If I've forgotten others, please know the oversight is regretted, and if you let me know I will include you on future editions.

We *are* the product of intelligent design.

— or —

We *are not* the product of intelligent design.

One of the above statements must be true.

— and —

Truth is not changed by your theory or mine.

Table of Contents

Foreword

Recently I was told about an experiment in which an event was staged by actors before a group. Only one member of the group, the "mark," was ignorant of the experiment. Afterwards, the group leader explained the event in a way wholly inconsistent with what the audience actually observed. When polled as a group, the mark always voted with the group even though his observation did not agree with the erroneous explanation. Only rarely did the mark buck authority, trust his observation and disagree.

When I was studying to be a professional geologist I was told that just as the "doctrine of organic evolution is universally accepted among thinking biologists, so also the geosynclinal origin of the major mountain ranges is an established principle of geology." I was never comfortable with either claim because I could not reconcile them with my own observations of the data. At the time I figured my intellect was not sufficiently competent to keep up with "thinking" geologists, so I switched my emphasis to the practice of law. Ten years later I was surprised to learn that geosynclinal theory was being abandoned for plate tectonics, an explanation that resonates well with my personal observations of the structure of the earth's crust.

Due to phenomenal technological advances within the past few years we are now observing bio-complexity that will take the breath away. Although the observed data is wholly inconsistent with neo-Darwinian theory driven by random mutation, the resistance to change is extraordinary. Although skeptics are no longer actually burned at the stake, their reputations and careers often are. Why? Why is it OK to accept a change to plate tectonics, but not to intelligent design? Why must science institutions go to such desperate lengths to suppress any challenge of evolutionary theory? Why are geosynclinal skeptics allowed to stand, while Darwin skeptics must remain seated?

The Cave Painting explores this odd situation in a fascinating allegory about whether certain historical events are best explained as the product of an intelligent cause or of only material natural causes. Although observation, logic and intuition lead to a conclusion that a painting is really a painting, in the story this rational inference is not allowed. The disallowance seems absurd. How could anyone conclude that a painting is merely the product of material causes? This allegorical absurdity reflects the brilliance of this work. The reader is constantly asking how this plot could be so absurd. Yet, today we are confronted with a paradigm that is just as absurd. Is it not even more absurd to argue that observed evidence of biochemical languages and messages used to build nano-machines having functions and attributes far more sophisticated than do the most brilliant of human inventions, are not possibly the result of intelligent causation?

Yet this work actually raises a much deeper question: Why? Why is it so important for institutions to suppress the rational basis for belief that life is a design and not an occurrence? Why must science argue that the idea of design is based wholly on a myth while the idea of no-design is rooted in rationality? What is the purpose and effect of this institutional doctrine?

The reader will find *The Cave Painting* to be much more than a simple allegory. Actually two books in one—a work of fiction and detailed end notes—the work provides a healthy dose of fun combined with a serious assessment of all aspects of the debate: social, ethical, legal, philosophical, religious, and, of course, scientific. The fully documented end notes provide not only a thorough criticism of evolution but also detail the many scientific arguments for design. Therefore, the work should appeal not only to those who simply seek to be entertained, but also to those who wish to deepen their understanding on how best to answer one of the greatest questions of all: Where do we come from? *The Cave Painting* is a book that serious students of the intelligent design/evolution debate will want to read and keep handy as a ready resource.

John H. Calvert, J.D.
January 31, 2006

Introduction

Why is evolution defended with religious zeal by the mainstream scientific establishment? Why must Darwinism be guarded from critical scientific inquiry by heavy-handed political power? Why do some scientists selectively exclude certain lines of evidence relevant to origins? These and similar questions deserve an answer from those who claim to *know* that we are products of natural, unguided, purposeless biological evolution.

Of course the questions themselves surely seem pointless to many people, scientists included. Such questions may as well be suggesting new discussions on the nature of the solar system: why must the heliocentric model be guarded from critical scientific inquiry? Or possibly we should re-investigate the earth's shape: why selectively exclude evidence relevant to a flat earth? In fact, these very questions are often raised in attempts to ridicule any non-naturalistic explanation of origins. Such objections are easily made and, while logically irrelevant, they serve the purpose of casting non-naturalistic theories of origins in the same light as other past, outmoded theories. We may as well have another go at alchemy.

But is the suggestion that living organisms are purposefully designed really akin to suggesting the earth is flat? Do scientists know we are not designed with the same kind of knowledge that they know the earth is not flat? To suggest that scientists *know* that all living organisms are the product of purely unintelligent processes is to say that they *know* that all living organisms are not the product of intelligent design. Where does this knowledge come from?

Intelligent design theory offers a compelling alternative to the current creation story of Western culture: Darwinian evolution. After all, the search for intelligence is routinely conducted in other scientific contexts such as archeology and the Search for Extraterrestrial Intelligence (SETI).[1] An unearthed artifact that appears to have been formed by a person is unquestionably and immediately considered to have been so formed.[2] To do otherwise might make the science of archeology interesting, but nonetheless futile. Specified, ordered data received by the radio telescopes of the SETI project even slightly resembling that of the simplest strand of DNA would no doubt be hailed as a breakthrough discovery of intelligent life beyond planet earth.[3] Any attempts to attribute such signals to chance would be met with blunt, perfunctory criticism.

To be consistent, at some point mainstream science (if the term can be personified) must reconcile the mixed signals that it sends in its treatment of intelligent design theory. Either design can be detected and attributed to an intelligent designer, or it can't. If it can't, then we must stop fooling ourselves into believing that there were early human cultures any place we dig up what appears to be a designed object—unless, of course, the maker had the foresight to sign, and possibly date, the object. Even then the signature would be suspect. After all, a signature simply has the appearance of design—a trap for the unwary naturalist.[4]

Highlighting the irrationality of purposely denying the possibility of a designer for an obviously designed thing is the objective of *The Cave Painting*. Most evolutionists, no doubt, will object to being accused of purposely denying the possibility of

---

[1] *See*, End Note [1], page 232.

[2] *See*, End Note [2], page 233.

[3] *See*, End Note [3], page 234.

[4] *See*, End Note [4], page 236.

intelligent design of living things, but such can be the case only when evolutionists attempt to limit science *by definition* to exploring only "natural" causes for all effects.[5] Such an *a priori* boundary defining what "science" is and what it is not may be appropriate for certain fields of scientific endeavor such as applied science. But when extended to exclude all but purely natural explanations for the origin of life, such a boundary must be recognized as being required by an assumed constraint, not an inherent limitation. Thus, the refusal to even contemplate the idea of intelligent design in living organisms is purposeful, and it will ultimately be shown to have been irrational.

The Cave Painting explores the issues of the contemporary origins debate in a manner designed to both inform and entertain. Those familiar with the debate will no doubt recognize the issues, as well as prominent personalities; where possible, actual quotes are used. Virtually every aspect of the debate has an allegorical counterpart. Hopefully the reader will come away from reading *The Cave Painting* with an understanding of the limitations of scientifically imposed naturalism, the possibilities of intelligent design, and the ability to ask the right questions of both.

---

[5] *See*, End Note [5], page 239.

Chapter 1

Jenny looked on with anticipation as she watched classmates disappear one by one through the small opening in the ground before her. A clear sky rose high over the ground made hot in the September sun as each child tentatively entered feet first into the cool air of the rocky gap. Nervous smiles and laughter were exchanged between those standing anxiously above and those descending hesitantly into the unseen world below. The open field where Jenny stood was distinguishable from the surrounding area only by the temporary fence encircling Ms. Hart's fifth grade class, and the craggy breach now swallowing kids whole.

"The cave," as the kids referred to it, was a recently discovered subterranean cavern named Lavoux Cavern, after its accidental discoverer. Lavoux, a rancher, had lost a sheep down this opening, and had been frustrated at his inability to fill what he thought was simply a small, deep crevice. After a futile effort involving several loads of dirt and rocks, he realized that this "hole" must be an open fissure of some kind.

A friend of Lavoux's in the Agriculture Department at the local university—Lutrell University—suggested to a colleague in the Natural Sciences Department that he check out the fissure. The university sent Dr. Seeting, an associate professor of Earth Science and a paleogeologist, to investigate. Not only was he young and athletic, Dr. Seeting had distinguished himself in academic circles by successfully blending a passion for spelunking with a keen scientific interest in studying underground geological formations. In addition to authoring many publications, Dr. Seeting was a popular speaker and lecturer in the newly developing field of paleogeology—the study of the history of natural formations and geological features. He emerged from an initial exploration of Lavoux Cavern with a proposal to purchase the field. Lavoux, fearing there may be more fissures, gladly sold the field and the immediately adjoining land to Lutrell University.

Ms. Hart had serious reservations about bringing her class of fifth graders to the cavern. This was her first year teaching, and she had been extremely reluctant to agree to supervise her class on what appeared to be a very risky venture. She was persuaded only after she and the school administration received assurances from Dr. Seeting that it was safe. Even now, as she watched each anxious face vanish from view, she wondered if her first field trip might be her last.

In the short time since the discovery of Lavoux Cavern, Dr. Seeting and his team had made significant improvements to the steep, narrow entrance, cutting steps and handholds, as well as placing safety ropes where necessary. Once he was satisfied that even a child could enter the cavern safely, he persuaded Lutrell University to offer class field trips as a way to publicize the cavern and garner local interest. The effort tested Dr. Seeting's power of persuasion, as he had difficulty assuring the university's Board of Governors that the publicity and local interest would, in fact, be positive, and not result in unwelcome, negative public relations issues. Ultimately successful, he was now in the cave, helping his first group of children reach the bottom of the most treacherous portion.

Jenny waited patiently, often turning to scan the area around her. With its rocky terrain and tall grasses, it reminded her of her hometown. She had recently moved to "the yonder side of the mountains," as her father put it, after the death of her mother. Although only a long half-day's drive away, the "highlands," as it was sometimes called, seemed like a different world. Memories of her mother, as well as friends from home, left an empty longing. Even though it was only the first month of school, finding a close friend here appeared doubtful. As with most new kids, Jenny knew her classmates only by distant looks and an occasional exchange with an unsure emissary. Such exchanges were usually followed by remote group whispers that Jenny pretended to ignore.

Often described as gangly, Jenny was thin and tall. As she waited outside the opening to the cavern her straight blond hair was clearly visible above her classmates' heads. The coveralls the children had been given to wear reached only slightly below her knees, a circumstance that added to the awkwardness of being new. But now she breathed deeply, enjoying the smells of the country and thinking about home. She waited until all the other kids had entered the dark opening, and then she slipped in right before Ms. Hart.

As the light receded above her, Jenny shivered slightly. Imagining how it must have felt for the first person that entered this hole in the ground, she wondered why one would do so. She had heard of Lavoux's sheep; she wondered if any people had ever fallen into this mysterious darkness. The thought of finding bodies made her shudder again.

Jenny carefully avoided the person below her as the group slipped, slid, and stepped to the bottom of the steep entrance. Sounds of kids helping kids down pierced the darkness that was otherwise disrupted only by the light beams from each child's helmet, what the children called "miners' hats." As Jenny looked below her, the lights appeared in the darkness like party lights strung up on a windy day. Dr. Seeting could be heard below speaking instructions in a hollow voice to those at the bottom.

Once inside and on more level ground, the group of kids followed Dr. Seeting as he labored to crowd himself through one narrow gap to another in a slowly descending, snaking route into the shadowy darkness. Because it was the first time a group of school children had been allowed into the cavern, Dr. Seeting went slowly and made frequent stops to count heads.

Ms. Hart's anxiety returned as she cautiously brought up the rear of the winding light path, often out of sight of Dr. Seeting. Before long she and the others were brought to a huddled stop on a flattened ledge, evidently in a great void. Nearby dripping sounds indicated a pool of water somewhere to their left; otherwise, the silence was broken only by the rustling of shoulders trying to get closer to the cave wall to their right.

"Boys and girls, stay back," Dr. Seeting almost whispered, as he herded the group away from an apparent drop-off into the darkness. "This is where we will begin to see some of the most wonderful examples of natural cave formations ever discovered," he continued, with keen anticipation. "You are about to see something that, until recently, no human being had ever seen before." With that, Dr. Seeting turned on a very large floodlight he had carried in, washing much of the blackness away with a wide funnel of light.

"Wow!" was the first exclamation, followed by an unexpected silence as each child's head moved with the slowly moving beam of Dr. Seeting's light. Ms. Hart was relieved to see that the imagined drop-off was only a sloping grade into a great, flat pool of water, perfectly still except for the repeating pattern of ripples caused by several steady drips from above. Jenny stood with the others, captivated by the eerie landscape as the flashes from cameras revealed in shifting shadows what had known only darkness for ages.

"We are standing beside what I have named the Great Lake, an underground reservoir that probably extends far, far underground, most likely fed by underground streams," Dr. Seeting explained in a hushed tone. "No one knows how deep it is or how big it is, or …" he paused for effect, "what lives in it!" he said while quickly grabbing a child nearby. The children laughed, their eyes large with hopeful apprehension.

"Look over there," Dr. Seeting said, shining his light across a glassy-smooth inlet of the water. Light shone on what appeared to be a large pile of gravel. The kids strained to see the top, hidden from view. "Rockslide," Dr. Seeting said. "It's interesting only because it's so unusual to see in a cavern like this, but it's not nearly as impressive as what we are about to see. Let's continue."

Dr. Seeting's great light led the way as he moved off into the darkness, now along relatively flat, level ground. Ms. Hart, glad for the even footing, nevertheless wished for a larger light because she continued to find herself out of Dr. Seeting's sight behind the last child in line.

The children were more talkative now as fear began to yield to excitement and their lights seemed to penetrate more of the darkness. Soon, in a cluster again, their eyes were following Dr. Seeting's light as it shone across an array of large, glistening, hanging formations, some reaching from the ceiling to the floor of a great expanse before them. Cameras clicked and flashed as Dr. Seeting's light came to a stop on the largest of the rippled columns.

"Stalactites," Dr. Seeting said matter-of-factly. He silently gazed on the forest of carrot-shaped forms with the apparent satisfaction of a farmer looking at a new crop. He clearly saw more than did the kids, or even Ms. Hart, for that matter. He noticed the small, orange-tinted "soda straws," old in time but young in growth, patiently changing over time into full stalactites. In addition to the stalactites, stalagmites could be seen on the cavern floor, each one built up from an eternal source of mineral-rich water slowly depositing the dissolved mineral on the floor. The entire scene spoke of eternity, of irrepressible orderly change measured not in years but in eras.

The noise of the students growing restless startled Dr. Seeting back to reality.

"Notice the small, tubular formations here," Dr. Seeting motioned with his light to the nearest soda straws. "These are called soda straws ..."

"Soda straws?" the kids interrupted. "Can we have some soda?" one child giggled. The others laughed as they pretended to drink a soda through imaginary straws like the small tubes they saw hanging above them.

Dr. Seeting continued, "Soda straws are baby stalactites. Slightly acidic surface water has seeped into this cave and dissolved the soluble rock over hundreds of thousands of years to produce the hollow tubes of calcite," he said, forgetting momentarily that he was talking to fifth graders. Dr. Seeting stopped short and tried again, "Basically, drops of water carrying tiny solid particles of a substance called calcium carbonate precipitate, uh ... leave behind, some of the solid particles before dropping to the floor. Over time, these tiny particles build up to become the soda straws you see here. And, over even more time, these soda straws grow into stalactites as you see here," he said as he slowly moved his light up and down a particularly large stalactite.

"Now," Dr. Seeting said quietly as he knelt down on one knee, "the really fascinating thing about these formations is that they formed themselves."

"What do you mean?" Jenny asked. The others were also interested as they stared in fascination at the stalactites and then back at Dr. Seeting.

"What I mean is this," Dr. Seeting explained. "Each of these formations," his flashlight again swept the cavern before them, drawing heads and eyes along with it, "formed over long periods of time without any people around. There was not even a person to watch, until we came along." Nodding heads waved light beams up and down the darkness as the kids began to understand and share in Dr. Seeting's amazement.

"You see," said Dr. Seeting, "what we see here is the result of blind, natural processes. These purposeless processes never had us in mind, but here we are!" The kids' lights shone on Dr. Seeting's face as they listened, sharing his enchantment. "Yes," he repeated, "here we are. And I don't know about you, but nature's work is something I find very beautiful. And it happened here without anyone even knowing about it."

Stalactites and stalagmites weren't particularly beautiful to Jenny, but the idea of seeing something that was made in complete darkness with no expectation of ever being found intrigued her. The thought of nature producing such interesting arrangements of rock and minerals captivated Ms. Hart as well. Seeing that her class of 10- and 11-year olds also seemed to be fascinated thrilled her even more. Maybe now they would find new interest in her science class. She may even have each of them write a report on how cave formations come into being, she thought.

At this point, Dr. Seeting, wanting to ensure full appreciation of the natural beauty around them, asked that each child turn off his or her light. With each click the darkness crept in slightly, still held back by Dr. Seeting's large floodlight. Once he turned his light off, however, the group gasped in startled delight. The cavern was completely, utterly dark. The kids tried putting their hands in front of their faces in vain attempts to see something.

Dr. Seeting waited for the nervous laughter to subside, then said in a low voice, "This is how this cave looked before we came, and this is how it will look after we leave. This cave has been without light since its formation." The children occasionally yelled at each other to stop stepping on toes, and they often inadvertently hit one another; any movement seemed to interfere with an unseen classmate.

"Shhhh," Ms. Hart quieted the kids.

"It was in this darkness that the rockslide happened and the stalactites were formed," Dr. Seeting said, still in complete darkness. "But they are not the most fascinating formations found in this cavern."

He stopped speaking and waited as the kids nervously moved about slightly, each wishing they could turn their lights back on. Ms. Hart wondered why the long hesitation. She imagined Dr. Seeting deep in thought as the seconds seemed to tick by in minutes.

"Now," Dr. Seeting finally continued, "we are going to see the most remarkable natural formation anyone has ever seen. So remarkable," he paused, "it actually looks like a picture! Do you want to see it?"

"Yes, let's go!" the kids said in unison, knowing that the lights must necessarily come on before they could go anywhere or see anything. Dr. Seeting, turning his light on again, instructed the others to do the same, and to follow him. The group wound between more narrow gaps and crevices, more than once having to help each other up and over wet, slippery crests and ridges. Soon the group came into a large, open cavern, apparently free of any dripping water as indicated by the virtual silence when the group came to a stop. The dry air was cool and welcoming; some of the students wanted to sit. Many of the kids looked around, their relatively weak lights searching the cave about them, but they were not able to make out any particularly remarkable formations. In fact, they could hardly make out walls in any direction.

"Over here," motioned Dr. Seeting, turning off his large light.

As the group followed him, the air was filled with the hollow sound of echoing footsteps. Dr. Seeting led the group to one side of the cavern where he stopped and set his light on the ground such that it faced a cave wall before them. He stood and waited for all the kids to gather about him, their lights slowly swarming toward the wall. Soon the whole class was seated or kneeling around him, facing the wall of the cave. With evident enjoyment at the thought of what they were about to see, Dr. Seeting clicked on his light and adjusted the beam such that it illuminated a significant portion of the stone mass in front of the group.

"Look." Dr. Seeting stood back with his head slightly tilted.

At first the kids didn't see anything. Ms. Hart also strained to see any significant natural formation in the light's glow. Dr. Seeting waited patiently.

"I see it!" one of the boys shouted, bolting forward.

"Don't touch!" Dr. Seeting called out, moving suddenly to intercept the boy. Drawing back, the excited boy pointed out what he saw. One by one, each of the kids began to see it as well.

Ms. Hart saw it, too; and, like the first boy to notice it, wanted to go touch it to see if it was real. Instead, she and others stared in awe as a barrage of flashes sent lightning about the cave.

Jenny's camera clicked and whirred as well, capturing the faint but distinct lines and markings that, together, "looked amazingly like a hunting scene" as Dr. Seeting told them. They all agreed the "man" appeared to be hunting. Jenny was captivated by the image of a man running, both legs bent at the knee, the back leg stretched more than the other, feet in the air. One arm was raised with fingers wrapping a spear, while the other stretched forward, pointing ahead. His hair blew behind him, as if by the wind; his eyes focused intently in the direction of his outstretched arm. The ground stretched underneath as a long horizontal line. Jenny could distinguish at least three colors, some brighter than others.

As the kids took this in, Dr. Seeting was plainly eager to show them even more. Excitedly, he said, "Wait until you see this." Slowly moving his light to the right caused all the heads to turn in the direction in which the man was pointing. "Look here," he pointed this time. "What do you see?"

On the wall before the staring faces were what appeared to be animal figures, this time less recognizable, but clearly distinguishable. The animals overlapped, making it difficult to see individual features, but each had a large head with long horns. It appeared that they were all the same type of animal. On some four legs were visible, and at least one was lying down. The one lying down even appeared to have something like a spear sticking into its neck.

"We call markings such as these that look like recognizable things 'specialties,'" said Dr. Seeting. "Each different type of figure, like a 'man' or an 'animal,' is a related, but distinct specialty. In this case, the specialties look like a painting, but they are not paintings, they merely have the appearance of paintings. They are, in fact, only *apparent* paintings.[1] These specialties are, as I mentioned, the most remarkable formations found in this, or any, cave, anywhere. Their discovery has captivated the attention of scientists worldwide, and has stirred serious debate. Many scientists are simply unable to comprehend how such an arrangement of colored pigment could form naturally to look like a picture. In fact, I am currently putting together a team of scientists to study this pattern to try to determine how it got here. Maybe one of you will grow up to be a scientist and help me."

The kids all nodded excitedly at the thought of being a scientist like Dr. Seeting. Ms. Hart nodded approvingly, herself impressed by Dr. Seeting. Some of the kids started considering various scenarios to explain the pictures, asking Dr. Seeting what he thought about each. He pretended to take each seriously, usually commenting at some point in his response that there is no evidence of people having ever been in the cavern.

Jenny stared, her small circle of light moving slowly across the wall in front of her. She followed the lines of color, black and red, with yellow shading on the animals. The man's eyes looked straight at one of the horned animals.

Soon most of the students became bored with the figures and were milling about, ready to move on. When Ms. Hart indicated that she thought most of the kids had seen enough, Dr. Seeting called out, "OK, time to go. We need to go before our batteries run out."

At the thought of once again being engulfed in the blackness of the cave, each child quickly turned in the direction from which he had come, eagerly waiting for Dr. Seeting to lead the way out. Dr. Seeting listened to the conversations behind him as he slowly started back. The kids were much more talkative now, and he found their chatter amusing as they continued to talk to each other about the marks on the wall, and how they might have originated. He had not gone far when a voice above the others caught his attention.

---

[1] *See*, End Note [6], page 246.

"Wait!" A girl's voice echoed about the cavern.

Dr. Seeting turned to find Jenny, still facing the wall, her tiny light continuing to follow the lines of color on the cavern wall. While the other kids nervously fidgeted, Dr. Seeting set his light down and answered, "Yes?"

"I have a question," Jenny said, turning to Dr. Seeting with a look of wonder and confusion. "Just ... what ..." She thought for a moment, "*Who* painted this picture?"

Dr. Seeting smiled.

Chapter 2

"Let's go," Dr. Seeting shouted as he picked up his light and turned it back toward the way they had come. The kids eagerly followed, anxious to get back to sunlight.

Jenny remained in front of the wall, still fascinated, her eyes once more carefully following each line of the hunting scene so graphically depicted before her. The apparent use of colors for shading, the convergence of lines, the placement of highlights, all intrigued her, and reminded her of an art class she had taken the previous year. She could hear the receding echoes of shuffling footsteps behind her as she took one last picture.

The flash brought Ms. Hart's startled attention. Alarmed that one of her students had been left behind, she whirled around as she called for the group to stop. It did not surprise her to find that the lone light she could see lingering by the wall belonged to Jenny. Ms. Hart rushed Jenny to the line of kids waiting with Dr. Seeting while loudly scolding her for being so absent minded. "You stay with me," she said, keeping a hand on Jenny's belt as they followed Dr. Seeting out of the cave.

Going out of the cave at the opening proved much harder than entering. The narrow tunnel-like crevice seemed steeper going out, and in some places rocks and dirt rained on the children in the rear as the children ahead of them clawed their way up. Finally, Dr. Seeting asked that everyone wait where they were as he worked his way up and pulled himself out into the clearing above. He had anticipated this problem and had brought along some rope, knotted in intervals, which he could let down into the almost vertical opening from which they were attempting to exit.

"Guide this down!" he yelled, lying on his stomach and peering into the rocky opening. "Keep it going!" he shouted as he fed more and more rope into the darkness. The children below passed the end of the rope by until it had reached the bottom where Ms. Hart stood.

"OK!" Ms. Hart shouted.

"OK, OK," went a chorus of voices back up the rope.

Dr. Seeting tied the other end of the rope to the bumper of the nearby school bus. He gave it a couple of tugs and then turned back to the opening. "Now, come on up." He was on his stomach again, hands cupped to his mouth.

One by one, Dr. Seeting grabbed each hand as it appeared, lifting each child out and onto his feet. The sunlight above was blinding, and the warm air, the rustle of the long grass, and the buzzing of insects all seemed magnified in contrast to the cool darkness of the tranquil cavern below. Finally Dr. Seeting pulled out a struggling Ms. Hart, who, once out gave Dr. Seeting a look of friendly irritation as she bent over and started shaking her head, trying to get the dust and dirt out of her hair. Dr. Seeting made one last call down the hole before pulling the rope out with an apologetic grin in Ms. Hart's direction.

Dr. Seeting counted children while Ms. Hart checked off a list of names. Once the head count was confirmed, Dr. Seeting slid a heavy wooden cover over the opening to, as he put it with a wry smile, "keep sheep or kids from falling in." Then Dr. Seeting brushed himself off and went about gathering up lights, rope and other equipment he had brought, while Ms. Hart rounded up all the kids, herding them onto the bus. Dr. Seeting and Ms. Hart eyed each other often with a look of wonderment, enjoying the kids' excited bantering. The cave was a hit.

After counting kids once more, Ms. Hart settled back as the bus rolled past the temporary fence and rumbled out onto the two bumpy ruts of the dusty access road. Dr. Seeting's van followed behind the dust of the bus as both made their way back to the school. Ms. Hart finally relaxed, thrilled at the response she was seeing and hearing among her students. The chatter continued all the way back to the school. Even the boys, normally aloof and difficult to engage, excitedly talked about the cave, speculating as to when they might go back.

Near the school, Ms. Hart stood and faced the back of the bus. "Kids," she started, waiting for quiet, "kids, listen. When we get back I want us all to go back to our classroom quickly because Dr. Seeting has agreed to stay for a few minutes to tell us some more about the cavern and answer any questions we may have. This would be a good time to thank him as well," she said with a look of "or else" attached to it.

"Yes, Ms. Hart," they all said in unison in a teasingly sarcastic response.

Once in the parking lot, the children poured out of the bus and went into the building, drawing questioning looks in the hallways as they noisily paraded in their dirty coveralls to their classroom. Dr. Seeting entered soon after, to the welcome of a super hero.

"Kids," he said, "what you have just seen, very few people have ever seen. You have seen the hidden treasures of nature, the patient, unending result of thousands of years of imperceptible change. The beauty, the splendor, the ..." He paused as he tried to think of the appropriate words. "Well, it's just astonishing that such magnificence was made in the hidden darkness of the earth. I've been spelunking—that is, exploring caves—for years and I've never seen any natural formations like those we all just witnessed." After that, he looked about the classroom. "Questions?"

Jenny's hand shot up.

Because they had been on their way out at the time, Dr. Seeting had asked Jenny to wait for an answer to her question in the cavern. Knowing what was coming, he pointed, "Yes, uh ..."

"Jenny." Ms. Hart said, looking about.

"Yes, Jenny." Dr. Seeting was upbeat.

"Who painted those pictures?" Jenny asked directly, with open wonderment.

Dr. Seeting sighed and looked up as if he was thinking. "Very good question, Jenny. However, it doesn't have an easy answer. Your question, in fact, points out the difficulty we scientists encounter when we see, and try to explain, natural formations. Our explanations can be based only on evidence—evidence to prove our theories."[1]

Jenny wasn't sure if that was good or bad, and the rest of the class seemed confused.

"Let me explain," Dr. Seeting continued. "You see, science, natural science, can consider only *natural* explanations for physical features of the world around us.[2] Natural explanations must be based on the physical evidence—what we can see, hear, smell and touch. This is the only way data and information can be gathered, tested and confirmed. It's important that theories can be tested independently and repeatedly to confirm various hypotheses."

Most of the class was attentive, but not understanding. Dr. Seeting had to remind himself again that these were fifth-graders. He looked at the class with a serious tight smile as he thought for a moment. Looking back at Jenny he said, "You see, Jenny, in this case, as hard as you or I may find it to believe, there is simply no evidence that people have ever been in Lavoux Cavern. Indeed, it seems impossible that people could have ever been there. Just consider how hard it was for us to get in and out. And it does not appear that the opening we used had ever been used before, except by that sheep!"

The kids laughed as they sat nodding, absorbing the explanation with interest.

"Therefore," Dr. Seeting's voice took on the tone of a lecture, "in keeping with the spirit and rigors of scientific inquiry, we can theorize only natural explanations for the figures

---

[1] *See*, End Note [7], page 259.

[2] *See*, End Note [8], page 271.

we saw on the wall of the cave, based on the evidence. And, again, there is no evidence of people having ever been in Lavoux Cavern. To simply resort to a 'someone painted them' explanation is convenient, but it ignores the facts, and is, in fact, an unprovable theory. We can simply have no knowledge of any such imagined person."[3]

"You mean *nobody* made those pictures?" Jenny asked in astonishment.

"That's right," Dr. Seeting said. "Even to use the words 'painting' or 'picture' is somewhat misleading, because the figures we saw are 'pictures' only because our brains see them as such. To use the term 'painting' infers a painter, and there is no evidence of any painter having painted the figures we saw. Without any evidence of people having ever been in Lavoux Cavern, it is simply naïve to assume that humans made these forms. So rather than using the term 'picture,' which infers some singular position among natural formations, scientists prefer terms such as 'marks' or 'figures,' or, more precisely, 'specialties.' At best, the remarkable specialty we witnessed today is simply an *occurrence*; as I said before, only an apparent painting."

The children listened intently, but Jenny's face was not convinced.

Dr. Seeting tried again, "There are many examples of natural phenomena taking a recognizable form. Take, for example, mountain formations that look in the distance like a person's face. It's simply a coincidence that our brains see a face when what we are really seeing is just some craggy rocks formed by wind and rain."[4]

"And clouds," one child blurted out. "Sometimes clouds make animal shapes."

"Precisely," Dr. Seeting replied enthusiastically. "Clouds, mud puddles, tree trunks, even stars! In all the random, even

---

[3] *See*, End Note [9], page 274.

[4] *See*, End Note [10], page 278.

chaotic working of nature, it is inevitable that occasionally some recognizable shapes and figures are made."

"Then," Jenny continued, still not sure about the explanation so far, "just how did those so-called 'figures' get drawn, uh ... get on the wall?"

"Again, an excellent question," Dr. Seeting replied, mentally noting her characterization of the figures. "Since the discovery of the *figures*," he placed serious emphasis on the term figures, "on the wall of Lavoux Cavern, scientists the world over have put forth theories of how they came to be. My staff and I have even received a research grant to explore this question. We have some theories of our own, but I prefer not to discuss them at this time. What I can say for certain is that the figures appear to have been made slowly over long periods of time, and there were no people around at any time during their formation. This, at least, is scientific fact."[5]

"*Fact*?" thought Jenny. But before she could speak, Ms. Hart interrupted to ask if there were any more questions. After a few more minutes she announcement that Dr. Seeting's time was up; he had important work to do at the university. Dr. Seeting nodded, smiling, thanking the kids for the chance to speak to them. The kids gave him an enthusiastic round of applause, each one thanking him for taking their class to the cave.

"Before Dr. Seeting leaves," Ms. Hart said quickly, "I want him to know about our homework assignment."

A swell of "boos" from the kids followed. "Now you're going to take all the fun out of today!" one child shouted.

"Yeah, why do you have to ruin it for us?" said another, looking around and laughing.

Ms. Hart raised her hand to silence them. "You may like this assignment," she started. "We are all going to help Dr. Seeting."

---

[5] *See*, End Note [11], page 282.

The "boos" were replaced by clapping and cheering as the children excitedly started asking questions. Ms. Hart again raised her hand for quiet.

"Do we get to go back to the cave?" one asked.

"Maybe," Ms. Hart said, mainly to keep their interest, "but our assignment is to help Dr. Seeting explain how the lines and colors we saw on the wall of Lavoux Cavern might have been made."

Dr. Seeting looked pleased. "Maybe one of you will be the next great scientist!" he said in mock seriousness.

"We've heard how scientists know that no people made the figures we just saw. And we learned that Dr. Seeting is actively investigating their origin. Each of you has two weeks to prepare a report with your explanation, your *theory,* of how those lines were made. Use your imagination, but think like a scientist. You will be graded on how well you can support your theory. We may even be able to convince Dr. Seeting to help grade our reports."

Dr. Seeting graciously agreed, and left to shouts of "thank you" from the classroom.

Ms. Hart was satisfied.

Jenny was confused.

The other kids were smiling.

Chapter 3

"All that's missing is the bright light," Dr. Seeting said sarcastically.

"It's a bright light that got us into this situation," said another angrily.

The small room looked like the interrogation room it had become as several men stood and others sat questioning Dr. Seeting. Although the meeting was intended to be friendly in nature, Dr. Seeting could not help but feel like he was being grilled. The men were his colleagues, but their longer tenure at Lutrell University made them impossible to ignore. One of them, Dr. Rich Hawkins, seemed particularly bothered that Dr. Seeting had taken a class of school children into the cavern and showed them the hunting scene. Others, including Dr. Ray Silva, were less bothered, but nonetheless interested in ensuring that future visits, if any, were handled with more forethought. A large, rotund man with a permanent frown insisted on smoking a cigar that put a light haze into the air. Moving his chair over near an open window that looked out over the quiet campus, Dr. Seeting sat annoyed. It was early on a weekend, and few students were up and around.

"We knew from the beginning that we had to handle this carefully," one of the men stated. "That's why I was against the field trip in the first place. If I had any idea you would show them everything I would have stopped it! You, of all people, should have known that we needed to think this through before pulling such a stunt."

"I know that; we did think it through, and I had the approval of the Board of Governors … for the *stunt,*" protested Dr. Seeting.

"You had approval to go inside the cavern, not to go showing off the cave figures. Those people on the Board are not scientists, but even they understand the ramifications of finding the cave painting. And they know The Rule! I'm wondering if you do," said another, glaring at Dr. Seeting.

At the mention of The Rule, a certain solemnity fell over the group. Rarely mentioned at all, most were surprised to hear the words spoken explicitly, even in this group. Usually invoked under only the direst of circumstances and even then only indirectly, most, including Dr. Seeting, considered its reference here to be premature.

Dr. Seeting looked upward and sighed loudly as he shifted in his chair. "Of course I do. But what are we to do? We can't pretend they're not there!"

"Yes! Yes! We can or, rather, we *could have* pretended they're not there!" scolded Hawkins. "That's the problem! What we *could have* done is no longer an option. We *could have* kept this within the scientific community. We *could have* waited until we understand the origin of the markings better before telling the world about them! We *could have* done more research. We *could have* involved more research scientists. We *could have* waited for some plausible theories to emerge before telling the world about the specialties. Now 25 kids and a teacher have seen the figures. Some even have pictures!" Hawkins continued in a bothered tone as he paced about the room. "Their friends, their families … It's only a matter of time now; those people will be all over this," Hawkins' voice trailed off.

The men sat in silence pondering the situation. Dr. Seeting was beginning to regret ever having come up with the idea for the field trip. He had originally planned on only taking the children to see the Great Lake, the rockslide, and the stalactites. It was only after the group was inside the cavern that he decided to show them the wall markings as well. He understood the implications of the discovery of the cave figures, but once inside the cavern and seeing the excitement and interest of the children, he couldn't resist showing them everything. Even though he knew that revealing the cave specialties would lead to questions about origins, he also knew that there was a logical, natural explanation for the markings. Why not simply tackle the problem head on? he thought at the time. Dr. Seeting was lost in thought when the silence was interrupted.

"How do you plan on controlling the situation," the smoker puffed, leaning with a pointed finger. "You know we have to control this. The National Board of Education is very interested in what's going on here. These things have a way of getting out of hand you know. The Board tries very hard to avoid controversy in these matters."

"There's no need to control anything, Manny," Dr. Seeting shot back, knowing that, in fact, the information did need to at least be monitored. "Let's face it; the facts will win the day. We were right about the sun, weren't we?"[1]

"Whose son?" asked Manny, looking around as he blew smoke upward with his chin in the air.

Dr. Seeting looked at him with a cutting look of astonishment. "Not *whose* son. *The* sun, the s-u-n," he said, looking away in disgust. "We settled that controversy didn't we? And we did it by facing the facts head on, without apology!"

"Son, sun, whatever. But look what he had to go through," said the large man. "Trumped up charges, a charade of a trial. They shamed him before the whole world. And he had the facts to back him up!"

"We have facts also," snapped Dr. Seeting.

"And we're in charge now, don't forget," Dr. Hawkins added, looking around at the group. "At least we have that going for us.[2] But that's the only thing at this point."

---

[1] *See*, End Note [12], page 287.

[2] *See*, End Note [13], page 294.

Dr. Seeting looked at Dr. Hawkins. "I'm telling you, Hawkins—all of you—there is no evidence of a human painter. Period. I don't know what you are all worried about."

"But you confirmed that the kids have an assignment to try to explain the figures," the smoker continued. "And they will talk. And they do have pictures, right? How can we be assured that all will be done in accordance with the Board's guidelines?"

"I was very clear about the truth of the matter—that there is no evidence of human origin of the cave specialties of Lavoux Cavern. No evidence! Believe me, I've spent more hours in that cave than anyone else, including any of you, and there is no sign of intelligent life down there—now or ever. Do you think that any of those kids will come up with evidence? They're *fifth* graders, for goodness sake!" Dr. Seeting sat back confidently.

"Someone was doing research in the rare book room of our library," Dr. Silva said calmly. "Apparently it was one of your young scholars."

Dr. Seeting looked surprised. Everyone sat silently, considering this information.

"The rare book room?" Dr. Seeting finally asked. "How did a fifth grader know about the rare book room? *I've* never even been to the rare book room."

"You should go sometime," Manny grumbled. "You might be surprised at what you find there."

Dr. Seeting's and Manny's eyes met in a brief stare long enough to communicate the seriousness of the matter.

"Not that it matters, but I still wonder how a fifth grader knew about the rare book room," stated Dr. Seeting. "And isn't the rare book room restricted? How would a student even get in there?"

"Who knows? She was with a parent or something," Dr. Silva said. "The point is that these kids have been given an assignment that involves research. My question is, why not expect someone to end up in the rare book room? We can't be too careful with these things, at least not for the near future until we can develop and test some theories."

"That's why it is unconscionable that you would even consider such a field trip, and with *cameras*!" Dr. Hawkins sat with his head in his hands as the others nodded solemnly.

"So tell us more about this assignment," Dr. Silva said, trying to shift the energy in the room to be more positive. "Whose idea was that, and what exactly does it involve?"

"Does it matter?" Dr. Seeting asked. "They're just fifth graders!"

"With pictures!" Hawkins again groaned.

"So what?" Dr. Seeting demanded impatiently, standing up to get closer to the window. "It is very clear that the figures on the wall are a natural formation. I know that. You know that. These kids know that. Their teacher knows that. The assignment is simply a challenge to teach them to think creatively. What can be the harm in that?"

"The harm is not in these kids knowing anything, or thinking anything, creatively or not," started Dr. Hawkins. "You know what the potential harm is—once the word gets out."

"You are all blowing this way out of proportion," Dr. Seeting said. "I mean, the facts are what the facts are. Can we control that?"

"Can we control that?" Dr. Silva asked, turning to Manny.

"We can try, at least until more information becomes available," said the smoker as he leaned back with his cigar in the air. "Does Kozak know?"

"Kozak knows."

"Kozak knows?"

"Yes."

"The Rule?"

"Yes, Kozak knows!" Dr. Seeting repeated, now becoming irritated.

"Well, just make sure he knows where we are on this," Dr. Silva said calmly, looking intently at Dr. Seeting.

The smoker nodded, smiling as he put out his cigar with deliberately long twists in an ashtray.

Chapter 4

"Jenny!"

Jenny had almost made it out the door of the classroom before Ms. Hart called her back.

"Jenny, may I speak with you for a moment?"

Sure, thought Jenny, *recess could wait for a few minutes*. She walked back and sat by Ms. Hart's desk until she and Ms. Hart were alone in the room with the sound of kids playing outside. More kids were noisily making their way down the hall, heads bobbing by the classroom windows as they went. Ms. Hart got up and shut the door and returned to her desk, sitting slowly while pushing her hair behind her ears.

"Jenny," she started with some hesitation in her voice, "it's about the rough draft of your report."

"My report?" answered Jenny as she looked over a stack of papers to see her rough draft in front of Ms. Hart. Jenny had put extra time into this report, sensing it was her first chance at the new school to do well on a major homework assignment.

"Yes. Let's talk about this for a moment. The assignment, you will remember, was to propose a theory, that is, an explanation, for how the figures we saw on the wall of Lavoux Cavern came to be."

Jenny shifted nervously, sensing that her rough draft was in some way deficient. She had made every effort to make it neat and legible, even though it was called a *rough* draft, she thought.

"Let me get right to the point," Ms. Hart said, looking down at a stack of papers on her desk. "Even though it is clear that you put a lot of thought into your report, and it is fine in almost every respect, I can't accept it, and I certainly can't let you present your theory to the class."

"What?" Jenny asked, surprised.

"Maybe I should have made it even more clear, but the assignment was to propose a *scientific* theory to explain the cave figures," Ms. Hart explained. "I know you are only in the fifth grade, but I'm trying to get you to think like a scientist."

"But ..." Jenny started.

"Furthermore, I've discussed this with our principal, Mr. Kozak, and he has assured me that I can't let you present your explanation."

The thought of Mr. Kozak frightened Jenny. He was a large, stern man, and to a fifth grader he was downright intimidating. Rarely smiling, he usually just patrolled the hallways like a battleship, spreading kids in his wake. Occasionally he would look down with a scowl at a particularly slow-moving child. *Why was he concerned with my report?* Jenny thought.

"You see," continued Ms. Hart, looking at Jenny, "I know that for some reason you think that people made the marks on the wall of Lavoux Cavern. That's OK; it's fine for you to have individual ideas about such things and I'm not going to interfere with your personal beliefs. But in this case to let you present your personal opinion in class would be in complete violation of our guidelines for teaching science.[1] If you insist, we'll have to take this up with Mr. Kozak for his approval."

Jenny suddenly wished to be outside. She glanced about in a nervous effort to make time move faster.

"This notion ..." Ms. Hart paused; she was trying to be gentle. "This notion that people painted the figures ..." She hesitated again. "Well, this is not a scientific theory. You remember what Dr. Seeting said. A scientific theory must be based on facts that can be confirmed independently by others.

---

[1] *See*, End Note [14], page 295.

You don't have any of those kinds of facts. Simply appearing to be painted by people is not enough to say people actually made the painti—the *apparent* painting. You remember about mountain formations and cloud formations, don't you?"

Jenny nodded. A few more heads bobbed by the window, some looking in. *Probably wondering why she was in trouble*, Jenny thought.

Ms. Hart glanced toward the door and continued, "These things—mountains, clouds, and such—have natural explanations that can be checked, and the explanations have been shown to be true. We're trying to find a similar explanation for the markings on the wall of Lavoux Cavern. But there is no way to prove that people could have, much less did, make these wall figures. Dr. Seeting has assured us that there is simply no evidence of a person having been involved in the making of these figures."

Jenny sat motionless, occasionally looking up at Ms. Hart, not sure what to say. She felt awkward again, and vulnerable. She looked around the room once more, as if she might find someone to help. The room was quiet except for the slowly ticking clock and distant sounds of kids playing. Jenny thought she occasionally saw shadows of someone standing around the classroom door.

"And, what is this about a *legend?*" Ms. Hart had a wincing look on her face.

"It's something my dad suggested," Jenny explained, straightening up in her chair. "I found information about it in the library."

"Your dad? The library?" Ms. Hart asked.

"Yes, I explained the assignment you gave us to my dad. I even showed him the photos I took in the cave. When I told him that I thought someone painted the pictures on the wall, he agreed. He even suggested that the picture might be the image of the person who painted it, but he said he couldn't prove it." Jenny looked hopeful.

"Your dad?" Ms. Hart repeated. "Is your father a scientist?"

"No, he works at the lumberyard; he drives a truck. At least he did; he's been sick lately, and can't work much any more." Jenny's eyes were bright. "My father said that he remembered hearing about people called ancient people who lived in this area long ago, including on the land around Lavoux Cavern. He said he thought those people had a legend about underground people."

"*Ancient* people? *Underground* people?" Ms. Hart was both intrigued and getting impatient. "And you found something on this at the library?"

"Well, actually, my Aunt Sema helped me find it; she lives with us. First we looked at the city library, but we only found information at the university library, in a special room. I think it was called the rare book room, or something like that."

"And you found a book about underground people?" Ms. Hart asked, surprised.

"Two books. Both had information about an ancient tribe that was indeg—, indegen—"

"Indigenous?" Ms. Hart helped.

"Yeah! *Indigenous* to this area many years ago. This tribe, they say, had a legend about people that lived underground, maybe somewhere around here. They even had a special word; I think it's 'ee-goo-ya' or something like that. It's in my report ..."

"I saw that," Ms. Hart interrupted, slightly grimacing.

"Yeah, and the word means 'people of the underworld,' I think," Jenny said excitedly.

For a moment Jenny was hopeful that maybe Ms. Hart would change her mind and let Jenny present her theory after all.

The bell rang moments before a rush of movement could be heard outside as students began returning to class. Jenny sat expectantly, realizing that she had missed recess. Ms. Hart looked at the clock, shuffled the papers in front of her, pushed back her hair, and looked at Jenny.

"Jenny," she said softly, smiling, "I want you to think of a real explanation for the figures on the cave wall, not one that relies on myths.[2] It may be fun to entertain ideas about 'cave people' or to imagine someone making 'paintings' on the wall, but I'm trying to get you to think about facts, about reality. I want you to learn how to think like a scientist. Can you do that for me?"

"I guess so," Jenny replied, shoulders sinking at the thought. "But it will be pretty hard to think of how a painting could paint itself."

"Not a *painting*," Ms. Hart stressed. "Think of it not as a picture, but simply as marks on a wall, random marks making a pattern, an interesting pattern, even a remarkable pattern, but a mere accident nonetheless." Kids were now huffing back into the classroom, noisily finding their desks.

"OK?" Ms. Hart asked with a hopeful smile. Jenny gave an unsure nod. Ms. Hart pointed Jenny to her seat.

Jenny ignored stares as she made her way back to her desk. Ms. Hart gave a knowing glance to the door where Mr. Kozak stood, now looking in the window, and then sighed as she looked at the busy classroom.

Mr. Kozak smiled.

---

[2] *See*, End Note [15], page 297.

Chapter 5

Ms. Hart's classroom buzzed with excitement when at last the day came for the students to give their reports before the class. The entire morning was devoted to the presentations, and most of the students had arrived early and were busy reviewing notes and making last minute changes. Jenny sat quietly, detached from the commotion around her, the noise sounding distant and muffled. She had finally managed to come up with a theory that met Ms. Hart's approval. But the few days that she had to develop her idea together with her own indifference to her proposed explanation made it difficult to prepare a report she was proud to present. As she sat amid the din of activity she yearned to be back home. She wished she had never visited Lavoux Cavern.

The classroom had been cleaned and decorated in preparation for Dr. Seeting's presence. A large banner welcoming him hung across a wall opposite the doorway. Each of the kids had written a note of thanks on the banner, and it was to be given to Dr. Seeting before he left. On tables in the back were clay models of caves with stalactites. Some of the models were part of elaborate dioramas complete with lakes, rockslides, stalactites, and, of course, multi-colored pictures. Hung around the room were various drawings and posters the children had made. Most of the pictures were crude attempts at recreating the hunting scene on the wall of the cavern.

Dr. Seeting arrived to a small celebration. After being greeted by all the students, he squeezed into one of the extra desks near the front. He sat in anticipation, not looking the least bit uncomfortable. Since the day of the visit to Lavoux Cavern two weeks before he had frequently thought about the students' assignment. It was an easy decision for him to agree to come back and help Ms. Hart grade the students' work. He increased everyone's level of excitement when he told Ms. Hart, right in front of the class, that he fully expected to find a solution to the mysterious cave markings, right here in her classroom.

One by one students went to the front of the class, nervously standing with their note cards. Each student presented an explanation for the cave markings; some even had poster board visual aids. The first student enthusiastically suggested that the lines were burned onto the wall by volcanic activity. Another suggested that birds had flown into the cave; finches, the student ventured. The finches, possibly carrying berries or seeds in their beaks, and not being able to see where they were going, repeatedly flew into the wall, making a small mark where they hit each time. Most of the students referred to the marks as a "picture" or a "painting," each time prompting Ms. Hart to jot down some notes. Dr. Seeting also made occasional notes, sometimes even nodding.

Jenny sat nervously, wishing she could just skip her turn. Her few note cards seemed of little help, and she thought about saying she was sick. She was only vaguely aware of the other students' reports, occasionally noting that another student seemed to struggle with the assignment just as much as she had. Her stomach fluttered, and she wanted to go to the restroom when Ms. Hart looked up and said, "Thank you. Now, let's see … Jenny!"

Jenny thought she saw Dr. Seeting attempt to sit up straighter upon hearing her name. Ms. Hart again looked toward the door. Jenny almost knocked her desk over getting up, and it seemed like ages as her heavy legs slowly made their way to the front of the room. She turned to look at the class. Dr. Seeting had an expectant look; Ms. Hart tilted her forehead with a hard stare.

"Um," Jenny was surprised to hear her own voice, "I am going to present my thery, my *theory,* on how the marks that look like a picture got on the wall of Lavoux Cavern." She was relieved that she had gotten out her first sentence without stumbling over too many words. Her breathing was coming easier now, but she wasn't sure what note card should be on top. At the word "picture" the pens scribbled again.

"I think that the marks got there when water got into the cave. If the cave flooded, the water level could be pretty high up on the wall. The water could have had dark objects floating in it, like tree bark, that could stick to the wall." She paused. Ms. Hart looked relieved. Dr. Seeting even appeared interested.

As Jenny continued, her words echoed about her head as if in the great cavern that held the pictures she was visualizing in her mind. Her eyes could not find the right note card and she thought she might faint. She couldn't seem to say the right words, and when she spoke her normally lively voice came out low and monotone. "When the water level got higher or lower the tree bark might stick in different places. Sometimes the tree bark might stick on the wall and parts of it might soak into the wall in a pattern that only looks to us like a person or an animal."

Dr. Seeting leaned forward, beaming.

"I think that is how the lines that look like pictures got on the wall of Lavoux Cavern." Jenny ended quietly, never looking up from her hands as they gripped her useless cards. She was not sure if she had any more notes, but her mind was blank, and she was fairly certain that what she had just said was the end of her report. She quickly took her seat.

Dr. Seeting shifted to a relaxed position.

Jenny held back tears.[1]

Ms. Hart, glancing toward Dr. Seeting, smiled.

---

[1] *See*, End Note [16], page 299.

Chapter 6

Jenny ran past Alex and into the room, throwing the door shut behind her. Papers flew as she madly searched her backpack. Alex, her dorm mate and best friend, opened the door and stuck her head in.

"Is everything OK, Jen?"

"Yeah, sorry," Jenny grinned, with a fleeting look in Alex's direction. Alex noticed Jenny's flushed look, but she knew that if there was a real problem Jenny would tell her. The two had been friends since the sixth grade, when Alex was Jenny's first, and almost only, friend. They had been inseparable in high school and had each chosen Lutrell University primarily because it was the only university that had accepted them both. Like sisters, they had learned to rely on each other for support. The two made a good pair; Alex's easygoing disposition often offset Jenny's intensity. Alex lingered long enough to give Jenny an "*Are you sure?*" look and then closed the door.

Jenny finally got her Earth Science book out of her backpack, and threw it open on her desk to page 446. Her long hair fell straight down on each side of her face as she stared down in disbelief at the photo she saw there, in a chapter entitled "The Nature of Science." She looked up from her desk to three worn photos on her bulletin board. Quickly pulling down one of the photos, she laid it side by side with the picture on page 446. *Unbelievable,* she thought to herself. The colors were slightly different; the picture in the book was of much better quality. Nevertheless, there it was, a picture of the "remarkable natural markings," as the book called them, of Lavoux Cavern.

Many years had passed since Jenny saw the figures for the first time with Ms. Hart and her fifth grade class. Now well into her freshman year at Lutrell, Jenny was still tall compared to her peers. Her blond hair framed a face that was friendly and engaging, often displaying a look of wonder as it did that day in Lavoux Cavern. But her slender frame and soft-spoken manner hid a fierce aggressiveness that she could hardly conceal when challenged on ideas with which she disagreed. Looking at the photo on page 446, Jenny wished again, as she often did, that she had vigorously challenged Dr. Seeting in Ms. Hart's fifth grade class.

Jenny sat and leaned back in her chair, holding her photos up in the light. She saw, as she always did, a work of art. Over the years she had often wondered about the artist. Although she never visited Lavoux Cavern again, she had followed the research of Dr. Seeting with interest. Not only was he now a full professor of Earth Science and Director of the College of Natural Sciences at Lutrell, he was something of a local celebrity. His theories on the "miracle of Lavoux Cavern" had made him one of the world's most prominent naturalists. Each summer he and a team of earth scientists would enter Lavoux Cavern and stay for weeks. Their findings often got published in leading scientific periodicals. He was the principal author of her freshman Earth Science text, now open before her.

Seeing the photo in her textbook brought back a torrent of memories. The September heat above the cavern. The cool, moist air inside. The narrow, rocky opening through which she half dropped, half slid into the cavern. The absolute immobilizing darkness of the cave. The great, bright light of Dr. Seeting.

Jenny quickly scanned the pages explaining the figures, quite a few pages, she thought. Dr. Seeting was quoted throughout. Much was said about "facts" and "theories" and "the nature of science." Jenny slowed down and read one particular passage that spoke of the "persistent imaginations of a few" that continue to believe in "legends" and other unscientific explanations for the natural formations found in nature.

Legends, Jenny thought. *A few.* Maybe this refers to the legend of the underground people. *Apparently,* she thought, *some people believe the legend.* She continued reading as the text explained that some people continue to believe that the cave figures were made by people despite the lack of any evidence that people had ever visited the cavern. To believe that people visited the cavern and that a person made these marks for some purpose, possibly even as art, could be, according to the book, "useful for some as a social convention," but otherwise "has no basis in science."[1]

Jenny's heart raced as she traced with her finger the lines of the hunter's legs in mid stride, the hair flying back, the raised arm with the spear. *Unbelievable,* she thought again. She looked at the animal on its side, a straight line extending from its neck. "A remarkable example of progression," read the caption under the photo. *Could it really be?*

Jenny read more about the theory of progression, which was responsible for the remarkable cave formation, according to her book, by a process called natural selective inclusion. A naturalist named Seth Neihil first published the theory of natural selective inclusion, which relies on small incremental changes that are naturally selected for inclusion into various forms called specialties.[2] According to Neihil's theory, over many "generations" of events single dot or single line figures are gradually changed into larger, more complex figures as certain of the incremental changes are preserved from destruction. The book quoted Neihil at length, often referring to Neihilian progression as the current "best scientific explanation" for the various "apparently painted figures" making up the "so-called hunting scene." In fact, Neihil had stated that, even though the

---

[1] *See,* End Note [17], page 301.

[2] *See,* End Note [18], page 302.

markings appear to be designed, "There seems to be no more design in the variability of the wall figures than in the course which the wind blows."[3]

As Jenny slowly turned pages, she read a sidebar note profiling Dr. Seeting. She studied his picture, trying to decide if she would have recognized him from the person who led their field trip in the fifth grade. Although he had aged noticeably in the decade since her field trip, she thought, he was still quite handsome, with a broad smile that was unmistakably his. The sidebar noted his achievements, but it was the final line that made Jenny sit up straight, a quote that unexpectedly seized a lost memory: "What I can say for certain is that the figures in Lavoux Cavern appear to have been made slowly over long periods of time, and there were no people around at any time during their formation. This, at least, is scientific fact."

Dad, Jenny thought, *I've got to show Dad.* She quickly grabbed her keys, the science book, the photos, and sprinted out the door.

☙ ❧

As she drove, Jenny recalled the day she sat, totally dejected, in Ms. Hart's class after giving her report on how the cave art in Lavoux Cavern might have been formed. She never forgot how, even as a fifth grader, she felt belittled, betrayed and even bullied. That day her father had noticed her despondency and had pulled her aside to talk. He had never been one to compromise his convictions, and he empathized with her, at the time simply telling her something she never forgot: "You did what you had to do, but don't worry, the truth isn't changed by what anybody thinks about it."[4]

In the intervening years she resolved in her mind that she would never let someone else tell her how or what to think

---

[3] *See,* End Note [19], page 304.

[4] *See,* End Note [20], page 306.

again. Such an obstinate attitude got Jenny into trouble on more than one occasion. Her father had often implored her to be less hardheaded and to listen to those in authority. She grinned as she thought about the first week of seventh grade when her father had to go with her to the school principal to discuss Jenny's refusal to recite the school's creed in a group assembly. At the time, Jenny told them, she could not "conscientiously abide" by the school's creed, and, therefore, she simply could not recite it like a "zombie." Slightly embarrassed at being there and irritated at having to miss work, Jenny's father sat quietly in the school office as the principal admonished and Jenny trembled. In the end, though, her father stood beside her in that principal's office, with a firm hand on her shoulder, and told the principal that he supported his daughter's position. If the principal had a problem with that, her father continued, he could discuss it with the school board. Jenny smiled again at the thought.

One of the reasons Jenny decided to go to Lutrell was its proximity to her father, now bedridden in their house, cared for by her Aunt Sema, his sister. But now the short drive home seemed like an eternity. As she drove into the driveway she noticed Aunt Sema's car. *Good,* thought Jenny, *she will remember this as well.*

"Hey, Dad! Aunt Sema!" Jenny yelled as she came in the front door.

"Jenny!" two voices cried together from different rooms.

"What are you doing here, I mean, this time of day, girl?" Aunt Sema asked, giving Jenny a hug and pulling her into the living room where her father lay. Aunt Sema seemed always to be moving, but now her large frame stopped momentarily as she held a strong arm around Jenny. Jenny impatiently hugged Aunt Sema.

"Wait until you see this, both of you!" Jenny said as she pushed away and walked around the permanently opened sofa bed to her father. "You won't believe this!" Jenny quickly gave her father a kiss on the forehead, and then, sitting cross-legged on the bed beside him, practically threw the book in front of him.

"Whoa, now, Jenny," her father said, "I'm an old man now. You can't come in here throwing things at me." He gave Jenny a big smile as he held the book out at arm's length, struggling to read the cover.

"I'm sorry," Jenny said, half laughing. "But here," she said, handing the man his glasses, "look at this." Jenny opened the book and placed it in front of her father, impatiently tapping the page to direct his attention.

Jenny's father took a long look at the photo on page 446. The photo reminded him of that day when Jenny came home from school noisily excited at having visited Lavoux Cavern—and the day a short time later when she came home quiet and sullen. He thought back to those times as he stared at the picture. That period had been hard for him, having recently moved after the death of his wife. He regretted not having helped Jenny more during the first few months of school, and as he stared at the photo in front of him he was again reminded of that difficult time of their lives. Jenny wasn't sure if he was looking at the picture, or reading, or even dozing. She bent down to look at his eyes.

"Dad!" she said excitedly. "Dad, do you know what this is a picture of?" Aunt Sema, leaning over, eyes looking down her nose, gave a knowing "uh-huh" as if to say, "of course he does" before walking out of the room in her endless routine of chores.

The gray-haired man turned the book to look at the cover, looked again at the picture on page 446, and then looked at Jenny. "Looks like a photograph you took."

"It *is*," Jenny said, putting her own photograph beside that of the book.

Her father compared the photos, just as Jenny had done earlier. He then put the book down on his lap, removed his glasses, and lay back, his fingers pinching the bridge of his nose.

"Jenny," he started, with a tired voice, "I'm an old man now, and I can't get around like I used to when you and I were both younger. But I have a strange feeling that I will soon be going with you again to the principal's office." He turned toward Jenny, smiling. *His body may be weak*, Jenny thought, *but his eyes are as bright as ever*. She squeezed his hand and kissed him again on the forehead. Aunt Sema continued in and out of the room, humming under her breath, with a louder humming and a deliberate glare at the mention of the principal's office.

"Dad, this book is teaching that those figures, those *paintings*, were formed by nature, over time by a purely natural, unguided process, an idea called 'progression!'" Jenny took the book and began reading parts to her father. "Listen to this, Dad," she began. "It says here that scientific explanations are based on empirical observations or experiments."

"*Scientific* explanations, huh," her father said, understanding the emphasis.

"Yes," Jenny continued, "it says that scientific explanations emphasize evidence and use scientific principles, models, and theories." Jenny continued scanning the pages. "And here," she said, pointing, "it says that there is no evidence to suggest that a person could have made the markings in Lavoux Cavern. Scientists have been unable to empirically observe any evidence to suggest a person ever visited Lavoux Cavern ..." Jenny continued scanning the page. "Here it gives examples of the lack of evidence ... no footprints in the area ... the lack of accessibility ... no evidence of people living in the area at the time ... the age of the rock formations, including the markings themselves ..."

"Jenny," her father interrupted.

"Yes?"

"What are you going to do?"

"What am I going to do?"

"Yes," her father looked sympathetic, "what are your intentions? Why are you so worked up about this?"

Jenny thought for a moment and said, "Dad, am I the only one in the entire world who looks at these cave markings and sees a picture? A painting? A work of art?"

"I'm sure you are not, Jenny. You know that I agree with you." He reached over and picked up her science book, holding it up as if he were trying to estimate its weight. "But, Jenny, these are *scientists* who have studied this phenomena for years. Who are you ..." He hesitated before continuing, "I mean, who are *we* to second guess things like this?" Her father slowly placed the book back on the bed in front of Jenny.

Now Jenny stared in disbelief. "Dad," she said, moving closer and deliberately holding his hands as she looked into his eyes. "Dad, scientist or not, something is amiss here. How can someone look at this artwork and not immediately consider the artist? Doesn't the artwork itself speak of an artist? Do you really need the footprints of the artist to explain the art?" Jenny was agitated at the thought.

"Jenny, I'm just saying that maybe you should first consider the evidence. You're young; you just started college. Learn. Make a good grade. Then make waves."

"Dad!" Jenny burst out. "I *have* considered the evidence. The evidence is the painting itself. A painting requires a painter!" Jenny was on her feet now. "A painting does not paint itself!" Aunt Sema returned to the room, feigning a sudden interest in a bureau drawer beside the bed.

Jenny looked at Aunt Sema, took a deep breath, and sat once again on the bedside. She said softly, "I remember the first time Dr. Seeting's floodlight washed over the wall of the cave and the wonder that I—that *we all*—felt at seeing the images of people and animals on that wall." Jenny thought for a moment more. "I even remember," she said thoughtfully, "wondering about the person who made these pictures. What kind of person was this? Where did he come from? How did he live? Who did he live with? What kind of family did he have? Even after Dr. Seeting's light was moved the image of these paintings remained burned into my memory. I imagine the hunter walking the same earth I walk. I wonder if he and I would have been friends had we lived at the same time." Jenny looked at her own photos for a few moments, finally putting them back into her science book like bookmarks.

Jenny squeezed her father's hand once more, leaned over and with a piercing look said, "Dad, I saw the same picture in that cave as everyone else. And what I saw was obviously painted by a painter. Maybe I can't prove who painted it, and maybe we'll never know who it was. But what was the point of sending such a great light into such oppressive darkness, only to refuse to acknowledge the obvious?"

"The obvious?" her father asked.

"Yes, the *obvious,*" Jenny was on her feet again. "The truth is that the *painting* was made by a *painter.*"

"The truth?" her father asked.

"The *truth,*" Jenny practically shouted the word "truth," "is that a painting needs no more evidence of a painter than a watch needs evidence of a watchmaker![5] Is that so unbelievable? Why won't people believe that?"

Jenny's father watched her for a moment as she stood over him with fire in her eyes. Aunt Sema shut a drawer loudly.

"Some will believe," her father sighed, "some will believe. I'm pretty sure," he shook his head with a chuckle, "some are *about* to believe."

Jenny left with an over-the-shoulder good-bye.

Aunt Sema muttered something about the principal.

Jenny's father smiled.

---

[5] *See*, End Note [21], page 306.

Chapter 7

"C'mon, Jen, let go and have fun," Alex implored as she paced about their dorm room. "You've been hiding behind those books all week. You would think you're in college or something." Alex waited for a reaction, but Jenny never looked up from the small fortress of books on her desk. "Jenny, just look outside, will you?" Alex said, squinting out their window at the first sunny day in weeks. Even though people below clutched coats and scarves against the wind Alex was anxious to get outside and away from their dorm room.

Jenny ignored Alex's loud sighs and replied to any statements from Alex with a detached "uh-huh" from behind her wall of books. Finally, Alex walked over and closed the book Jenny was reading, keeping her hand on the closed cover as she leaned into Jenny's face.

"Alex!" Jenny cried, looking up surprised at Alex's friendly glare.

"Hello! Earth to Jenny, come in please," Alex said, continuing to lean stiff-armed on Jenny's book.

"Give me a few more minutes—this is important. Besides, it's cold outside!" Jenny said as she glanced toward the window.

"I would love to give you a few more minutes, Jenny, but I can't," Alex said in a completely serious tone.

"You want to but you can't, huh?" Jenny asked, smiling as she sat back with a sigh. "Why not?"

"Well, you see, Jenny, the library called. They want their books back," Alex deadpanned. Jenny looked at Alex as she continued, "They want to know when to send the truck over."

Both girls stared, Alex still leaning on the closed book, Jenny with a raised eyebrow of disbelief. After a few seconds, neither could hold back, and both burst out laughing.

"Look, Jenny, I've let this go on all week hoping it would go away," Alex said, motioning to all the books stacked up around the edge of Jenny's desk. "So now I want to know, what's going on with all the books, Jen?" Alex looked over the various volumes as if they were contraband. "And *science* books!" Alex said in derision, drawing back quickly as if suddenly afraid of leaving her fingerprints. "You already have one science book. Isn't that good enough?"

Jenny looked closely at Alex's face, wanting to pour out her thoughts. The two kept no secrets but Jenny had never fully confided to Alex about her fifth-grade experience at Lavoux Cavern and her report in Ms. Hart's class. Besides, Jenny had not entirely worked through her own thoughts and was not sure she could explain to Alex why the sudden, intense interest in science. In any event, she was sure that Alex would not understand. As good a friend as Alex was, she had little interest in discussing intellectual topics, much less controversies.

"It's nothing. We can talk about it later," Jenny said, trying to sound convincing as she reached for the book Alex had closed.

Alex snatched the book away and quickly went back to the window.

"Look, Jen," Alex said, as she held the book up in the window. "Look outside. It's sunny! Let's go out. I know it's cold, but we don't have to be out long. Just go with me to the Student Center. There is something there I want to show you—something you will find interesting, I promise. Anything is better than sitting in this dark room, reading …" Alex paused as she looked at the cover of the book she still held, and then continued with exaggerated scorn, "*On the Origin of Specialties.*"[1] Alex put the book down slowly, holding it out between her forefinger and thumb as if to keep from becoming contaminated. "Come on, Jen. I have a surprise for you at the Student Center—something you will really like." Alex walked back and tossed the book onto Jenny's desk.

"Some-*thing*? Or some-*one*?" Jenny asked suspiciously, suddenly realizing that Alex might be setting her up. Alex often tried in vain to get Jenny interested in a boyfriend. So far, the best Alex could do was arrange an occasional date. All through high school she had dragged Jenny to dances, concerts, pool parties, and even to weddings, always with an eye toward meeting boys. Often Alex's plans were quite elaborate. Once she convinced Jenny to ride her first roller coaster, promising to sit beside her. Jenny was shaking as she stepped into the seat, and then terrified when at the last second a stranger was sitting beside her. The ride attendants, informed of the ploy beforehand, ignored Jenny's pleas to be released from the security bar that was quickly lowered into her lap. By the end of the ride Jenny was clinging to the boy next to her, who had very agreeably placed his arm around Jenny. At the time Jenny was furious; she didn't speak to Alex the rest of the day. She eventually learned to laugh about it, but now she was always suspicious of Alex's ideas to go somewhere or do something new.

"Some-THING," Alex assured her. "I've given up on you and guys. What a waste of pretty. We should have each been born in the other's body. Why is it that you got tall and thin and I got short and dumpy? But this is not about guys, I promise. Come on, I know you will like it. In fact, I'm surprised you don't know about it. If you did you would be hauling me there."

Jenny studied Alex for a few moments. She wished Alex would stop imagining herself to be short or dumpy. In fact, Alex was shorter and heavier than Jenny, but she was in no way

---

1 *See*, End Note [22], page 307.

unattractive. Why she didn't have a better self-image always mystified Jenny. "You aren't going to give up, are you?" Jenny said with a laugh.

"Nope. Come on. We're going out to the Student Center for something hot to drink. We'll brave the elements for some coffee—double latte, on me. You can tell me all your problems."

Jenny leaned back in her chair and stretched. "What makes you think I have problems?" Jenny asked, trying to sound convincingly surprised. The sound of Alex yanking coats out of the closet was followed by Jenny's jacket flying into her lap. Alex quickly had her coat on as she started for the door.

"You always have problems because you take things so seriously. You need to lighten up, Jen. But that's another reason to go to the Student Center. We're going to see some others who think about serious things. If you like you can even discuss important topics with them." Now Jenny was puzzled, but interested. Alex was standing at the door. "Are you waiting for springtime? Am I going to have to put that coat on you? Come on, let's go!"

As Jenny walked out into the cold sunshine the chill awakened her as the sunlight closed her eyes.

"Wow! It is really bright out here!" Jenny exclaimed. Blinking and looking around, Jenny breathed deeply and realized that Alex was right—it was good to be outside.

"It's called sunshine. It's warm and bright; you'll get used to it," Alex said, starting off toward the Student Center. "You've just been hibernating in your dark cave too long." Jenny flinched, but let the comment pass. She held her hands over her eyes as she kept up with Alex who had picked up her pace.

"Why are we running?" Jenny asked. "I'm starting to get hot."

"I don't want you to miss any more than you already have," Alex replied without explanation and without slowing down. She reached out and grabbed Jenny's arm to pull her along.

Jenny noticed as they neared the Student Center that there appeared to be a large number of students outside in front of the building. On most days this would be a common sight, but on such a cold day it struck Jenny as unusual. It appeared that the students were grouped in active clumps—some large, some small. "What's going on?" Jenny asked.

"You'll see. You will like it," Alex grinned, still pulling Jenny by the arm.

The noise level increased as they approached the Student Center but no one could be heard distinctly. Not until they got closer did Jenny realize that each group of students was clustered about a speaker. Alex and Jenny stopped and took in the scene.

"It's called Open Forum," Alex said loudly, leaning next to Jenny's ear. "It happens once a year; it's organized by a group known as the Scholastic Community Advocating Magniloquence. Students can stand up and spout on any topic they wish to their heart's desire—or until dark, whichever comes first. What do you think?" Alex asked excitedly, rubbing her hands together.

Jenny did like it. Her eyes went from speaker to speaker. Some had signs and posters; most were standing on makeshift platforms—boxes, stools, and the like. It seemed almost every political or social cause was represented. Certain speakers seemed to draw more opposition than others. Soon Jenny was dragging Alex through the students from one speaker to the next. Occasionally a lone orator opined to no one but random students milling about. At one point Jenny and Alex had to squeeze through the convergence of two crowds, each attracted by opposing speakers who seemed to be addressing each other more than the students around them. As the girls pushed through the throng, Jenny shouted, "Who's that?" as she pointed to one of the groups.

"That's NORM," Alex said over the noise as she and Jenny emerged from the swarm and headed for the main door of the Student Center.

"Norm who?" Jenny asked as she stopped momentarily to look back at NORM.

"Not Norm *who,* NORM *what,* as in the National Organization of Men. They're also sometimes referred to as Pro-Voice," Alex said, turning Jenny back toward the Student Center. "Come on, I'm cold."

The girls entered the relative quiet of the noisy Student Center and found a table near a window in the coffee bar. Jenny waited while Alex left to order two cups of coffee. Out the window Jenny's eyes took in the action of the Open Forum. From where she sat, Jenny could clearly see the NORM group and even noticed several Pro-Voice banners.

Alex returned carrying the two cups of coffee. With jackets off they began sipping and looking around.

"What do you think? Didn't I tell you that you would like it?" Alex remarked. "It's unfortunate that it's too cold to stay out very long. I'm surprised so many people are out there now."

Pro-Voice signs bouncing around the air about the NORM group continued to attract Jenny's attention.

"Yeah, I must admit you know me pretty well. Why didn't I know about this?" Jenny asked. "If I had known about this I might have brought a box to stand on for awhile."

"This place is pretty busy inside, too," Alex remarked, looking around. "I hope we don't run into Chris—I think he likes me."

"Chris who?" Jenny asked, grinning. "Do I know him?"

"Stafford."

"Chris Stafford? The guy that helped you get your computer on line?"

"Yeah. Ever since then he's wanted to go out with me. At least I think he does; he's too much of a nerd to just ask me."

Jenny was laughing. Alex handled boys like a child plays with toys—eager but quickly bored. Rarely did a guy show an interest in her that was not returned.

"So what's wrong with Chris? I don't remember him that well, but he seemed nice enough to me," Jenny said, still chuckling.

"He speaks a foreign language," Alex replied, again looking around.

"Chris Stafford speaks a foreign language? What's wrong with that?"

"It's computereze. Computers are the only things that interest him," Alex said, glancing about. "Except you."

"Oh, please!" Jenny groaned. "Not that again!"

Alex often suspected that she was only a bridge to Jenny's island. More than once Alex avoided introducing boys to Jenny if she had any interest in them herself. While not usually requiring a conscious effort on her part, Jenny made it a point to show little interest in Alex's boyfriends.

"You always think they are after me."

"Not always. And sometimes I'm right."

"Well this time you are wrong," Jenny said, still grinning in disbelief. "He wouldn't know me if he saw me, and we never see each other. And if you keep looking around like a fugitive you're bound to attract his attention."

"You'll see," Alex said as she shifted around to look outside. The two sat quietly watching the Open Forum activity that could occasionally be heard inside.

"Alex," Jenny said after a few minutes, "tell me something. I know I'm from the other side of the mountains and all … the 'lowlands' as some people say. But I've been here a long time now and I've never heard of NORM. What does 'pro-voice' mean?"

Jenny looked at Alex who appeared to be slouching slightly to be less noticeable should Chris walk by.

"Do we have to talk about that now?" Alex asked. "I would rather talk about what's been bothering you lately."

Jenny looked surprised. "Bothering *me*?"

"Yes, *you*," Alex said. "Every night this week you've been holed up in our room behind your mountain of books, studying.

You haven't relaxed since you almost knocked me over running into our room last week. Even for you, such focused intensity is unusual. So, tell me, what's the deal?"

Again Jenny looked closely at Alex's determined brown eyes and considered telling her about discovering the picture on page 446 and about the cave painting in Lavoux Cavern. But again she decided that the time was not right. She wasn't even sure if Alex knew there was a cave painting, much less a controversy over its origin. A lull in the Student Center let in some muffled noise from outside.

"It's nothing, Alex," Jenny said quietly. "I just suddenly became interested in science, that's all. Who knows, I may become the world's next great scientist!" Jenny said facetiously.

Alex looked at Jenny for a few seconds as she waited for the truth.

"OK," Alex said, nodding slowly as she continued to study Jenny's face. "I don't believe you, but I understand. I just hope this isn't a permanent change in you. Or it's going to be a long four years."

The two girls sat silently for a few minutes, occasionally turning toward a swell of cheers or clapping from outside. When it became clear that Jenny was not going to explain more, Alex sighed and said, "NORM is a national men's group pushing for rights that would permit husbands to legally divorce their wives at any time in their marriage, for any reason. It's called 'termination.'"

Jenny thought for a moment in disbelief, sure that Alex was trying to be funny. "Termination. You're kidding, right?"

Alex motioned outside. "Termination on demand. Go listen for yourself."

"Termination of a marriage? Like divorce? For any reason? And what does 'pro-voice' have to do with it? And what if the wife objects?" Jenny reeled off the questions before Alex could answer.

"Any reason," Alex replied. "Even if the marriage is simply inconvenient for the man, NORM advocates unilateral marriage termination—the wife has no voice. Because the men want to have the only voice in the matter, they are *pro-voice*."

Jenny wanted to laugh. "You aren't serious! If this isn't the wildest idea I've ever heard!"

Alex again gestured and motioned outside, saying, "I didn't make this up. These men want to have the only voice in deciding whether to terminate or not, and they believe it's their right."

"They want to have a voice in the matter?" Jenny repeated, finding the idea incredulous. "To *terminate?*" she questioned quietly to herself as the word began to register and she turned to look out the window. She closed her eyes, surprised at the sudden waves of indignation beginning to roil within her as she considered the life-destroying implications of termination, especially for children. Having experienced the loss of a parent, she wondered why anyone would willingly choose such a condition. Her eyes shut tightly, silently wincing as she recalled her father's grief at the loss of her mother, and how he still says he would give anything if he could just restore that innocent life again.

"And they want the *only* voice in the matter?" Jenny asked, turning back to Alex. "Aren't there two voices to consider before terminating a marriage? What about the wife?"

"See that other group facing NORM?" Alex pointed. Jenny had noticed that NORM and another group seemed to be particularly at odds with one another. She also realized that those two groups were the two she and Alex had pushed through to get to the Student Center.

"That's the Pro-Wife group," Alex said. "Like their name says, they represent the interests of the wives, they're anti-termination. But so far it looks like the High Court will rule against them."

"Against them!" Jenny was astonished. "Why the courts? The very idea! How could anyone support such a crazy idea?"

"Support?" Alex asked, as if it were a silly question. "NORM needs very little popular support to win its court case. It need only convince a few black-robed judges that termination is a fundamental right. For NORM, it is a privacy issue, and like all such matters NORM believes that termination should fall under the umbrella of protected rights associated with personal privacy. But termination has support. Another national group called 'Planned Marriages' supports NORM, for example, and believes that termination should be an unrestricted right of all men."

"Planned marriages? *Planned … marriages*?" Jenny repeated slowly, emphasizing each word. "Alex, how could an organization with a name like that support the exact opposite of marriage? How could it support the killing of a marriage?" Jenny thought for a moment as she looked outside and noticed for the first time a "Pro-Wife" sign. "It seems to me that if Planned Marriages is for termination, and even backs Pro-Voice, then it actually supports neither planning nor marriage," Jenny said as she finished her coffee. "It should be called '*Un*-planned *Un*-marriages!'" Jenny declared.

Jenny had finished her coffee and was slowly picking at the rim of her Styrofoam cup as they watched outside. She looked across the table into Alex's cup. Alex knew what Jenny was thinking.

"Look, Jenny, I knew you would think this Open Forum thing was a cool idea, but it's too cold to stand out there now. Let's just sit in here and watch. We can grab some of their literature on our way back."

"Come on, Alex. You're the one who brought me over here. It's not that cold. I just want to hear some of this for myself. I'm not going to make a scene, Al," Jenny promised, smiling at Alex as Alex shook her head. "But I've got to find out more about termination, especially this Pro-Voice stuff. Don't worry, I won't embarrass us."

"You won't embarrass *us* because *us* isn't going out there," Alex said, sitting back in her seat. "I'll sit right here and watch. If you need me, just wave or something," Alex grinned.

"Chicken," Jenny said as she stood up and put on her coat. "I won't be gone long. Don't leave without me."

Alex watched through the window as Jenny first approached the Pro-Wife table and struck up a conversation with a woman sitting under a large, colorful quilt that was draped over her head. The woman held the quilt under her chin such that only her face showed, her breath visible as she spoke. It was apparently an informative exchange, with Jenny alternating nods with disbelieving headshakes, her breath also visible with each word. Occasionally Jenny would turn as particular people were pointed out on the NORM side. Soon Jenny left the colorful quilt and worked her way into the NORM crowd. The NORM speaker was in the middle of a speech that included an appeal to the practices of other civilized countries and sophisticated cultures.

"For too long our termination rights have been denied to us!" the speaker shouted. "Even today there are those who would deny us our voice to terminate. Such people claim to be 'pro-wife,' but, in fact, they are really anti-termination and anti-voice!"

At this point a loud exchange ensued between the Pro-Wife and Pro-Voice groups. Alex could see Jenny trying to get close enough to the NORM speaker to ask a question. In all the years Alex had known Jenny she had never seen Jenny shrink from confrontation. Alex found Jenny's quick wit and questioning style amusing, even though in most instances she wondered why Jenny would care. But something about Jenny's passion had always appealed to Alex. A heart-level honesty radiated from Jenny, warming their relationship, as Alex never found Jenny to be duplicitous in any way. When challenged on ideas, Jenny's style was to ask pointed questions. The "right questions," as Jenny would say, are the quickest way to the truth. If you ask the right questions, Jenny would point out, it hardly matters what the immediate answer is—the point of the question is to get people to think. Like insect bites, the question might not have an immediate effect on the listener, but later the itch will demand a response. Now as Alex watched Jenny's breath coming quicker she even considered going out to join the fun.

When the noise died down and the NORM speaker was at a pause, Jenny shouted, "Excuse me!" Alex saw Jenny's breath go up with her hand and the speaker's face look down. "Excuse me," Jenny continued, "I'm sorry if you have already covered this, but do you really support a man's right to have the only voice in ending a marriage?"

"Our voice is our right!" the NORM speaker shouted over Jenny's head. "Termination is a right! Pro-Voice supports man's right to …"

"You didn't answer my question!" Jenny shouted. Now a larger crowd was beginning to gather around Jenny such that Alex could see her blond hair only sporadically. "I asked if you really support a man's right to have the only voice in terminating a marriage."

Without looking directly at Jenny the NORM speaker answered loudly, "I suppose that you, obviously a woman, think you should decide what's best for me!"

"Are you going to answer my question?" Jenny asked. "Or are you afraid to answer truthfully?"

The quilt stood up in interest.

"A man's voice is the only voice that matters," the speaker said. "That's why we've asked the courts to end legislation that restricts our right to termination …"

"Wait!" Jenny's heart was beating faster. "You mean that a man can ask a woman to marry him, get married, and then, if he finds the marriage inconvenient, he can simply dump his wife in the trash bin behind the town hall?"

Loud applause came from the Pro-Wife side. Jenny stood directly in front of the NORM speaker, hoping the commotion would not be used as a distraction to ignore her. It was apparent, however, that he welcomed Jenny's challenge, and after a minute continued.

"Are you married?" he asked.

"Does it matter?" Jenny retorted.

"Only because you are a living example of why we must win in court!"

The NORM guys cheered. Even Jenny had to laugh. She had long ago learned to take *ad hominem* remarks in stride. The NORM speaker continued, "Were I to be seduced by your beauty and make the unfortunate mistake of marrying you, I most certainly would want the choice of remedying the situation cleanly and safely."

The NORM people clapped and cheered again.

"Well, in that isolated case, I might welcome your decision!" Jenny shot back. After some laughter from the growing crowd, Jenny added, "But aren't you overlooking the fact, sir, that there is another voice in addition to yours in a marriage—another voice crying to be heard? Can you really simply ignore this other voice ... this person ... this *life*?"

Alex could barely see Jenny through all the heads that now surrounded her and decided that based on the interest apparently being generated, this might be something worth hearing. She slipped on her coat and headed outside.

"The courts will rule on whether it's a person. What we support is the husband's voice in the termination decision. You may be anti-voice, but you have no right to force your will on us—you cannot legislate morality![2] And don't think termination is an easy decision for us. Nevertheless, sometimes we must make the decision to remove their names from our lives—and let the chips fall where they may!" The speaker shouted.

Jenny bristled at the words. *Remove their names? Let the chips fall?* She stared hard in disbelief, studying the man's countenance for any sign of familiarity. Wanting to lash out at his insolence, she held her emotions in check and forced herself to focus on the main issue. How could there be any question that a wife—a wife created by the act of marriage by the husband—is

---

[2] *See*, End Note [23], page 308.

not an "it," but is a person? Alex was approaching the crowd now and heard Jenny's voice cut in above the others.

"So, then," Jenny started, "you would say that there is only one person whose voice is relevant in this decision?" Without waiting for an answer Jenny continued, "Do you deny that there is another person whose voice, were it to be considered, might have a very different opinion in the matter?"[3]

"It is for the courts to decide ..."

"NO! I'm asking you! Do you, yourself, as a person—a human being—deny that there is another person—another human being—whose voice, if it could be heard, should be considered before simply choosing to kill a marriage through termination?"

For the first time the NORM speaker seemed confounded as he stared angrily at Jenny. The few seconds of hesitation lingered uncomfortably like a heavy mist in the cold air. It seemed that most listeners were eager to hear the reply. Jenny stood resolutely, until finally taking advantage of the prolonged quiet to speak in a voice at first low and then rising to a shout.

"Your silence rings with the shrill sound of selfish ambition and fills the air with the stench of a self-centered conceit that would go so far as to destroy the life of one whose very station in life is due to your voice in the first place!"

Clamor and commotion again surrounded Jenny as the Pro-Wife group cheered and the Pro-Voice group jeered. The NORM speaker began a monologue about termination rights and the courts, but Jenny heard nothing as she worked her way quickly out of the moving bodies.

Others were now openly challenging the NORM representative as Alex found Jenny emerging from the din, her face still flushed.

"Well, *that* was fun," Jenny said, laughing nervously as she pulled her jacket close, feeling the cold once again. Alex took

---

[3] *See*, End Note [24], page 309.

her arm as the two turned toward the dorms while Jenny filled Alex in on the details of her exchange.

"What was that about chips falling?" Alex asked with eyebrows raised in astonishment.

"You heard that?" asked Jenny, as the two looked intently at each other.

"Yeah, do you think ...?" Alex started.

"No. I don't. I think he's just full of bombast and bravado," Jenny said.

"But, you know, Alex, I still think ..."

Jenny continued discussing the merits of termination as they walked. Alex listened patiently, glad to be with Jenny but privately wondering, as always, how Jenny could be so openly passionate about some things. For her part, Alex was finding that the college experience was helping her to be more open-minded and tolerant of all ideas. Who was she, she thought, to say what is right or wrong, good or bad, truth or falsehood?[4]

"Alex, is it just me, or is the world changing? Sometimes I feel like an oddity in an odd world." Jenny looked at Alex as if she expected Alex to answer. "Do you ever feel that way?" Jenny asked.

"Everyone feels odd sometimes, Jen," Alex answered. "The key is to be open-minded and tolerant of other people's ideas."

Jenny respected Alex, but she had noticed that Alex had changed over their first few months in college. There was a time when even Alex would have thought Pro-Voice was wrong and a bad idea, and it surprised her that Alex didn't react more strongly to it now. After experiencing a bad divorce with her own parents, it seemed Alex would be against unrestricted termination.

"I know, Al, but ... this Pro-Voice/Pro-Wife controversy ... this Pro-Voice stance ... this idea of unrestricted termination as a legal right ... is the craziest idea." Jenny paused for a

---

[4] *See*, End Note [25], page 312.

second. "Well, it's *one* of the craziest ideas I've ever encountered."

Alex listened without speaking as the two walked slowly arm in arm in the cold. Experience had taught Alex that just listening is what Jenny needed. Rarely did Alex have an interest in Jenny's intellectual ramblings, much less any thoughts to add. In fact, she often did find Jenny to be odd. In the past it was Jenny's oddity that appealed to Alex; she found Jenny's peculiarities to stem from candid openness, not obtuseness. But lately Jenny's oddness seemed to be a result of adhering to old-fashioned notions that seemed out of place on a college campus. Even after all these years in the highlands, Alex thought, Jenny seemed to harbor certain old-school lowland ideas that ran counter to the contemporary thought she found so refreshing at Lutrell.

As the two girls continued toward their dorm the sound of the Open Forum ebbed in the distance where, in the still active multitude, a colorful quilt stood motionless, with eyes straining to follow the two girls into the distance. When Jenny and Alex were no longer in sight the quilted face turned away and sat again, smiling.

Chapter 8

"Are you sure this is it?" Jenny asked Alex. Jenny had seen the signs as they entered, but she was certain she had never seen this place before.

"Sure I'm sure, I've been here. Twice," Alex said, matter-of-factly as she slowly entered the parking lot.

"Twice?" asked Jenny. "You never told me that. So you've seen the painting?"

"The *painting*?" Alex asked with sarcastic emphasis.

"Yes, the *painting*," Jenny said, mocking Alex with an exaggerated tone, wagging her head.

"What I saw was nature exhibiting change over time—it's called progression," Alex responded in a truly disinterested tone. "Anyway, let's go, Jen. This was your idea and I don't want to spend all day here. A nice day like today shouldn't be spent in a hole in the ground." Alex got out of the car and started toward the main building.

It was Jenny's idea. She had contacted Dr. Seeting's office at the university to arrange an appointment with him to discuss his book's explanation of the cave figures. Surprisingly, he remembered Jenny and had agreed to meet with her right away, even suggesting that they meet at his field office at Lavoux Cavern. Jenny had persuaded Alex to accompany her for moral support. Although not particularly interested in Lavoux Cavern, Alex had agreed to come and back Jenny up. It was a mutually beneficial arrangement—Jenny needed Alex for encouragement, and Alex enjoyed watching Jenny "in action," as she described it.

Many weeks had passed since Jenny first discovered the pictures on page 446. Recently she had finally discussed the issue with Alex, who, predictably, found Jenny's interest difficult to understand. It was the first time Jenny divulged all the details of her class field trip, the recess-time conference with Ms. Hart and the oral report in front of Dr. Seeting. Other than expressing amazement that Jenny would truly believe that a person had drawn the figures on the wall of Lavoux Cavern, Alex simply listened patiently and tried to appreciate Jenny's current preoccupation on the topic. Both agreed that it made sense to talk to Dr. Seeting now and try to come to some resolution, because the topic of progression was soon to be discussed in their science class.

Jenny stepped out of the car and looked around in astonishment. This was not the Lavoux Cavern she remembered. The dirt road leading in was now paved. It led to a large parking lot complete with bus parking and shuttle service to an imposing, glass-walled building named The Neihil Research and Science Center that welcomed visitors with large lighted signs indicating current attractions.

"Are you coming, Jen?" Alex called back after realizing that Jenny was still at the car.

Jenny stepped slowly over perfectly manicured grass to a sidewalk, inching her way toward Alex. Late winter rains had greened the lawns and the unusually warm weather had even coerced some blooming flowers out of the ground. "This place used to be an open field," Jenny remarked, gazing about her. "There was just a hole in the ground, not even a hole, just a little crack of rocks and dirt. You could hardly see it." Jenny stood looking down with her arms outstretched toward the ground at an imaginary hole before her.

Alex walked back and grabbed Jenny by the arm, pulling her toward the front door. "Things change, Jenny," Alex said. "Come on! I've never seen anyone so enamored with a parking lot!"

Cool air rushed past the two as the glass doors opened automatically before them into the air-conditioned lobby. A large, circular information desk to the right was staffed with blue-shirted kids. A flashing red number above their heads indicated the next tour to leave from Gate 2.

"Wow," was all Jenny could utter. She stood staring, wide-eyed at the expansive lobby.

"They've changed it since I was here," said Alex without irony. "I think they used to wear red shirts."

The floor below them was transparent in places, showing lighted cracks and crevices in the ground below. Alternating male and female voices over a loud speaker announced the current showing in the theater, along with prices and discounts available. Jenny thought she smelled popcorn coming from a gift shop over to the left where T-shirts hung in the doorway. On one was screen-printed the familiar hunting scene with the words "Nature's Miracle."

Jenny walked over to the information desk and asked one of the blue shirts where she could find Dr. Seeting. After some questions back and forth, as well as a phone call with several looks at a huge clock over the desk, the girls were given visitor badges and pointed to some doors marked "Staff Only." A click unlocked the doors as Jenny swiped her badge through the reader. Once past the doors, the sounds changed from loudspeaker announcements to music, the smells from popcorn to new carpet and paint.

Before they could get halfway down the hallway, Jenny noticed Dr. Seeting walking towards them. She recognized him from the picture in her science book; he had aged significantly since the field trip in Ms. Hart's class. He was smiling, but her heart raced nonetheless. She suddenly wanted a drink of water. Alex was a step behind Jenny saying, "There he is" in a singsong voice under her breath.

They met with outstretched hands. "How are you, uh Ms …"

"You can call me Jenny," squeaked a thin voice.

"I'm Alex," said Alex, all smiles.

"I remember you," Dr. Seeting said, looking closely at Jenny. "But you've changed quite a bit!" The smile never stopped, and he seemed genuinely glad to see her. "I actually think of you often; in fact, you have contributed to my success."

Jenny could not speak, but her face gave an expression of "I what?"

"Yes," Dr. Seeting turned. "Come and let me tell you all about it." He started briskly down the hallway.

Jenny was not so keen on hearing how she helped make Dr. Seeting successful, and she was not prepared to do much listening; she had hoped to do most of the talking. But he was talking, and she couldn't, so on the three went toward his office, with Dr. Seeting chatting all the way. At the end of the hallway the girls were ushered into a room that looked more like a museum than an office.

"Make yourself at home, look around," Dr. Seeting said, motioning around the room. "I'll be right back." He left quickly, passing one of his assistants on the way out.

"Can I get you something to drink? Water, juice?" Dr. Seeting's assistant asked as Jenny looked around for a place to sit. *What luck*, thought Jenny. Alex didn't hear; she had found an in-wall fish tank and was bending over, tapping at the bottom.

"I'll have a glass of water," Jenny squeaked again. The assistant waited a moment for Alex, still tapping and oblivious, and then left. Dr. Seeting returned as quickly as he had left.

"Please, have a seat." Dr. Seeting motioned to some leather chairs surrounding a low wooden table. The water came out of nowhere. Jenny took a quick drink, trying to be discreet. They both sat for a moment as Dr. Seeting waited for Jenny to get comfortable.

"How many years has it been?" asked Dr. Seeting, looking at Jenny as if she were an old friend. Jenny didn't remember the blue eyes that now she imagined were looking right through her. His hair was graying more than the picture in her book showed.

"I was in the fifth grade," said Jenny, finding her voice, "and now I'm a freshman—at Lutrell. Alex and I are roommates." Both of them looked around at Alex who was examining a shelf full of rock samples.

Dr. Seeting turned back to Jenny. "So, you have continued your interest in science, I see." He was relaxing, one arm over the back of the chair, legs crossed.

"Well, yes, and that's why I wanted to talk to you today," Jenny answered. "I've been reading a lot lately. Alex thinks I'm a little crazy, actually," Jenny said nervously. Alex looked up briefly; she had found a jigsaw puzzle of the "miracle of Lavoux Cavern" and was busy putting it together.

"Let's talk, and then I'll show you around," said Dr. Seeting. "Have you been here since the new building was built?"

"No. The last time I was here it was just a big hole in the ground," Jenny laughed. Dr. Seeting shifted as he smiled knowingly. "Actually, it was just a *little* hole in the ground," Jenny continued, laughing again and surprised that Dr. Seeting was so easy to talk to. "I remember you putting that flat board over the hole to keep people from falling in. I hope it worked!" They both laughed as a quick shiver went through Jenny at the thought of accidentally ending up in the cave.

"Yes, it worked. No more sheep ... or kids!" Dr. Seeting smiled. "Although I was lucky for that lamb—no telling where I'd be today if it hadn't shown me the way."

Dr. Seeting sat back, looking closely at Jenny as he thought back to the day of her class visit. "Yes, I remember that field trip. Didn't I have to tie knots in a rope to get you kids out?"

"Yeah, and even then I wasn't sure if we would make it," Jenny said, laughing again.

"Well, there's no knotted rope anymore. We have a funicular now. It takes 50 people at a time down and back in padded-seat comfort." Jenny didn't know what a *funicular* was, but she nodded with an impressed expression, imagining a big train car like those used in subways.

"So, what did you want to see me about today?" Dr. Seeting asked, rubbing his palms together and leaning forward with interest. Jenny pulled out her Earth Science book. Dr. Seeting's eyes opened wide in fake astonishment, "Hey, you have my book! You didn't find a mistake, did you?" Dr. Seeting asked in fun.

"Actually, I think I did," Jenny said, regaining some composure and remembering why she was there. "Maybe not a mistake, but, maybe," she fought for the words, trying to remember how she was going to bring up the subject, "maybe I just need some clarification."

"Maybe you do, maybe it's a mistake," Dr. Seeting was upbeat, giving Alex a "who knows" look. Alex had finished the puzzle and was looking around wondering how Jenny had gotten a glass of water. "Show me what you have concerns about," said Dr. Seeting, emphasizing the word concerns.

"Here," Jenny said, handing the book over to Dr. Seeting. "In the chapter called 'The Nature of Science' on page 446 is a picture of the picture on the cave wall."

"Only a picture because we see a picture," Dr. Seeting interjected cheerfully. "To nature it's just random markings in the dark. At least it was until we lit it up," he said, beaming.

"Well, that's just it," said Jenny tentatively, "don't you think that one alternative, one *scientific* alternative that is not mentioned in your book, is that a person, not nature, made the picture?"

Dr. Seeting thought for a moment. "Yes ... and no."

"Excuse me?"

"Yes, it is an alternative explanation, but, no, it is not a *scientific* alternative. You know the difference, don't you?" Dr. Seeting asked with a serious look.

"Well, I've read what you say in your book, about science relying on evidence that can be empirically observed, and all that," Jenny replied.

"That's right, Jenny," Dr. Seeting said with a solemn look on his face. "You see, people believe things for a host of reasons, some valid, in the sense of being related to reality, and others, let's say, not as valid. What science tries to do is sort out what is fact and what is not. To do that, we, meaning we scientists, can't go on hunches, or intuition. We look for independently verifiable evidence. So you, or anyone else, for that matter, can believe that a person made the apparent painting we see here in the cavern, and I say that's great! Really! But I have to question you if you say you believe that based upon scientific reasoning."

"Why is that, exactly?" Jenny pressed. "Why isn't a painting evidence itself of a painter?[1] Dr. Seeting was impressed by the conciseness of the question, one with which he had wrestled many times.

"Because there's a rul--, a *method* of science!" Dr. Seeting said, seeming to struggle with his words. "Good science requires rigor with respect to the methods of proposing and testing hypotheses."

"I understand that," Jenny started. "And it seems that your method permits only the 'it is not a painting' hypothesis. But what about the "it was painted by a person' hypothesis? Why is that not 'science' also? It seems so intuitively obvious to me."

Dr. Seeting listened patiently as Jenny continued to question.

"Such a hypothesis is even consistent with what your book says, the part about making and testing hypotheses. Isn't an inference of design just as much a scientific hypothesis as the inference of chance?"

"Look, Jenny," Dr. Seeting said slowly, troubled by the word design, "I would also like to believe that a person painted these apparent pictures. Really, I would. It would be easy. It would greatly simplify my research. As I understand it, there's even a legend about 'underground people' or some such thing that is consistent with such a view of 'design,' as you say. But I have been tromping around this cave for almost 15 years, and

---

[1] *See*, End Note [26], page 315.

the only footprints I have ever seen are my own. There is no indication of a fire ever having been built in the cavern, there are no tools in the cavern, there are no remnants of clothing, no animal bones. There is simply no trace, no *fingerprint,* if you will, that a person had ever been in this cavern until after old Lavoux started throwing rocks and dirt into that hole. Not only that," Dr. Seeting was getting somewhat animated, gesturing around him, "there is no evidence to suggest that there were any people in this entire region until relatively recently, much too soon to have made such primitive *paintings,* as you call them. In short, there is simply no scientific evidence for the existence of a person who could have painted pictures on the walls of Lavoux Cavern."

Jenny sat thinking, not sure how to respond. Dr. Seeting looked at her earnestly, then sighed and tried again.

"You remember the stalactites, don't you?" Dr. Seeting asked. "Do you think that a person made the stalactites? Of course not. Such a thought would seem absurd, I'm sure, even to you. But there is really no difference. Nature, functioning slowly over time in the dark and unnoticed can gradually make very ordered structures such as stalactites—and wall markings. The only difference is that we know with a greater certainty how stalactites are made; that is, we have a testable theory, an explanation that can be tested by observation, even repeated in the lab. With the wall markings we may lack a complete theory, but that is the work of scientists like me. The question that concerns scientists is not *whether* the markings were made by natural processes, but *how.*"[2]

Dr. Seeting sat back in his chair. Jenny's mind was spinning. She was thirsty again.

Dr. Seeting continued, "The role of a scientist, in one sense, is to not be fooled by the appearance of design in nature. As my colleague Dr. Hawkins said so eloquently, the study of Earth Science 'is the study of complicated things that give the

---

[2] *See,* End Note [27], page 316.

appearance of having been designed for a purpose.' The cave painting, as you call it, is just one of many, many things that appear to have been designed."

Dr. Hawkins! Jenny thought. He was her Earth Science professor, and she had hoped that he would be amenable to discussing alternative theories. *Were there no scientists that believe that a person painted the cave painting?* Jenny looked closely at Dr. Seeting, sensing that there was some other reason that he was so fervently opposed to even considering a human painter.

"Not only that," Dr. Seeting straightened up at a new thought, "I have supervised extensive experiments to determine the age of the rock walls of Lavoux Cavern, as well as the pigment from the markings, and the ages are consistently similar and much older than accepted history would indicate that humans were in this region."

"But wouldn't it seem OK to at least offer as a *theory* that a person made the paintings?" Jenny asked weakly.

"Why?" asked Dr. Seeting sharply, almost cutting her off. "Why? To make some people feel better about themselves? To authenticate their psychological wish projections? Or maybe to personalize the unknown forces of nature in a misguided attempt to give some certainty in an uncertain world?[3] To do so would be tantamount to belief in fairies!"[4] Jenny thought Dr. Seeting was mocking someone, but she wasn't sure who it might be. "Or, maybe," Dr. Seeting finished curtly, "so we can align ourselves with the *legend*."

At the word "legend," Dr. Seeting stopped smiling momentarily, as if bothered by the thought. Jenny noted the second reference to the legend. Regaining his composure and his momentum, Dr. Seeting continued. "Look, Jenny, and Alex," he said, noticing that Alex had become bored with other distractions and seemed to be taking interest in the conversation at this point, "we can discuss this forever, but basically we are

---

[3] *See*, End Note [28], page 317.

[4] *See*, End Note [29], page 319.

discussing the difference between science on the one hand, and other very valid disciplines on the other hand. These other disciplines, such as history, which the legend might inform, have their place, but they are separate and distinct from science."

Dr. Seeting looked at his watch. "Why don't we take a look around and I'll show you some of my latest research, and don't forget, I want to tell you how you helped me in my work, Jenny."

Dr. Seeting got up and motioned toward the door. Jenny took a last drink of water before heading out of the office. Out in the hallway, Dr. Seeting led the way, walking briskly and asking over his shoulder, "Do you know why this is called the Neihil Research and Science Center?"

The girls looked at each other. Alex gave Jenny a "*don't you*?" look and then elbowed her as she whispered under her breath, "How can you disagree with this guy?"

Without waiting for an answer to his question, Dr. Seeting explained, "It was named after a late colleague of mine, Dr. Seth Neihil, probably the world's greatest naturalist. He is credited with publishing the most comprehensive theory on the origin of specialties, particularly the apparent painting of Lavoux Cavern."

Dr. Seeting swiped his badge through a card reader and opened a door that led the girls to a line of people pointed toward Gate 1. He took them to the front of the line, nodded to an eager blue shirt, and in a few minutes the three were sitting in the front seat of a funicular, Jenny supposed, headed to the "Great Room," according to the recorded information piped into speakers above them.

They rumbled slowly down a track through a steep, lighted tunnel; Jenny didn't remember having to walk or climb down as far as they seemed to be going now. In about a minute of listening to both a recorded description of the Great Room as well as Dr. Seeting's unceasing explanations, the three were out and standing in the cavern that Jenny had first visited during Ms. Hart's class field trip.

Once again Jenny found herself gazing all about her as if suddenly awaking from a long sleep in a strange place. She remembered darkness punctuated by little orbs of light from the kids' 'miners' helmets." Now the cavern was lit up like a great ballroom, the floor smooth and dry, colored lights highlighting particularly interesting rock formations.

"Over here," Dr. Seeting gestured. The sounds of footsteps softly echoed about. Although many people were in the cavern at the time, it was remarkably quiet, as if everyone were instinctively speaking in undertones. Jenny looked at the wall in the direction they were heading. There, behind a large piece of Plexiglas, was the cave painting.

"Looks a little different now, huh?" grinned Dr. Seeting. Jenny quickened her pace, coming closer and standing in awestruck wonder. It was more beautiful than she had remembered.

After a few minutes Dr. Seeting said, "Come over here, I want to show you something." He walked over to a series of exhibits that made a sweeping arc, following the contour of the cavern walls. Before each exhibit stood quiet visitors listening to headsets explain to them the secrets of nature's handiwork.

"You get a personalized tour," said Dr. Seeting, stopping in front of a life-sized portrait of someone Dr. Seeting described as "the eminent natural paleogeologist, Seth Neihil." Jenny recognized the picture from her Earth Science book. She always thought Dr. Neihil appeared unhappy, with his long gray beard and bushy eyebrows pushed forward by a jutting forehead.

"This man came to believe, as all scientists do, that everything around us, including the remarkable figures here in Lavoux Cavern, could be explained in purely natural terms. What made him great is that he was the first to publish a workable theory to explain *how* the natural markings we see there," Dr. Seeting pointed back to the Plexiglas, "might have originated. I'm sure you will find his theory very interesting," he continued, looking at Jenny. "In fact, you played a role in the development of the theory!" Jenny, recalling her reading about natural selective inclusion, tried to make a connection that might indicate her role. "Here," Dr. Seeting said, pointing past some kids with headphones.

A field trip? Jenny thought.

"This diagram does a good job of summarizing Dr. Neihil's theory. We all, I mean all scientists, believe that the markings that make up the apparent paintings, that is, the specialties we see here, are a product of progression, which is simply shorthand for 'change over time.' It is clear that all natural formations are made little by little over long periods of time. But what we lacked was a workable *mechanism* to explain how progression might work in the case of creating new specialties like the wall figures here in Lavoux Cavern. That's where Dr. Neihil's theory of progression comes in. It is, as Dr. Hawkins has noted, 'a bewilderingly simple idea.' It involves the preservation of small, random permutations over time, by a mechanism Neihil referred to as 'natural selective inclusion.'"

Jenny was familiar with the theory; she had read it over and over in her science book. She could imagine it working for something like stalactites, but she could not understand how it could possibly make a work of art.

Dr. Seeting motioned to his right, "Look at this series of diagrams. On the first is a random smudge, maybe just a dot. How did the dot get there? By magic?" Dr. Seeting was being sarcastic. "No, of course not. This dot—it could have been a line or almost any other mark—was put here by nature randomly by some event. If this dot was preserved, another dot or line randomly placed by another event could be preserved nearby, effectively modifying the first dot or line. This process of events with modification could continue over many generations of events until the collection of randomly placed dots and lines makes a recognizable figure—a picture, as you would say."

"So, you really think that the cave painting here in Lavoux Cavern, *this* cave painting," Jenny pointed over to the cavern wall, deliberately using the term painting, "is a result of random wall markings simply lining up in the right order?"

Jenny tried to make the proposition sound absurd, but she knew that Dr. Seeting would not be moved.

"That's right, Jenny." Dr. Seeting said calmly. He wasn't moved, and he seemed to be surprised that Jenny could not grasp such a simple idea. He wondered what motivation she might have for not readily embracing the only theory accepted by scientists such as himself.[5] Although there was no hint that she was superstitious or prone to believe myths, he could only surmise that she nevertheless lacked the ability to distinguish wishful thinking from scientific inquiry. As visitors shuffled around them, he looked closely at Jenny and continued, "Over long periods of time, of course. You see, given enough time, and the selective preservation of certain marks and colors, it is not hard to suppose how such figures could progress from one specialty to another, from a simple dot or line to something we perceive as a recognizable object."

Jenny looked at Dr. Seeting with equal amazement at his unshakeable certainty. The confidence in his voice, and his unwavering conviction that progression—chance and natural law alone—was solely responsible for the picture she could see behind him, caused her momentarily to doubt her own thoughts. She sensed his genuine concern that she come to understand, and she found herself drawn by his confident assurance. His eyes were locked to hers as she glanced over his shoulder at the eyes of the hunter bearing down upon his prey. Immediately she stepped back, looked again at Dr. Seeting and said, "So really it's just chance, pure and simple, that made this work of art?" Again, Jenny tried to make the proposition sound preposterous.

"Chance plays a part, Jenny," Dr. Seeting said, seemingly tireless in his refutation of what he referred to as "that old canard." "We must have some process for fresh variation of lines and spots of color. Marks must be put on the wall somewhere in an ever-changing variety. But the *randomness* of mark variation must be contrasted with the non-randomness of selective

---

5 *See*, End Note [30], page 321.

inclusion, the other side of the process. It is not *necessary* that variation be random in order for natural selective inclusion to work. Selection can still do its work whether variation is directed or not. Emphasizing that variation *can* be random is our way of calling attention to the crucial fact that, by contrast, selective inclusion is sublimely and quintessentially *non*-random."[6]

Jenny listened closely. This was the first time Dr. Seeting had said something that, in her opinion, got to the heart of the issue. How can a random process produce such order, such *meaningful*, non-repetitive order? He explained it to her again, and she paid attention to the emphasis on *necessary* and *can*.

"So, are you saying that the entire process is *not* random, but is actually *directed*?" Jenny asked.

Dr. Seeting shifted slightly, looked annoyed for the first time, and started to say something before Jenny first said, "I mean, a simple two-step process in which one of the steps relies on random chance, luck if you ask me, renders the whole process chance-dependent, right?" Jenny was pretty sure she was on the right track, and was feeling confident as she continued. "If I proposed to you a two-step process in which the second step was a non-random automatic payment of cash, but only after you first successfully shot an arrow into the center of a distant moving target while blindfolded, would you pay me for a try? Would you ever get rich?"

Dr. Seeting was impressed with Jenny's spontaneity and the acuteness of her remark. He smiled at her, still unmoved. "Jenny, maybe it's just semantics. But it's ironic that this emphasis on the contrast between variation and non-randomness of selective inclusion has led people to think that the whole theory is a theory of chance."

Ironic? Jenny thought. *It's not ironic; it's common sense.* Again Jenny was surprised at Dr. Seeting's unshakeable conviction, but she decided to try one more tack. "OK, but just

---

6 *See*, End Note [31], page 322.

how did the spots, the first specialties, get there in the first place?"

Dr. Seeting smiled broadly, sensing that he was finally making progress. "Well, they didn't just spontaneously generate—as some kind of idiopathic cave phenomena. It simply took an event as Neihil's theory proposes. In fact, there are many theories involving various events; and, believe it or not, that's where you helped," Dr. Seeting said, turning to Jenny. "Do you remember your report in the fifth grade?"

Do I remember? thought Jenny. *It was only the worst day of my life*. "Sure," said Jenny.

"You proposed a rather unique event for getting the first dot of marking on the wall—water!"

Jenny remembered, all right. But she had no intention of the idea being taken seriously. In her estimation the finch beaks were more likely to explain the markings on the wall.

"Suppose water flooded this cavern." Dr. Seeting said. "It's certainly possible; much of this cave is an underground reservoir. We can go see the Great Lake later if you like. Water could have entered this portion of the cavern many times over the past million years."

Jenny listened intently. Alex, who had already made the round of exhibits and had returned, was looking around impatiently for the nearest exit. The thought of a sudden flood made her nervous.

"I never forgot that concept, *your* concept, of water," Dr. Seeting continued. "Really, it was very clever. I proposed several scenarios using water myself, but it took the genius of Dr. Neihil to put the pieces together."

Dr. Seeting pointed to the first diagram again. "Suppose this cavern became flooded with water. And suppose the water carried in a dark colored substance, maybe berries that had burst in the water, maybe dark clay deposits, maybe even tree bark!" he said brightly, turning to Jenny. "Now suppose, as well, that the walls of this cavern had an affinity for this substance, maybe a chemical attraction, maybe a metallurgical affinity, you know, like magnetism, such that the dark colored substance stuck to the wall. And suppose that, after the water receded, the color on the wall remained in certain places, and *voila!* You have the start of a cave painting." Dr. Seeting looked pleased.

A lot of supposing is going on, Jenny thought. She was looking ahead, following each exhibit. One dot, another, a line, a blotch, and on and on.

"The real key with respect to wall figures seems to be that over time, with many floods, variations in the markings occur. That is, after each event, each 'generation,' so to speak, there is a slightly different combination of lines and colors on the wall. These variations are the *permutations* I mentioned earlier, certain of which are only selectively included to be preserved as permanent marks, thus the term, *selective inclusion*, such that the marks remain only in selected patterns. Just as a human painter might select certain colors and lines in a painting, nature *naturally* selects for inclusion small, incremental random permutations to produce a figure that looks like a painting. The exact means for accomplishing this is still being investigated. Some lab tests have successfully demonstrated simple patterns using magnetism, for example."

Jenny stood leaning slightly forward, examining a large photograph showing a team of white-coated scientists posing beside a large stone having on it a series of fuzzy parallel lines. *The Stanley-Harold Experiment,* as it was titled, purportedly demonstrated how the "building blocks" of cave specialties could have originated naturally.[7]

"Over thousands, even millions of years," Dr. Seeting's arms were spread, "the selected marks line up, lines are formed, and soon you have random lines and marks that look to us like a picture. It's really very simple. In fact, what surprises me is that we don't see more 'pictures' everywhere."

"Shouldn't we see some partial pictures, then?" Jenny asked, remembering the objections Neihil himself raised in his

---

[7] *See*, End Note [32], page 324.

book *On the Origin of Specialties*. "I mean, we should see lots of lines and blotches around, shouldn't we?"

Dr. Seeting was again impressed by her quickness, answering, "Yes, but that's the amazing thing, we *do* see intermediate forms, partial 'pictures,' if you will. Look over here."

Jenny peered through a small Plexiglas opening in one of the exhibits to a lighted portion of the cave wall behind. On the wall was mounted a stainless steel arrow pointing to a cluster of small blotches that appeared to be deposits of charcoal. Dr. Seeting turned a handle that rotated a clear overlay in front of the blotches. On the overlay were lines drawn to show how, like a constellation of stars, the blotches could make a figure that resembled an upside down spider. Further rotations of the clear overlay showed how additional dots and lines could make the spider appear to be descending from a web, or even being eaten by another insect. A final turn of the overlay showed the spider on a web suspended between two tree branches.

"You see," continued Dr. Seeting, "these marks are all around us; we just don't see them because our brains simply don't see them as 'pictures.' But that," he pointed into the small Plexiglas, "is the start of one of your paintings. Come back in a million years and it may look like a flower, or a computer, or something not even invented yet. And your 'hunter' may be wearing a hat!" Dr. Seeting gestured enthusiastically, pointing back to the large Plexiglas. "And speaking of hats, take a look at this," Dr. Seeting said as he walked quickly back to the big Plexiglas.

"Look here," Dr. Seeting said, pointing to the ankles of the hunter. "I think you would say these are the ankles, right?"

Jenny nodded.

"Well, what, I ask you, are these?" Dr. Seeting was pointing to a short line crossing perpendicular to the line of each leg at each ankle.

Why identical lines on both ankles? Jenny thought.

Dr. Seeting stood grinning, as if the short lines were proof positive that the figure could not have been designed. Thinking Jenny might be missing the point, Dr. Seeting continued, "What's with the extra, useless body parts?[8] I don't know about you, Jenny, but I don't have any parts bulging out at my ankles. Don't you think that if a person painted this, he or she would have at least gotten the body parts right? And look at these funny-looking 'bulls.' It looks like they have wreaths around their necks. Is that the work of an artist, or a lowly greasepaint monkey? It appears to me, Jenny, that if these figures are the work of an artist, this 'artist' is a curiously inept cobbler of species."[9]

Dr. Seeting finally stopped. Jenny felt overwhelmed. She wanted to leave and wished Dr. Seeting had another appointment or some other important business to attend to. She couldn't help thinking about the organization of the hunting scene, the colors used, the perspective, the movement. It just seemed impossible to her that such a composition of matter could have come about by chance, no matter how long it had to do so.

"Yeah," Jenny said slowly, trying to think of something as they walked back to the exhibits, "I see what you are saying, but still … I can't see … what are the odds of this happening like you say? What are the *chances*?" Jenny found her voice again.

Alex, standing behind Dr. Seeting, rolled her eyes and motioned toward the exit, mouthing, "I'm ready to go."

"Chance is a funny thing," Dr. Seeting answered. "The occurrence of almost anything unusual happening is improbable in hindsight. Suppose I had a deck of cards and I were to lay out 20 cards in succession, laying them face up in a row on a countertop here." Dr. Seeting pointed to a nearby counter, motioning to an imaginary line of cards. "What are the odds that I could pull these 20 cards out in that order?" he asked, waving

---

[8] *See*, End Note [33], page 325.

[9] *See*, End Note [34], page 327.

his hand over the imaginary cards. "The odds are high; you might even say they're astronomical. But I just did it. You saw it. And it took only a few seconds. I could do the same thing with all 52 cards, and if you were to calculate the odds of that particular sequence you might be tempted to say it was impossible. But you couldn't say it's impossible; it just happened. So, to talk about the odds, or even to try to calculate the odds of a present reality occurring is somewhat irrelevant—*it happened*, regardless of the odds.[10] The same is true with the cave markings. Imagine I had not one deck of cards, but millions of decks, and millions of years. Given so much time, the 'impossible' becomes possible, the possible probable, and the probable virtually certain. One has only to wait: time itself performs the miracles."[11]

"Miracles?" Jenny asked.

"It's just a figure of speech, Jenny," Dr. Seeting said, showing some signs of exasperation. "What other options are there, really? Should we just wink and say it painted itself? Should we treat it as an acheropite image of sorts—and turn this research center into some kind of *sancta sanctorum* for deluded pilgrims?"[12]

Jenny felt besieged as she tried to take in all Dr. Seeting was saying. It became clear that whatever argument she could muster would be met with a seemingly indisputable counter argument. She was tired, confused, and decided it was time to leave. She thanked Dr. Seeting and pretended to be surprised at the time, saying she and Alex had to get back to Lutrell for a study session.

"Remember, Alex?" Jenny asked.

"That's right, Jen, we better go." Alex feigned surprise as well. "What's the fastest way out?"

"This way," Dr. Seeting answered. "Just follow me."

---

10 *See*, End Note [35], page 329.

11 *See*, End Note [36], page 331.

12 *See*, End Note [37], page 338.

The three started toward a gate marked "Return," past more blue shirts, and got back on the funicular as it channeled more information about items in the gift shop.

"Believe me," said Dr. Seeting as he made a quick check of the seat padding and sat down, "I can understand your desire to simply say a person painted the marks on the wall down there. But as a scientist I must assume that a person never entered this cavern to paint such a picture. I urge you to consider the evidence. Because even though you are a very bright student, you still have a lot to learn. Consider Neihil's theory and you will find that the concept of progression has great explanatory power. For example, as we discussed, it can explain stalactites—over time, each miniscule amount of mineral deposited is added, by natural selective inclusion, to the whole, eventually forming a stalactite. No purpose, no planning, no *person* involved."

At the top the three got out into the artificially cool air of the lobby. Announcements, popcorn, and blue shirts filled the air once again. Alex walked a step ahead, hoping Jenny would follow without being distracted.

"Dr. Seeting," Jenny turned to speak as Alex again rolled her eyes and threw her head back in exaggerated hopelessness, "What if ... what *if* a person did paint the picture down there. Wouldn't you, wouldn't *science*, want to know that?"

Jenny and Dr. Seeting stood facing each other as Alex paced back and forth behind Dr. Seeting, who looked tired as he stood with his arms folded in front of him.

"Science doesn't, and I don't, for that matter, *want* to know anything other than the facts. That's what science is all about—the facts and generalizations drawn from those facts."

"But what if," Jenny tried again, "the *truth* was that a person painted that picture? If it were the truth, or even if it was possibly the truth, wouldn't that make a difference? Shouldn't truth be the aspiration of reason?"

Jenny's face was turning red. Alex had seen this before and decided to study the posters in the gift shop window.

"Are you, Dr. Seeting," Jenny's voice was rising as she looked directly at his eyes, "ultimately interested in knowing the truth, or suppressing the truth?"

"The *truth*!" Dr. Seeting blurted. "What is truth? The *truth* is that I, and you, for that matter," he leaned in with emphasis, practically touching Jenny with a pointed finger, "can know for sure only what we can see, hear, smell, taste, or touch. No more, no less.[13] That, unfortunately for people like you, is the constraint honest people must accept if we are going to have any progress in this world, much less arrive at *truth*!"

Dr. Seeting realized some people were looking and managed to step back, pretending to straighten his collar. He surprised himself with his outburst, and he knew he would regret the personal attack and the implication of Jenny's dishonesty. Intellectual honesty was actually something he couldn't help but admire in Jenny. Alex, still examining the posters, braced for the inevitable.

"All I know," started Jenny, her face becoming redder, "is that *truth*," she almost shouted the word, "is not changed by what you or I think about it. Either a person painted the picture in this cavern or he didn't, but the fact of its origin is not changed by your theory or mine … or Dr. Neihil's! Something doesn't make sense here, and I aim to find out! I intend to learn the TRUTH!"

The last line came out a little too loud, turning heads and even prompting two blue shirts to walk toward the two. Jenny turned and stormed toward the automatic doors so fast that she had to stop abruptly until they opened far enough to squeeze through.

Dr. Seeting stood staring after her, blue shirts at his side.

Alex quickly skipped after Jenny, smiling in delight.

---

13 *See*, End Note [38], page 339.

Chapter 9

Alex wanted to burst out laughing but didn't say a word as the car approached the rising gate at the exit. As they went under a "Please Come Again" sign, Alex glanced sideways at Jenny, who was turned facing out, her forehead practically up against the window.

"I guess you told him," Alex said, suppressing a laugh. "You were in fine form this time, Jen," she said shaking her head and laughing. "Wow, talk about a wind bag!" Alex laughed again.

"We're missing something," Jenny said quietly, still staring out the window.

"We're missing English, smarty pants," Alex said. "Look at the time!"

"There's a piece to the puzzle that's missing," Jenny said, slowly turning to Alex.

"I'm not missing any puzzle pieces; you saw me finish that puzzle of a picture that wasn't painted by anyone."

Alex turned to grin at Jenny. Jenny was staring past Alex as if in a trance.

"Besides, I don't really care about that painting thing," Alex said. "That's your problem."

"There's something missing ..." Jenny repeated quietly to herself. "Like some unwritten rule or something."

"Come on Jen, drop it," Alex urged. "It's just a bunch of colors on a wall. Who cares? Besides, you had your chance to tell Dr. Seething what you think."

"Seeting, See-TING," Jenny said.

"Whatever. He looked like he was seething to me."

"Alex, did you notice how he kept mentioning a legend?" Jenny asked.

"Oh, please, Jenny, *drop it*."

"Did you?"

"Ugghhhh, what is this about, Jenny? Is this about revenge of the fifth grader, or something?"

"It's about truth, Alex," Jenny replied sarcastically. "Something you probably don't ever think about unless you're playing 'Truth or Dare.'"

"Who cares about truth? I mean, really, Jenny, who cares if a little man made your precious little drawing or not?" Alex laughed at the thought.

"Drop me by my dad's house," Jenny said, turning quickly to face Alex.

"What? That's ten miles out of my way, and *we* are supposed to be in English class in exactly 17 minutes."

"It's not that far," Jenny pleaded. "Believe me, it's important."

"Oh, I'm sure," Alex replied sarcastically, turning on a side road that went by Jenny's father's house. "I'm sure truth is hanging in the balance."

The car had hardly stopped before Jenny was out on the driveway. "Thanks millions, Al," Jenny waved as she shut the door. "And take good notes!" she yelled through the closed door as Alex sped away.

☙ ❧

"Jenny?" Aunt Sema's voice called through the screen door.

"Yes, Aunt Sema, it's me. Sorry to show up unexpectedly, but it's important."

"Come in, girl," Aunt Sema hugged Jenny. "You know you can come here any time. Of course, you should be somewhere else, and I'm sure it can only be trouble that you're here, but I won't ask." Aunt Sema stood blocking the narrow entryway, waiting for Jenny to tell her why she was there.

"Where's Dad?" Jenny asked, looking past Aunt Sema's shoulder.

"He's where he always is," Aunt Sema answered, wondering why Jenny would ask. "Asleep."

"Asleep?"

"Like a baby."

"Not for long."

Jenny squeezed past Aunt Sema into the room where her father lay.

"Dad," she shook her father gently. "Dad, wake up."

"Jenny? I hope this is a dream, because I sure don't want to get up right now." Jenny's father turned over slowly, first opening one eye and then the other.

"Dad," Jenny said impatiently, "come on, wake up. I have to tell you something."

"Something that can't wait?" Her father lifted his head, looking around to get oriented. "Are you getting married?"

Aunt Sema squelched a laugh. Jenny pushed a pillow under her father's head.

"No, Dad."

"Then what is it?" asked her father, straightening up slightly. "Sema, may I please have a soda?"

"Get comfortable, Dad." Jenny pushed another pillow behind his back. "How's that?"

"It would be nice to be standing, but this is alright for now," her father said. "Now, what is it that's so important as to wake the dead?"

"Dad, listen. Do you remember, ..." Jenny was interrupted by Aunt Sema returning with a glass of fizzing soda and some crackers.

"You want some, girl?" Aunt Sema asked with a frown.

"Yeah, I'll have some, *girl*," Jenny said playfully. Aunt Sema gave her a sideways look as she left the room.

"Dad, do you remember back in the fifth grade when you told me about some ancient people and about a legend of the underground people?" Jenny asked expectantly.

"Yes, they are referred to as the Ancient People, but ..."

"Tell me more." Jenny was beside him on the bed now.

"More?"

"Yes, what do you know about the Ancient People?"

"We used to live around some," Aunt Sema said as she came back in the room with more fizzing soda.

"*You* know about the Ancient People, also?" Jenny asked excitedly.

"Well, I don't know how much I know about them, but I know *of* them," Aunt Sema said. Jenny's father nodded as he listened.

"I can't say as I know too much about them either. I just know what my father told me, and I knew one or two that lived around us, and I had heard about a legend," her father said, thinking hard as he tried to remember.

"A legend? You know about a *legend*?" Jenny asked in astonishment.

"Hold on, Jenny," her father held up his hand. "I don't really know that much. I've probably already said more than I know."

"How can I find out more? Particularly about the legend. I *really* need to find out about the legend."

Jenny noticed Aunt Sema suspiciously wide-eyed, with raised eyebrows. She looked back at her father.

"Do you remember where we lived before we moved here?" asked Jenny's father.

"Yeah, pretty much," Jenny pretended as she thought hard. Her memories were few and mostly involved elusive images of her mother.

"Do you remember Mr. Hunn, the man who lived across the street with about ten kids?"

Jenny again thought hard; she vaguely remembered the name. "You used to throw rocks at his barn roof, until one day he ran you off with his dogs."

"Oh, you mean Attila!" Jenny suddenly remembered, a shiver running up her back at the memory of the barking dogs. "That was his name?" Jenny asked, amazed. "I mean, *Hunn* was his real name?"

"He was a nice man, and one of the Ancient People," her father continued, taking a long sip of the soda.

"You're kidding," Jenny's mouth was open in disbelief. "Well, if he was ancient then, he must be petrified by now." Jenny laughed until she realized that she was laughing alone. "Do you think he's still there?" she asked in a suddenly serious tone.

"Maybe. There were quite a few of them over the mountains," said her father. Aunt Sema agreed with a nod. "They kept to themselves, mostly. Nice people, however; good to have as neighbors."

"I want to visit them," Jenny said. Aunt Sema let out a loud sigh, as if to say, "What now?"

"Really, I want to visit the Hunns as soon as possible," Jenny pressed.

"I don't even know if he lives there anymore. But some of his family is probably still there. Sema, where're my maps?" her father asked, reaching to a side table among a stack of papers and magazines.

"Why don't you just drop in?" Aunt Sema asked sarcastically. "We haven't seen that man in over ten years! He wouldn't know Jenny from me, unless she was to throw rocks at his barn roof."

"Sema ..."

"Here's one of them," Jenny's father said as he unfolded a large, well-worn map from a stack Aunt Sema tossed into his lap as she stood over the bed, clearly annoyed. "This one will get you over the mountains." After a few more minutes he found another, smaller map and unfolded it. "Here's our town, and, look, where we lived is still marked." Jenny's father handed her the two maps. "It's a long way to go alone. But I suppose if you want to go I can't stop you."

Jenny grabbed the maps and jumped up. "Aunt Sema, will you go with me?" Jenny asked, hugging her across the shoulders.

Aunt Sema twisted out of the embrace and started out of the room. "No, girl. I don't even want to know what you have in your mind right now. Sometimes you worry me, girl. Anyway, I'm not going anywhere; I got work to do right here. Who do you think is going to take care of your father?"

Jenny looked at her father, wishing she could take him with her. His eyes met hers, his mind also momentarily considering the possibility of accompanying her, before reluctantly succumbing to reality.

"You can find it, if you look," he said. "You remember what our house looked like?"

"Sort of," Jenny replied, knowing that she had only fuzzy memories, at best.

"Here," her father said, pulling out a box of pictures from a nearby end table. After rummaging through for a few minutes, he pulled out a photo. "Look," he said, "here it is. This was right after we repainted and had the new chimney put in. Best chimney in town!" Her father turned the photo over. "And, in case you can't find it, here is our address. Just ask anyone."

"Thanks, Dad," Jenny kissed her father. "Can I borrow the car to get back to school?"

"Borrow the car? Where's your car?" Aunt Sema hurried back in with a broom in one hand and a trash bag in the other. "You'd better not have it for long," Aunt Sema said with a scowl as she handed Jenny the trash bag.

"Don't worry," Jenny said as she pushed the picture and map into her purse. "I'll have to bring it back to pick you up, remember? We're going over the mountains!"

Aunt Sema swept at Jenny's feet as she ran out the door. "Have your bag packed, hear?" Jenny said, looking back. A loud crash sounded as she tossed the trash bag into a metal can and ran to the car.

Aunt Sema watched her drive away, shut the door, shook her head, and smiled.

Chapter 10

The large oak conference table looked as heavy as the faces of the men and women sitting around it. At one end sat Dr. Silva, now the Department Head of the College of Natural and Applied Sciences. Dr. Silva, a nationally known paleogeologist, was instrumental in transforming Lutrell University into one of the nation's premier research universities. In particular, under his guidance Lutrell University led the nation in Earth Science education and research. A successful author of popular books and essays on the topic of progression, Dr. Silva also sat on the board of the National Institute of Sciences, the NIS, a prestigious group of the nation's top scientists. He seldom called department-wide meetings, preferring instead to delegate and manage in a "hands off" style. Something must be wrong, everyone thought. Budget cuts were on most people's minds.

To Dr. Silva's right was Dr. Seeting, sitting comfortably, elbows on the table, reviewing a stack of papers in front of him. Dr. Seeting was Dr. Silva's colleague, both at Lutrell as a professor and at the National Institute of Sciences, where Dr. Seeting had recently been elected president for his second term. The two men were longtime friends, and both were influential in and out of the academic setting.

Around the rest of the table sat professors, assistant professors, lecturers, and other department heads. An overhead projector stood ready at the head of the table, even though Dr. Silva rarely used it. For most this was their lunch period and many waited impatiently for Dr. Silva to start the meeting. The older professors had brought work to do; the younger ones wished they had.

"OK," Dr. Silva said, standing and looking at his wrist, although there was no watch there, "this shouldn't take long. Could we all sit down and we'll get started."

Conversations around the room slowly subsided as people found their seats. Someone got up to shut the conference room door. The room became quiet except for the sound of papers shuffling and muffled voices from students playing in a courtyard two stories below.

"We are well into the academic year here at Lutrell, and I want to congratulate each of you on a job well done." Dr. Silva paused, and then added, "So far." There were a few forced chuckles. Dr. Silva smiled and continued, "Dr. Seeting has notified me, however, of a potential issue we should all be prepared to deal with." He looked around as if he had forgotten what the issue was. "As you know, each year we have tried to integrate progression into more and more of our science curriculum. And, each year," he sounded as if it had become monotonous, "we have some students who want to challenge progression, particularly Dr. Neihil's theory."

"Is this about the picture in Lavoux Cavern?" a voice came from the back.

"Yes," said Dr. Silva, somewhat irritated for the interruption and the characterization of the wall figures.

Everyone groaned. It seemed that each year the professors had to deal with a few students who insisted that the wall markings looked like a painting to them. These students tended to take up valuable class time with various theories to explain who the painter was or might have been. Invariably it seemed that the discussion devolved into an argument about the validity of The Legend.

Dr. Silva looked at his wrist again, this time rubbing it as if that might make the time appear. "This time it's different. I'll let Dr. Seeting explain."

Dr. Seeting stood slowly, papers in hand, looking over his reading glasses at the group. He looked up at a large white screen as he turned on the overhead projector. Without putting any transparency on the projector he began talking.

"As you know, Lutrell University is considered one of the best universities in the land, primarily due to our proximity to Lavoux Cavern. I don't have to remind you that Lavoux Cavern has provided almost every one of us with unprecedented opportunity for research and publishing." Dr. Seeting glanced around the room before continuing. "The prestige we all enjoy is a result of strictly adhering to The Rule—the *rules*—of scientific principles and avoiding the urge to make simple cause and effect assumptions."

The mention of The Rule didn't go unnoticed with a few. Dr. Seeting tried to cover, not wanting to make that the issue for this meeting. Dr. Silva started to interject something about the scientific method when he was interrupted.

"We know all this; get to the point," another voice came from the table.

"It's different this year," Dr. Seeting said quickly, with a hard look in the direction of the voice. "I've had the opportunity to discuss the origin of the Lavoux Cavern wall markings with one of our freshman students. She's smart, she's informed, and she is asking all the right questions." He paused for a moment, and then, as if remembering something important, continued. "That's good, of course; we want students like that. We just need to be uniform in our response to such questions. She is in Dr. Hawkins' first year science course. Dr. Hawkins and I have discussed the situation in some detail. I would like to go over some general comments, however, by way of reminder for the rest of us."

Dr. Seeting put a transparency on the overhead projector. Most in the room recognized the transparency from previous years.

"Here are some suggested ways to handle questions as they come up. First," Dr. Seeting began, to the sounds of papers and people being repositioned, "we must ensure that our students know the difference between myth and science. Students must be assured that there is nothing wrong with believing myths. However, we must educate students as to the critical difference between myths and science, including the ways of knowing presented by the two, and the questions they answer. We must stress that myths can speak to purpose and values, both of which are very important, of course, and these are areas that are not for science to inform. Science simply provides the only reliable way of knowing about the real world."[1]

Dr. Seeting paused and looked about the room. He noticed that most of the people around the table either were not listening or were writing far more than he was saying.

"Uh-hum," he cleared his throat loudly. Some of the younger professors looked up and moved some papers around. "Again, we need to assure our students that myths have their place, and it is OK to believe myths and legends. In fact, many prominent scientists believe myths.[2] But when it comes to explaining the origin of natural phenomena, these scientists rely on empirical data and not on personal beliefs."

Dr. Seeting waited on the chance that someone may ask a question. Most in the room looked bored as they politely listened. Several of the people at the table wished they were in the courtyard.

"Second," he continued loudly, moving the transparency up, "we need to be able to quickly deal with this fact/theory question. Basically, keep it simple. Theory does not mean 'guess,' and progression has been observed so many times that there is no longer a compelling reason to question it. In this sense, it is a fact.[3] Stress that scientists no longer question whether progression occurred, but only how it occurred."

---

[1] *See*, End Note [39], page 341.

[2] *See*, End Note [40], page 342.

[3] *See*, End Note [41], page 344.

"Third," Dr. Seeting looked up but didn't bother waiting for eye contact, "we need to stop the rumor that many scientists reject progression as an explanation for all natural formations. This is simply not true. Be aware of quotes out of context, and stress, if necessary, that the quotes come from scientists engaging in the healthy debate about *how* progression occurs, not *if* it occurred."

"Fourth, and finally," Dr. Seeting said, forgetting to push the transparency up, so that number four could not be seen, "we need to be prepared to deal with the question of 'gaps' in the record of cave wall figures—the supposed absence of missing links between specialties. I had to deal with this misconception recently," Dr. Seeting said, perking up with a fresh thought. "Basically, while it's no longer a trade secret that we don't see as many intermediate forms as we might expect, we do see some transitional marks, such as those demonstrated in Lavoux Cavern.[4] Students need to understand that many changes may happen too quickly to leave transitional marks."

Dr. Seeting paused at this point, fully expecting a comment from Hawkins to initiate an unwanted diversion into a side conversation. Hawkins and Silva had a long-running public disagreement on just how gradual progression must have occurred. While Hawkins was a staunch proponent of strict Neihilian gradualism, Silva had published his own theory—interrupted stasis.[5] More consistent with observed wall markings, Silva theorized relatively long periods of stasis interrupted by relatively short periods of rapid change. Hawkins let the statement pass.

Never missing a chance to push Lavoux Cavern, Dr. Seeting continued, still eyeing Hawkins and Silva, "And, of course, many marks don't survive. We are quite fortunate that the conditions in Lavoux Cavern appear to have been perfect for preserving a large number of marks from extinction."

---

[4] *See*, End Note [42], page 346.

[5] *See*, End Note [43], page 348.

The noise in the courtyard became noticeably louder, signifying a class change.

"Are there any questions?" Dr. Seeting asked as Dr. Silva stood up. The room was silent except for the shuffle of papers being gathered up, binders shutting, pens clicking.

"It's very important for the education of our students that we stand unified on this topic and be consistent in our response to questions on this subject. By the way, remember that I have copies of this little booklet." He held up a brochure and spoke louder as the room became restless. "Put out by the NIS, it's called *Teaching About Progression and the Nature of Science*. You may find it helpful in dealing with common misunderstandings."

The sound of chairs rolling back and the stacking of papers indicated the will of most. Dr. Hawkins stood with Dr. Seeting and Dr. Silva at the front as the others left the room. While they were standing there, James Sinclair, a graduate student teaching assistant, approached.

"Dr. Seeting, may I ask you a question?" Sinclair asked.

Dr. Seeting looked around, "Sure. Here?"

"This is fine. Maybe the others want to hear as well," answered Sinclair, glad for the face time this opportunity provided.

All three looked up at a big clock over the door and then agreed to listen.

Sinclair pushed his glasses up on the bridge of his nose and then started earnestly, "I've been thinking, wouldn't it be easier if we just let the students, I mean, only the students who wanted to, um ... if we just let them explore the idea, so to speak, that a person made the natural markings in Lavoux Cavern?"

The three men stood motionless. "Are you serious?" asked Dr. Seeting flatly with an expression of disbelief. Sinclair fidgeted, shifting some papers from one arm to the other.

"Well, it seems to me that if we would just let them have some fun, so to speak ... because, their ideas can't be supported.

Let them present whatever evidence, so to speak, they have, let them get it out of their system. Surely their ideas can be countered easily with scientific reasoning. By letting them try to support their ideas, we can hang them with their own rope, so to speak." He tried to smile as he pushed his glasses up again.

Dr. Seeting put his arm around Sinclair's shoulders, drawing him off to one side, dismissing Dr. Silva with an "I'll handle this" look. Dr. Hawkins followed the two men over to a window overlooking the courtyard below.

"Listen, Sinclair," started Dr. Seeting. "We've thought about your idea before, but it doesn't make sense. Think of it this way: what if one of your students decided that he or she wanted to explore the idea that the earth is flat?"

Dr. Hawkins laughed, hand to mouth and turning away.

"Or what if one of your students insists on the notion that the sun revolves around the earth? Are you going to let them 'explore' that idea? You see, Sinclair, we are here to educate, not to indulge."

"But this idea of a person … or something, making the markings in the cavern, a picture, so to speak, is it so far-fetched?" Sinclair's voice faltered a bit.

Dr. Seeting's voice was low and firm, "Listen, Sinclair, and listen closely." He looked around; Hawkins got closer. "There is only one bit of evidence that even remotely supports the idea of the markings in Lavoux Cavern having been made by a person." He paused, waiting for Sinclair to guess.

Sinclair suddenly realized the enormity of his error. In an effort to appear conciliatory he had unwittingly made the worst mistake he could possibly make; he had run afoul of The Rule. As thoughts of career-ending measures shot through his mind he regretted letting his desire for "face time" cloud his better judgment. Even in his short tenure he knew that he was suggesting the one option that was off limits on the topic of the historical origin of the markings in Lavoux Cavern: the suggestion that they were made by a human for a purpose. Although not entirely sure why The Rule was enforced so ruthlessly, he nevertheless knew better than to suggest traversing it, and now wished to be anywhere but here, with eyes glaring down upon him.

"The Legend?" Sinclair offered weakly.

"That's right, The Legend. And do you know what The Legend is? Of course you do; it's a bunch of made up baloney from some peculiar people. These naïve people have a tradition that says they originally had dominion over the whole earth. If they had their way they would re-take this land and have us all believing their superstitious claptrap. And it's not just idle desire, Sinclair. There are activists among them, some of whom just happen to have a standing petition before the High Court claiming this is their land! Do you understand?"

Sinclair's hesitation spoke for him.

Dr. Seeting got closer to Sinclair's face and continued, "It means, Sinclair, that if The Legend is true, then the Ancient People have a legal claim to this land. If The Legend is true, we lose everything, *everything*, Sinclair. Lutrell University—gone! Lavoux Cavern—gone! Your *career*—gone!"

Dr. Seeting spit out the last line with special emphasis. With another look around, Dr. Seeting looked Sinclair in the eye, "Sinclair, if The Legend is true, then, thank you, that's it, we are *finished!*" He paused and then added sarcastically, "So to speak."

Sinclair shuddered.

Dr. Seeting straightened up, looked around, and patted Sinclair on the shoulder. "Understand?"

Sinclair nodded quickly.

Hawkins smiled.

Chapter 11

"Alex?"

"No."

"What?"

"No."

"No, what?"

"No, I don't want to see your cave pictures, no I don't want to hear your latest theory about cave men, no I don't want go with you to Lavoux Cavern, no, *no*, NO!"

"How did you know I was going to ask?" Jenny asked, astonished.

"Because that's all you think about these days, Jen. Give it a rest!" Alex rolled over on her bed. "Look at this," she said, shaking a newspaper. "It says here that the Earth is going to explode in less than six months. Now, *that's* something to worry about."

"Alex, I want you to go see my childhood home."

"No."

"*What?*"

"No. That's not the way I want to spend my first day of a long weekend. Besides, I've known you since the sixth grade, and you've never asked me to go see your old house. The only reason it seems so important now must have something to do with your stupid stick figures in Lavoux Cavern. I can't imagine what the connection is, but I'm right, aren't I?"

Jenny was surprised. *Am I really that obsessed?* Jenny sat still, sullen. She couldn't lie to her best friend.

"I can't make the drive alone. It would take all day to get there and back, maybe two days. I've contacted my old neighbors and everything," Jenny pleaded. "I've got to go today. Are you really going to make me go alone?"

"Take Chris," said Alex, shaking the newspaper to the next page as she lay on her back.

"Chris Stafford?" Jenny asked.

"Yeah, he's smart, and he's into geeky stuff. He would probably find your conversations engaging."

"He's a *computer* geek, Alex. He hardly knows me. I don't even know where he is today. How am I supposed …"

"He'll be here, don't worry," interrupted Alex.

A knock at the door prompted a stare of disbelief from Jenny.

"He told me he was coming right over," Alex confided with a smirk as she jumped up and grabbed a sweater. "Don't just stand there, open the door for your friend."

"Hey, Chris, uh, come on in," Jenny stammered, surprised at the cheer in her voice.

"I was just leaving," Alex said, quickly throwing on the sweater. "Sorry, Chris, I forgot—tennis lesson today."

Alex left Chris standing in the doorway. Jenny thought about closing the door with an apology, but today was the only day for weeks she would have time to visit her old neighbors.

"Hey, Chris, what are you doing today?" Jenny found more cheer in her voice.

"I thought I was going with Alex somewhere," he replied with a confused look out the door. "I didn't know she played tennis."

Neither did I, thought Jenny.

"I've got an idea, have you ever been over the mountains?" Jenny asked.

☙ ❧

The sun was still low in the sky as Jenny's car headed into the foothills of the highlands. Chris had never been over the mountains and agreed to go as long as they could make it back that night. He had plans to go visit his folks over the long weekend. Jenny was surprised at how interesting Chris could actually be, but she tried not to let on too much about the real reason they were going to her childhood home. Most of the conversation centered on the classes they were taking, how they liked various professors, and other small talk.

"This was your house?" Chris asked, picking up the picture Jenny had tucked in with the map.

"Yeah, at least that's what my Dad said. I can remember a lot about it, but it's different than I imagined. It seemed so much bigger. Have you ever noticed how things of your childhood look smaller than you remember them being?" Jenny asked.

"Yeah. I know what you mean. I remember this drainage ditch behind our house. When I was young it seemed like a river with steep, imposing banks that stood like a moat between our yard and our neighbor's. Now I look at it and wonder if it's the same ditch!" Chris said, smiling. "It looks like a little trickle; I could jump over it. What used to be a formidable obstacle to cross as a child looks like minor obstruction to me now."

Jenny laughed. "I guess that's the way life is! What seems like an insurmountable difficulty when you first encounter it often seems insignificant with a little time and growth."

Chris nodded and the two continued discussing their backgrounds. Eventually Jenny was surprised to find herself telling Chris about her mother's death. Up till then she had never discussed the details with anyone except Alex. But Chris seemed genuinely interested and concerned as Jenny told how her mother had died during the pre-mature birth of her little brother.

"My dad," she looked at Chris, smiling as she recalled, "from the moment they knew, he used to come home each evening and talk to my mother's belly. We would all laugh, but he was beside himself over the thought of having a son. He used to say, 'I know you can hear me in there!' Somehow he knew it was a boy; he even had a name picked out. As the time neared he made ready a room." Jenny paused as she mentally reminisced. "He even carved the name over the door—right in the wood!"

Jenny was quiet for a while.

"What happened?" Chris asked softly.

"I'm not sure," Jenny replied. "I was only in the fourth grade at the time, and my father never talks about it. All I remember is my mother being in the kitchen and having sudden pains so that she couldn't even stand. It must have been a weekend because we were all home, but my dad was outside. I tried to help my mom, but soon she was sweating all over and almost crying as she doubled over holding her stomach. I was so frightened; I knew it must have something to do with the baby. I ran out and screamed for my dad. Everything happened so quickly—my mom crying, my dad rushing her to the car. I was left home alone."

Jenny glanced at Chris, whose look said, *You what*?

"I was used to being home alone, but this time the hours seemed like days. Late that afternoon my father returned alone. His face was changed. I knew something was very, very wrong; but he didn't say anything, and I was too afraid to ask. Soon friends began coming over with food. Everyone was crying, and before I could overhear what happened my father took me into my room. We both cried. He told me not to worry, and we both cried some more."

Jenny had to stop talking. Chris was still.

"Is that one reason you moved across the mountains?" Chris asked after a few moments.

Jenny looked over, wiping her eyes. "Sorry," she said, trying to force a smile.

"It's OK," Chris said gently. "We don't have to talk about it."

"Yeah," Jenny continued. "My father," she paused as she looked over to Chris, "he had a hard time of it. Something inside him died also … and I'm not real sure he's ever really recovered. I often hear him comment on how much he would give to have her back again. And him."

"That must have been a hard time for you, too," Chris said.

"About a week later," Jenny continued quietly, "I woke up in the middle of the night to the sound of a loud scraping noise and something that sounded like someone groaning. I've never been so scared in all my life. I could hardly get up the courage to run to my dad's room. When I did I found he wasn't there and I became more frightened. I peeked around the corner to where the sound was coming from and saw my dad standing on a chair, using a chisel to scrape wood off over the door where he had carved my brother's name. I stood trembling as I watched for a few minutes, undetected by my father. He worked feverishly removing the name as splinters flew and wood chips fell to the ground. He was in such anguish, crying in agony and repeating 'my son … my only son' over and over again. He didn't stop until the name was completely removed and there were wood chips all over the floor. I never saw anything like it, before or since. The memories still make me shudder," Jenny said, with a quick shiver. "Soon after that we moved over the mountains to live with his sister."

"Wow," Chris said. "And this is the first time you've been back?"

Jenny nodded.

"We've lived with my Aunt Sema ever since," Jenny said, straightening up and taking a deep breath. "She's been a real life saver for us."

Jenny wanted to change the subject. The two sat in silence for a few minutes. After a quick stop for gas and something to eat they were back on the road, once again chatting as they wound up the steepest parts of the mountains.

"So, going back for the first time in years," Chris mused out loud. "What are you going to do when you get there?" he asked casually. That was the question Jenny didn't want to answer, but knew she must.

"Chris, I'm going to be honest with you," Jenny said tentatively. "I'm not exactly sure why I'm going to my old home, and I'm not sure exactly what I'll do when I get there." *That was a good start, at least literally true,* Jenny thought. *So far Chris seems OK with the whole thing.*

The thought suddenly struck Jenny that Chris might regret coming if he knew the true reason why she was making this trip—he might even want to go back to Lutrell, and they were getting farther away by the minute. She waited a few seconds, hoping for a diversion into another topic of conversation. When the silence became awkward, she decided it was only fair to tell him the real reason for their trip.

"Chris, I'm going to be honest with you, *again*," she laughed. "I'm going to my old hometown because I want to talk to some old neighbors my parents knew who are," she hesitated, "Ancient People."

Jenny waited for a reaction, but Chris simply listened quietly.

"Ancient People." Jenny repeated, "I want to go talk to some; they were my neighbors. I want to learn more about them."

This time Jenny stopped talking, shoulders tensed.

"Cool," Chris said, straightening up with a look of expectancy. Jenny relaxed.

Shortly after midday they were winding down the mountains into the lowlands and onto the road that led them to Jenny's hometown. Chris drove now, while Jenny navigated from her father's map, although she surprised herself when they got to her hometown by remembering a few turns without consulting the map. Eventually they turned onto her old street. It was a small road that was once paved, lined on both sides by small farms. Most of the farms looked like they were no longer being worked. There seemed to be more mailboxes than houses.

"Look for a yellow house with a nice chimney," Jenny said. "I'll look for an address on one of these mailboxes."

The two drove up and down the road twice and didn't see a yellow house or the house number.

"Maybe we are on the wrong road," Chris said, leaning over to look at the map. "Maybe we are in the wrong town!" Chris joked. "Are you sure you can see the house from the road?" Jenny hadn't thought about that.

"Look!" Jenny said as Chris instinctively slowed the car. "See that chimney?" Jenny held up the picture for Chris to see.

"I think that's it," he said, "but the house is green." Chris drove slowly on as Jenny looked across the street.

"That's it!" she shouted.

Chris looked around, "How can you tell? You're looking the wrong way."

"The barn! The barn with the metal roof!" Jenny shouted. "It's the barn I used to … well, never mind. Turn here."

"Here?" Chris pointed right. "But your house is to the left."

"I know, but turn into this driveway, and drive up to that house on the right; I think they were our neighbors. Yes, this should be the Hunn's house."

The car led a small cloud of dust past the familiar barn and came to a gravelly stop beside another car near the front porch. As the dust lazily passed over and around their car, Jenny looked closely at the small house, with neatly painted shutters on weathered siding that seemed to merge seamlessly with the ground. She was pretty sure she had never seen this house before. Breathing in and out deeply, she sat in the still car, looking intently at the front door of the house, which suddenly seemed very uninviting. She heard dogs barking. *Do it now*, she thought to herself, *do it now*.

"Let's go," she said as she opened her door and stepped out. Before she could get to the steps, a man opened the front door and stood silently on the threshold. Jenny stepped up onto the porch, the man watching her with his head slightly turned like he was going to speak to someone behind him.

"I'm here to see Mr. Hunn," Jenny forced the words out.

"I'm Mr. Hunn," came a voice from a face with no expression. *Boy, he sure is young looking,* thought Jenny. The man stood still, looking at Jenny suspiciously and squinting past her at Chris.

"I'm Jenny ..."

"Wait here," Mr. Hunn said as he abruptly turned back into the house, leaving the front door open. Jenny was suddenly worried and wanted to leave. She looked back at Chris to suggest he stay with the car and be ready to drive away quickly. Before she could say anything, however, a much older man came out the door. He was bent over and appeared as if he might topple at any moment as he made his way out onto the porch.

"Jenny! I'm Mr. Hunn," he said in a voice much stronger than the body looked. *At least this one is smiling,* Jenny thought. He looked over Jenny's shoulder. "You brought a friend?"

"Yes, but first, *you're* Mr. Hunn?" Jenny asked, studying his wrinkled face. "Who was that other man?"

Mr. Hunn smiled as he motioned with a crooked arm for Chris to come up to the porch. "I'm 'old Hunn,'" said Mr. Hunn as he leaned against the porch railing. "That was my son you saw before. His name's Andrew; some people call him 'young Hunn.' I must admit he's not too sociable these days; times have been hard for some folks here lately. He means no harm."

Mr. Hunn paused as he looked Jenny and Chris over with a steady gaze from a slightly wobbly face.

"You're here a little earlier than I expected," said Mr. Hunn. "But that's OK; we're just glad to see you back. Of course, I wouldn't have recognized you. The last time I saw you, you were only this tall," Mr. Hunn motioned a shaky hand down to a level that seemed to Jenny to equate to about two years old.

"We got an early start," Jenny laughed. "I was anxious to see you as well."

"Enough talking out here; come on in, both of you," said Mr. Hunn. The group walked slowly through the still-open front door, which Mr. Hunn shut behind them. Jenny and Chris made their way through the darkened entryway, finding themselves in a tired living room from which, apparently, a knick-knack had never been removed. *How does she keep it dusted?* Jenny thought.

"Is Mrs. Hunn here?" Jenny asked.

"Margaret? Oh, yes, she's out back; she'll be here in a moment. Like I said, we weren't expecting you quite so soon. Can I get you something to eat or drink?"

"Water is fine," Jenny and Chris both replied, still standing in the neatly cluttered room.

Mr. Hunn set off for the kitchen; the crooked arm and stooped posture making it appear as if he were using an invisible cane. Jenny watched him totter away, resisting the urge to reach out and steady him. Once she could hear that he had made it to the kitchen, she glanced about and walked over to an aged settee, sitting carefully, wondering if it might not hold her. Springs creaked as she slowly sat, leaning forward. Chris looked around the baseboards. "I'll bet they don't get broadband out here," he mused. Jenny could hear children through an open window to the backyard. Probably grandchildren, maybe great-grandchildren, she thought. The sound of a noisy refrigerator came off and on, and Mrs. Hunn was obviously preparing egg salad for dinner.

Mr. Hunn came back in carrying drink glasses jingling with ice. This time Jenny felt certain that she had to help him to avert a disaster. She stood up just as Mrs. Hunn entered from the backyard, stepped around Mr. Hunn, and reached out her hand.

"Well, see here," she almost shouted, "if it isn't Jenny and …" she searched Chris's face.

"And my good friend, Chris Stafford," Jenny finished her sentence, glad that she had used the word "good." The three exchanged handshakes and smiles.

"Well, welcome to our home. Don't mind the mess; grandkids, you know," Mrs. Hunn winked, pointing with her thumb to Mr. Hunn, who had stopped amid the commotion, still holding the drinks, jingling nonetheless. "Come on in and have a seat," Mrs. Hunn motioned, almost pushing them back into the living room. She took the drinks from Mr. Hunn, holding them until Jenny and Chris were seated.

Jenny returned to the creaky sofa. Chris sat down beside her, patiently looking interested as Mr. and Mrs. Hunn talked about Jenny's father, Aunt Sema, and how much things had changed since the death of Jenny's mother. Soon, realizing that an hour had already passed and the noise of children in the backyard seemed louder, Jenny approached the subject of her visit.

"Mr. Hunn, Mrs. Hunn, can I talk to you about something? Something personal?" Jenny didn't like the way it came out, but there it was. The old couple looked at one another and shrugged.

Mr. Hunn grinned, "Sure! Just don't ask Margaret's age!" Jenny forced a smile, glad for the humor but still uneasy.

Jenny looked at both of them carefully for what seemed like minutes and decided the best approach was direct. "I want to learn more about the Ancient People," she blurted quickly. "I heard … my dad told me … well, he *thought* that you might be, or were …"

"Ancient People," Mr. Hunn stated matter-of-factly, smiling. "That's right; we are Ancient People, the whole family, about 30 of us these days."

Wow, thought Jenny, *they really did have a lot of kids.*

A slamming door interrupted the living room as a group of about five kids ran through the house, from the back door to the front and out again.

"Everyone?" Jenny asked, apparently surprised. "Even the kids?"

Mr. Hunn chuckled. "Yes, the whole lot of us. We're all Ancient People." He noticed Jenny's look of astonishment. "We look pretty normal, don't we?" Mr. Hunn laughed quietly again, looking around as if he expected more kids to come noisily through the room. Mrs. Hunn went to shuffle the children into another activity. "You see, we, meaning the Ancient People, aren't ancient in the sense of being old. Ancient refers to our heritage, our tradition. We are a people having an ancient common ancestry. There aren't a lot of us anymore, but there is a remnant. We're around, here and there."

"Here and there?" Jenny asked. "How can you tell?"

"We can usually tell each other," Mr. Hunn said, "even though we look just like everyone else. There are even some at Lutrell University, I'm sure."

Jenny and Chris looked at each other in disbelief. *At Lutrell? No way*, Jenny thought.

"What makes you Ancient People, then?" Jenny paused. "I mean, besides ancestry. Could I be an Ancient Person, for example?"

Mr. Hunn thought for a moment, smiled, and said, "Yes, but it's not so easy to explain. First, you have to be chosen." Mr. Hunn hesitated a moment. "Maybe we can discuss that later. Let me tell you about the Ancient People."

This was what Jenny was waiting for, so she pulled out a pad and pen. "Do you mind if I take notes?" she asked.

"Not at all," Mr. Hunn said, relaxing in what was once an overstuffed chair.

"Where to start," he mumbled, looking into the air. "The Ancient People are a people without a land. This land used to be ours, but it is no longer, so we are here, but not here, so to speak."

Jenny was writing and not understanding. Chris shook his head, "What?"

Mr. Hunn smiled. "We have an oral tradition that has been passed down from father to son for many, many generations. That's why we are *ancient* people," he said with a wink. "The tradition says that many years ago—long, long before your people came to this land—it was the land of the Ancient People."

"Wow," said Chris, genuinely interested. "So all this land used to belong to you, I mean your people, you know, the Ancient People?"

"Yes, our tradition says that this land was given to us, and one day it will be ours again—the lowlands, the highlands, and the mountains. For now, we are to occupy as strangers sojourning in a foreign country, faithfully passing the traditions from generation to generation. Relatively recently your people came, and if it were not for our warrior king we would have perished."

Mrs. Hunn had returned, sitting quietly as she nodded in agreement.

"A warrior king?" Chris asked, on the edge of his seat. "This is too much. A warrior king, for real?" Jenny was glad Chris was with her. "What did this warrior king do?"

"There is a legend about it," Mr. Hunn started.

"A legend?" Jenny perked up. "Did you say a legend?"

"Yes, Jenny, a legend, about our warrior king."

"Mr. Hunn," Jenny was scribbling, "is there only one legend, or are there many?"

"Just one." Mr. Hunn replied, "Many traditions, but just one legend. We usually just refer to it as *The Legend*."

The Legend! Jenny thought. She could hardly restrain her emotions. She thought that Mr. Hunn might hear her heart beating. *Deep breaths, deep breaths,* she thought.

"Are you ready?" Mr. Hunn inquired, looking at Jenny with kind eyes.

"Yes, will you tell us The Legend?" Jenny's fingers were sweaty, and she had to constantly dry them on her jeans. "But first, I'm curious, is this the story of the underground people?"

"Almost," Mr. Hunn replied, looking surprised that Jenny knew about underground people. "It's actually the story of the underground *person*." Jenny thought she might wake from a dream if she just moved. *Still, be very still,* she thought.

"The Legend ..." Mr. Hunn was again thoughtful. He paused for a few seconds, nodded slightly several times, and then started, "Again, long before your people were here, we were a people in this land. We hunted all over this side of the mountains and, occasionally, the other. At that time there were many animals to hunt. Tradition says that our warrior king himself once killed seven charging bison single-handedly. That's pretty good for a man who, according to tradition, wasn't much to look at—a little small and thin, actually."

"Hunting? Bison?" Jenny asked, visualizing the cave painting and trying to remember how many animals it depicted. *They sure don't look like bison, but maybe the painter was in a hurry.*

"We call them bison; in our language it is *impueue,* which is similar to a bison, but with much longer horns and a ring of coarse hair about its neck. They are no longer around."

Jenny was scrawling.

"Your people began to come in on the other side of the mountains in steadily growing numbers. We were threatened by your mounting presence, but we had great confidence in our warrior king. He was preparing us to confront your people; we were sure he would be victorious to rule on the throne of a peaceful kingdom. Our people followed our warrior king over the mountains to meet your people in battle. Several days before we were to go to battle, he went out at night into a large field with some of his council, his most trusted warriors. It was the new moon, the traditional start of the time for final preparation. They each had five days' provisions for food and torchlight, and markings of different colors—war paint."

"War paint?" Chris asked, enthralled. "Real war paint, like in the movies?"

"The war paint was also a tradition. Each day before battle, the warriors would add another mark and another color to their faces and arms." Mr. Hunn made motions of making stripes with his fingers on his face. "After five days of preparation they would be prepared both physically and mentally; and they would look quite fierce. But this time, something went terribly wrong, and the council returned the next day without our warrior king."

"What happened?" Chris asked, his eyes locked with Mr. Hunn's.

"That night he vanished," Mr. Hunn said thoughtfully.

"In battle?" Chris asked.

Jenny could barely keep up, sometimes forgetting to write as she searched Mr. Hunn's face for credibility, any hint that this was all in fun. His moist eyes were steady, and his tone serious as he continued.

"No, there was no battle," Mr. Hunn paused.

"His own men killed him?" Chris was impatient. Jenny wished Chris would be quiet and let Mr. Hunn talk.

"No, even more remarkable than that," Mr. Hunn continued. "It seems he simply disappeared into the blackness of the night. The next morning when the council came back into camp alone, they were discouraged and bewildered. 'He was there and then he wasn't,' was their testimony. 'It was as if the ground swallowed him whole.'"

Jenny couldn't write any more.

"Our people went into mourning. For three days and three nights we wept for our king. Each day search parties would scour the field in which he was last seen, but there was no trace of him. But," Mr. Hunn leaned forward slightly, "and this is the amazing part, on the third day he appeared among us—it was as if he had been raised from the dead."

"What happened?" Jenny asked. "Did he say?"

"He said he had been underground, in the belly of the earth," Mr. Hunn paused, "and that he had left his mark."

"His mark?" Jenny blurted. "What mark?"

"That," Mr. Hunn paused, "we don't know," he finished, sighing and sitting back in his chair. "Some suspect it was a shoe; he came back wearing only one shoe."

Jenny was momentarily motionless as her excitement drained quickly from her face. "A *shoe?*" she asked.

"We don't know; he never said. But he did return with just one shoe, which is why to this day he is sometimes referred to as 'intapao' which means 'man of both worlds' because he had experienced a foot in both worlds, above and below."

"A shoe," Jenny said matter-of-factly as she slowly wrote down a note.

"It's only one possibility. Our warrior king never talked about it. He just said he left his mark as a sign that he was victorious over the grave."

Shaking off the thought of the shoe, and realizing that the cave painting itself was fully consistent with and strong evidence for the truth of The Legend, Jenny began once again to get excited.

"Mr. Hunn, you don't know how much this means to me!" she said. "Let me ask you, is there any proof of The Legend? Was it ever written down, verified from two independent sources?"

"Now you sound like one of those newspaper people," Mr. Hunn smiled, pretending to be suspicious. "No. Ours is an oral tradition, but it is the job of the father to faithfully transmit The Legend, word for word, to the sons. I believe it is true."

"I do too, Mr. Hunn; believe me, I do, too!" Jenny gushed, surprised at her own enthusiasm. "Is there anything else? You mentioned that he saved your people."

"Yes, well, after his return from the dead, as they say, he led us back here to this side of the mountains. If we had continued with a military campaign we would have certainly been slaughtered, as we greatly underestimated your force. Upon our return he taught us that for now ours must be a secret kingdom, in which we live quietly among ourselves and in peace among you."

"I'll say," Jenny said. "I had no idea. None!" Chris concurred with expressions of surprise.

"So what happened to this warrior king?" Chris asked.

"That's another interesting part of The Legend," Mr. Hunn smiled. "The Legend, as well as our tradition, does not say. Some say he is still alive. Some even believe he will return."

Chris and Jenny looked at each other. Mr. Hunn could see their disbelief.

"That is The Legend, briefly," Mr. Hunn said. "He was the last warrior king, the last to wear the imbalal."

"The *what*?" Jenny and Chris asked in unison, looking at each other and laughing.

"The imbalal," Mr. Hunn repeated, noting that Jenny was again writing, "i-m-b-a-l-a-l. It was a ceremonial bracelet of sorts, designating the victorious king as the king of kings. It was worn about the ankles."

"The ankles?" Jenny practically jumped. "Both ankles?"

"Yes," Mr. Hunn was amused at Jenny's excitement, "both ankles. Tradition says that the imbalal of our warrior king was modified upon his return by his father as a special gift, to be worn as a promise never to be apart from our people again. It is said that they made a slight whooshing sound as he moved from place to place as a reminder of his presence; and it was reported that the sound became louder when the warrior king was moving quickly from place to place—the sound of a rushing mighty wind."

Jenny wrote furiously, flipping pages as if at any time an alarm clock would wake her. She quickly reviewed her notes, scanning down each page, making an occasional additional notation.

"And you say the Ancient People were hunters?" Jenny asked.

"Yes, at least we used to be. Now all I hunt is food in the refrigerator," Mr. Hunn said. Everyone laughed. "You don't need a spear for that!" he said laughing.

Jenny realized it was getting dark. "Mr. Hunn, you do not know how much this visit means to me. But I must get back to the university tonight. Can I visit again?"

"Of course, but don't you want to stay for dinner?"

"No, I really have to go." Jenny was too excited to sit. She wanted to jump up and scream. Chris was shaking Mr. Hunn's hand as Mrs. Hunn came in wiping her hands on her apron.

"Are you sure you can't stay?" she asked earnestly. "We're having egg salad."

"No, really, I have to go," Jenny practically pulled Chris to the door. Once outside, she pressed the keys into his palm. "Drive!"

As the car pulled out of the yard past kids playing in the dusk, Jenny was hitting her knees with her fists, saying, "I don't believe it; I don't believe it; I don't *believe* it!"

"What's not to believe?" Chris asked.

Barking dogs followed Jenny past the barn as Mr. and Mrs. Hunn waved from the porch, holding hands and smiling.

Chapter 12

Quietly entering her dorm room, Jenny was surprised to find Alex still awake and not alone. The surprise was mutual as the sudden interruption startled Alex and her friend as well—an awkward situation that Jenny decided to curtail by quickly closing the door and waiting in the hallway. She considered simply leaving, possibly going to stay with her father. But as she looked at her watch and waited the young man opened the door and left sheepishly, not even acknowledging Jenny's presence as she leaned with her back against the wall outside the door.

Jenny waited a few more moments, considering her options before again quietly entering the darkened room. Too disturbed to even consider talking to Alex, Jenny silently went about getting ready for bed.

"Jenny … I'm sorry …" Alex started.

"Alex," Jenny snapped, cutting her off. "I don't really want to hear about it right now." Jenny could not look at Alex and didn't want to talk to her at the moment. She was even wondering if she would be able to sleep in their room, thinking about what might have happened, or worse, what might have been happening. She wanted to give Alex the benefit of any doubts, but if such a benefit were not warranted, she preferred not to know, at least not right now.

"Jenny, I can explain," Alex tried again. "It's not what you think," she said in an unconvincing voice.

"It's well after midnight, Alex. Can we talk about it tomorrow?" Jenny said as she slowly pulled covers over her. She had her back turned to Alex, still not wanting to even look at her side of the room.

Alex sat on the edge of her bed, watching Jenny in the dim light of her nightstand. Embarrassed, she had not expected Jenny to make it over the mountains and back in the same day. She had also not planned on the evening's turn of events; the entire episode seemed to just unfold naturally from what started out as a friendly dinner date. Now she wanted more than anything to talk to Jenny, but she knew better than to press the issue any further this night. Slipping under her covers, she turned out her light and stared at the ceiling for what seemed like hours.

Both girls lay awake through much of the night. Several times Alex awoke, each time her mind vainly struggling for a good way to explain to Jenny what had transpired. Light through the window shade eventually fell on both girls sleeping. Not until late morning did Jenny get up, dress, and steal out of the room.

༄ ༄

The Student Center was practically empty and Jenny was able to get a cup of coffee and find a secluded booth without being seen. Being the second day of a long weekend meant that many students had left campus. Jenny was glad—it was one of the few times she had ever left her room without putting on makeup, or even brushing her hair, and she hoped to remain undetected as she thought through the events of the last 24 hours. The troubling thoughts of Alex effectively extinguished the excitement she experienced in having learned The Legend from Mr. Hunn. Her coffee was growing cold when she was startled by a touch.

"I thought I would find you here." Alex appeared as unkempt as Jenny as she walked by and sat across the table from Jenny. "Just coffee?" Alex said, looking at Jenny's cup and trying to make small talk.

Jenny was silent as she struggled with some words, any words, she could say. The two sat like strangers, Jenny's gaze alternating from the table to the window, occasionally stopping to study her Styrofoam cup as if she had never seen one before.

Alex finally broke the silence. "Nothing happened, Jenny," she said quietly, anticipating Jenny's thoughts. Jenny believed Alex, and though relieved, she also knew that the statement would more accurately express the truth if it had ended with the word *yet*.

After another long silence involving continued interest in her near-empty cup Jenny looked up at Alex.

"Alex, remember the promise we made in high school?"

"*High school?*" Alex asked, as if the subject were completely irrelevant. "Jenny, that was over a year ago! Maybe two years ago when we made that promise. A lot has changed since then."

"*What*, Alex," Jenny said sternly, glaring at Alex. "Tell me, what has changed?"

"*We've* changed, Jenny," Alex said, becoming more comfortable now that some words had been exchanged. "I see the world differently now, Jenny. Somehow what I used to think and believe doesn't seem so important anymore."

"I don't think we have changed in the same way, Alex."

Jenny looked at Alex and then looked away at some other students noisily walking through, hoping she didn't know any of them. The slowly rising din of dishes clattering behind them indicated lunchtime. The background noise was welcome—Jenny hoped no one would overhear their conversation.

"We've both changed, Jenny. It's not like you weren't out half the night with Chris!"

"What is that supposed to mean?" Jenny shot back angrily. "We were *driving home* in a *car!* And I was with Chris only because you wouldn't go with me yesterday. Why would you say such a thing? If you want Chris you can have him, but I can tell you this, he's too good for you! In fact, he was a perfect gentleman, unlike ..."

"Shhhh ... I'm sorry, Jen," Alex said defensively, looking around to make sure nobody could hear them. "I didn't mean that you and he ... well, you know."

"I found Chris to be delightful to be around, but if my being with Chris is somehow a problem for you I will never see him again." Jenny was leaning forward, whispering loudly.

"No, that's not the issue," Alex said.

"Then what *is* the issue, Alex," Jenny whispered angrily.

Alex looked away as she thought how to answer Jenny.

"I don't know what to think, Jenny. It just seems that college has opened my mind up to new ways of thinking—better ways. Even last night, what's so wrong with just having a little fun? I mean, we weren't hurting anyone."

Jenny stared at Alex for a long time. She was beginning to see in Alex's eyes now a change that she thought she had detected earlier, a change that in one sense Jenny approved, but in the end she knew to be dangerous. A new skepticism was emerging in Alex in which she questioned any authority and in which the ends justified the means. Jenny could appreciate the skepticism—she considered herself to be a skeptic. But she knew that a viewpoint that valued only ends without considering the means was trouble.

"Alex, that's about the dumbest thing I've heard you say. And you've said some dumb things. What would your father say if he heard you say such a thing?"

"My father is miles away, and frankly, I'm better off without him," Alex said curtly, surprising Jenny.

Alex's father and stepmother had moved several hours away after she started college at Lutrell. Since then Alex's contact with her father had diminished to only sporadic phone calls, usually when Alex needed money for something. Jenny thought hard as she considered Alex's words, and she wondered if she might think differently if her own father were inaccessible.

"Jenny, the beauty of college is learning to be free. Free from authority, free from the past. I feel like I'm finally liberated, and I like it!"[1]

"Alex, the thought of disappointing my father, the one responsible for giving me life in this world, is too much for me to even consider. That's not to say that everything he said is incontrovertible truth, but I think it would be wise to be very slow to jettison all you learned from your parents growing up."

Jenny could tell that her words sounded like foolishness to Alex, but she continued anyway. "Alex, there are some things in this world that are right and some things that are wrong. A prudent person knows the difference and behaves accordingly."

"Oh, please, Jen," Alex groaned. "Do you think I could disappoint my father? The one who 'gave me life in this world'" she mocked, before continuing, "and then left my mother for some tease he had to pick up from school because she didn't have her driver's license yet?" Alex laughed. "Supposedly it was for tutoring. Well, she must have needed some intense tutoring. No sooner did she turn 18 than *bam*, they were gone. You've met her—remember? She called you 'ma'am'!" Alex laughed loudly, slapping the table, as she recalled the time. "So just how is it that I might disappoint my father?"

Jenny winced at the thought and realized that bringing Alex's father into the conversation was a mistake. Although he always seemed to love and care for Alex, the divorce left her mother devastated and completely estranged. Alex had not seen her since. Jenny gave an apologetic grin and sighed as she recalled the difficult times they had shared during that period of Alex's life.

"Look, Jenny," Alex continued, "I know that maybe, somehow, certain things are wrong. I mean, I wouldn't kill anybody. But who's to say what is right and wrong for me? There is no 'ultimate authority' that can be invoked," Alex said

---

1 *See*, End Note [44], page 350.

in a voice ridiculing the very idea.[2] "It's just us, Jenny. Just us humans to sort things out. Nothing but a bunch of atoms in motion, Jenny, and you and I can either enjoy our chance with nature while we are alive, or we can fight against it until we go back to being lifeless matter. That doesn't mean that we have to understand everything about nature, but we can be sure that new discoveries will only enlarge our knowledge of the world around us."

Jenny was amazed at Alex's words. This was apparently something she had thought about and a topic on which she had a definite opinion—something very unusual for Alex.

"What is that, some sort of manifesto?" Jenny asked, eyebrows raised in a questioning expression.[3] "When did you get so interested in such intellectual ideas?"

"You can make fun, Jenny, and I know we disagree on this, but the world around us—call it matter, energy, substance, or what you will—exists independent of our minds, or any conceivable mind. We can make the world what we wish. And I wish to be free."[4]

Alex's words sounded like so much gibberish to Jenny, but it wasn't the first time she had heard similar sentiments parroted about. She was just surprised that they had found a perch with Alex.

"Don't you think you are making some ultimate assumptions? How can you know the truth of what you are saying?" Jenny asked, still surprised that Alex would even think such thoughts.

"The truth," Alex muttered, slumping back in the seat looking annoyed. "Listen, Jen, I know I'm not as intelligent as you, but I am learning something here at Lutrell. For the first time in my life I'm feeling smart. You probably can't understand that; but I, Alex, for the first time feel like I know something, like

---

[2] *See*, End Note [45], page 351.

[3] *See*, End Note [46], page 352.

[4] *See*, End Note [47], page 352.

I'm enlightened. I'm learning that we can make this a better world by emphasizing human fulfillment, freedom, and the dignity of the individual. I also know that there is no absolute moral code; we just have to adjust our ethical standards in each situation according to our judgment. As we learn to do this, emphasizing fairness, kindness, and tolerance, the world will be a better place."[5]

Alex sat back again with an air of sophistication. Jenny was dumbfounded as she slowly realized that for the first time she and Alex were disagreeing on something substantive, something that mattered. In the past Alex may have had an occasional serious thought, and she may have even challenged Jenny on various ideas, but Jenny could not recall Alex ever having expressed an idea with the apparent conviction Jenny now observed. The two sat for a few moments in silence before Jenny continued, leaning forward to speak.

"Why were you sorry last night?" Jenny asked quietly, looking kindly but intently at Alex.

"What?" Alex asked, sitting back straight.

"Last night. You said you were sorry. Remember? Were you apologizing because what happened last night was unfair, unkind, or intolerant? Those are the criteria, aren't they? Weren't you just adjusting your ethical standard to the situation? What's to apologize for?"

"I don't know, Jenny. I don't have all the answers. I guess it was unfair to you. You are still my best friend."

"So what's fair to you might be unfair to me?" Jenny asked, continuing to be gentle, but probing. "Who's standard are we to use?"

"Like I said, I don't have all the answers, Jenny, and I'm not going to be able to answer your questions, for sure," Alex said, smiling. She meant her statement as a compliment to Jenny. "But my knowledge is expanding every day. And science is explaining more and more things in exciting new ways—new

---

[5] *See*, End Note [48], page 353.

technology, new solutions, and we get to be a part of it here at Lutrell."

Jenny noticed the reference to science, which gave her a clue as to the basis for Alex's newfound views on life. She wondered who it was that Alex might be hanging around to pick up such ideas.

"Alex, I have a lot of questions of my own. But I'm not so quick to buy in to this 'science can explain everything' mentality. I'm not so sure science can explain everything. What about your conscience, for example? Can science explain that?"

"I don't know, Jenny. But I suppose if it can't, it will someday. Some things take time, but just because something is unknown doesn't mean it's unknowable."

"Alex, I'm not going to debate you on these points; you seem to have very definite thoughts of your own, and, honestly, I can appreciate that. But I'll tell you what I think. I think that there are not only things we *can* know, there is also something that we can't *not* know. Regardless of how enlightened we become, or how much we try to deny it, we can't not know it.[6] I'm not even sure what that *it* is, but it's real, it matters, and it has something to do with truth and absolutes. It's something beyond ourselves; we can't escape it and we ignore it at our peril. Our parents and our grandparents knew something about this, and while their ethics and morals may seem old-fashioned to you, I would be slow to abandon them with such enthusiasm."

Jenny watched Alex's expression, which was patient, but not receptive. Again Jenny realized Alex's ethical context was colored by her own father, but she knew that Alex would get the point.

"I appreciate that, Jenny. And you know I respect you. But this *it* you are referring to is exactly what I'm being freed from. The *it* is simply society's way of enforcing morals by threatening punishment. Threats of punishment are what my

---

[6] *See*, End Note [49], page 355.

father used to make me behave a certain way—his way. Well, now I don't have to do things his way anymore. Does that mean I'm doing something wrong? Who's to say?" Alex paused for a few moments, noticing more people coming into the Student Center.

"What are all these people doing here?" Alex asked, looking around. "Why didn't they go home for the long weekend? Why didn't *you* go home, Jenny; you just live a few minutes away," Alex asked turning back to Jenny.

"I thought about it, but I knew I would be returning late last night, and I didn't want to surprise anyone, coming in so late ... unexpectedly."

Jenny didn't mean for the words to have the added meaning, but she was glad they did. She sensed Alex was trying to change the subject.

"Alex, I breathe the same air you breathe here at Lutrell," Jenny said, wanting to make a few last points before leaving the topic. "This air is permeated with the idea that right and wrong are relative, authority is oppressive, and the scientific method is the only sure method of arriving at knowledge. But that doesn't seem right to me. I think there are universal value judgments that are inherent to us. I think proper authority is liberating, and I think that science provides one way but not the only way to knowledge. Can I prove these things? Probably not, but here's what I'm learning, Alex." Jenny waited to see if Alex was listening. "When it comes to conflicting ideas like ours, we may each be wrong, but *we can't both be right*. Do you understand what I mean, Alex? Some ideas are exclusionary with respect to others; not all ideas can be *right*.[7] And, ideas have consequences; good ideas have good consequences, and bad ideas have bad consequences. And, to be very, very honest, Alex, as I would be only to my best friend, I see you buying into these modern notions, and it bothers me. Don't be fooled by people who like to

---

[7] *See*, End Note [50], page 356.

pretend to think lofty thoughts." Jenny sat back, looking at Alex earnestly, hoping for a positive response.

"Look, Jenny, you are my best friend, and, if I can speak honestly as well, I see you as being too rigid, too … obstinate in the face of overwhelming consensus. Take this cave painting thing. I mean, what will it take to get you to see that it is simply the product of nature? Don't you think you are a little extreme insisting that a person painted it? It's that kind of extreme, unyielding, intolerant position that causes problems in the world, not peace and harmony. Why can't you just believe a person made the picture, and let others believe it was made naturally? There's room for all beliefs if we just show tolerance instead of dogmatism."[8]

Sadness swept over Jenny as she realized that there was a wide and widening gulf between her and Alex. Jenny knew that the gulf was likely to become a chasm and could ultimately become an abyss that permanently separated the two. She doubted that Alex realized or cared about such a possibility. For a few moments Jenny considered possible options, including abandoning any attempts at proving that the cave painting of Lavoux Cavern owed its origin to an intelligent being. At least she could stop being vocal about her views and simply quietly accept the status quo. Doing so at this point would likely ensure her friendship with Alex; failing to do so would probably drive a wedge between them to the point of hopelessly fracturing their relationship. The dilemma grieved Jenny, but she knew she could not compromise her convictions and resolved instead to be optimistic that the gulf could be bridged with kindness and respect, just as Alex had suggested. For now she decided to change the subject and move on.

"Alex, will you be patient with me as I sort out my thoughts on these issues?" Jenny asked, reaching across the table. Recognizing that Jenny was attempting to close a divide,

---

[8] *See*, End Note [51], page 358.

Alex accepted the challenge with a smile, squeezing Jenny's hand.

"Jenny, we've been together too long to let things like this come between us. We're both learning more than we've ever learned before, and we're both being challenged with new ideas every day. I think we'll work through them OK."

Alex and Jenny sat back and looked around. More people were in the Student Center now, and both girls were wondering how they could leave without being seen by anyone, particularly anyone they knew.

"Do I look as bad as you do?" Jenny asked with a laugh.

"You know you always look better than me, but I must admit that this time it might be a draw," Alex said, also laughing.

They decided to quickly leave together, heads down, and ignore any possible calls from others. After successfully leaving the building without being noticed, they slowed down, taking their time going back to the dorm.

"So, are you going to go see your father over spring break?" Jenny asked. Jenny knew that even in Alex's family situation the best antidote to the kind of enlightenment engulfing Alex was spending some time at home with family and friends—away from Lutrell.

"No, I've decided to spend the time with my friends this year," Alex said, as if it was the only natural thing to do. "I think we are going to go to the beach."

Alex was oblivious to the irony that apparently her "friends" didn't include the one she had just referred to as her best friend. Jenny waited for a moment to see if this might just be an oversight, before replying.

"Friends?" Jenny asked nonchalantly. "Anyone I know?"

"Oh, no," Alex replied. "No, these are some new friends you haven't met. But they're really neat people. They're part of a discussion group I found—a lot of upper classmen, even some grad students. We have had some great discussions. I've learned more from them than from some of my professors."

The girls were not so fortunate in avoiding other students as they neared their dorm. A group of people Jenny did not recognize was just pouring out of the doorway as Jenny and Alex approached. Surprisingly, they all seemed to know Jenny by the expressions on their faces. As Jenny tried in vain to place faces with names she realized that they were not looking at her, but at Alex. Apparently Alex had planned to meet them here and then they were all going "out for some fun." As the lively group huddled around a flattered Alex, Jenny realized that she must be looking at a lot of upper classmen and even some grad students. While Alex was caught up trying to explain why she was out and why she looked the way she did, Jenny quietly slipped away and into the dorm. As she reached the window on the second floor landing she paused to notice Alex still outside with her friends, all of them laughing and smiling.

Chapter 13

"Jenny!"

Alex deliberately slammed the door behind her.

"Wake up!"

"Alex," Jenny groaned. "Leave me alone. It's Saturday. Go play tennis." Jenny rolled over and pulled the covers over her head.

"Jenny, come look at our door!"

Jenny thought she detected a sternness that was not typical of Alex. Their relationship had improved over the last several weeks—Alex didn't bring her friends around and Jenny tried not to talk too much about the cave painting.

"Alex, I've seen our door, will you please go away?" Jenny said as she pulled a pillow down tight over her head to block out the noise.

"Jenny, this is *serious,*" Alex said as she yanked the covers off of Jenny.

"OK, OK, I'm up. What is it? Is something on fire, or what?"

"Worse," Alex said, going to the door and throwing it open. "Step out here and take a look."

"All I can say is this had better be good," Jenny muttered as she wrapped herself in a blanket and the two stepped into the hall.

Alex shut the door, and the two girls stood there, looking. Across the door and onto the walls someone had painted "Legend Lover" in large red letters. Paint ran down in long drips from the bottom of most of the letters. Beneath the words was a crude depiction of the hunter, lying on his back with a spear in his chest. Jenny tried not to look disturbed as her eyes were drawn like a magnet to smaller words under the hunter that read "Dead—The Man." She blinked and looked away. Others in the hallway had also noticed and could be seen trying not to look, speaking in whispers.

"Who did it?" Jenny asked, not knowing what else to say.

"Who did it?" Alex said in a raised voice. "Does it matter who did it? Obviously *someone* did it, Jenny; it didn't just paint itself! And my guess is that it was not one of our friends!" Alex was now shouting, her voice showing signs of strain Jenny had never heard before. Others began poking their heads out of their rooms.

"What is going on here, Jen? I mean, it's OK for you to have your little causes, but this is my room, too."

"Al, I'm sorry," Jenny started, glancing around and trying to speak quietly.

"Jenny, when is this going to stop? We're in college now; we're supposed to be having fun. Instead you've turned all serious, and now *this*!" Alex threw her hand toward the door.

"Alex, someone is just playing a joke," Jenny tried. "It's probably because today is the day we start studying progression in Hawkins' class."

At the thought of Hawkins' Earth Science class, Alex groaned in dramatic despair. She wished she had not decided to have any classes with Jenny, especially not science.

"I'm not going," Alex said matter-of-factly.

"What do you mean, 'I'm not going'?" Jenny mocked Alex's anguish.

"Just what I said, Jenny! I am not going because I know you. You're not going to be able to sit there without popping off with something brilliant that you heard from your cave cretins! And I am not going to sit there while you make a fool of yourself, and I get tarnished with the same … ," Alex paused, looking at the door in full frustration, "the same BRUSH!" she shouted, again gesturing toward the painted letters. Jenny had never seen Alex so disturbed.

"Alex, I'm sure there's a very reasonable explanation for this. Someone is just trying to be funny. We'll find out who it is and you'll see, it's just a joke. Come on, let's get away from the room for awhile, let's go get some coffee. Double latte, on me," Jenny tried to sound cheerful.

"Coffee?" Alex cried. "Someone has threatened us, and you want to go get coffee? I'm reporting this; I'm getting security."

"Stop GAWKING!" Alex yelled as she stormed down the hallway.

☙ ❧

Jenny went back into her room, trembling slightly under the blanket still draping her shoulders. *Who would do such a thing, and why? Surely someone is just having fun.* But Jenny could not shake the thought that the message was meant as a threat, and the idea frightened her. Even more troubling to her, however, was the notion of subjecting Alex to even the embarrassment of such an incident.

Searching her memory trying to think of who might have painted the message on the door, Jenny thought about the night before, when she and Alex had attended a meeting of the Student Ancient People. Following Mr. Hunn's suggestion, Jenny had checked a student activities bulletin board, and, sure enough, a small card gave information about a weekly meeting open to all. Alex went unwillingly, dragged by Jenny, who still was not sure what kind of people they might find.

Jenny and Alex were both surprised to find that the students appeared to be normal; they seemed like typical students at a typical meeting. Some of them they recognized, and a couple of them they even knew by name. A refreshment area on one side served as a convenient place for Alex to pretend to be interested. Jenny milled about, talking to as many students as she could.

As Jenny worked her way around the room she was surprised to find that two of the people present were not students, but professors! Another student pointed out two people sitting leisurely at a corner table, bantering with the students. Jenny had to wait a few minutes before getting their attention, but once they made eye contact she quickly slid into a seat at their table.

"You are Ancient People? And professors here at Lutrell?" Jenny asked bluntly, smiling but with a look of suspicion. Aside from the casual clothes and apparent youth, Jenny was mentally trying to comprehend actual professors being Ancient People. Up until tonight she doubted there were even any student Ancient People at Lutrell.

"Yes," answered one. "I'm professor Clark, Jim Clark, and this is professor Susan Welch." Professor Clark extended his hand as he reached across the table; professor Welch nodded with a smile. "We're academic advisors for the Student Ancient People. Is this your first time here?"

"Yes. I'm here with a friend," Jenny said as she motioned to Alex, who was sitting uncomfortably, holding her purse close and looking about charily. "We, or I," Jenny continued, "am interested in learning more about the Ancient People, and about the tradition of your Warrior King."

"Great!" professor Welch exclaimed, excitedly extending her hand. "Welcome to our group."

Leaning over the table, Jenny thanked the two, and knowing her time was limited, began asking questions. "What are your backgrounds? Are either of you PhD.'s? Do you believe in progression?" The questions poured out as Jenny watched their smiles get bigger. A pause allowed Dr. Clark to answer.

"Earth science, yes, and no," he said, looking into the air as he thought.

"Ditto for me, but my doctorate is in climatology," added Dr. Welch. "Jim's earth, I'm wind. All we need now is fire," Dr. Welch laughed as Dr. Clark rolled his eyes.

"Wow!" exclaimed Jenny. Turning to Dr. Clark she asked, "I've got to know, how do you teach about the cave painting? Do you use Dr. Seeting's book? Do you teach that the cave painting was made by nature alone?"

Dr. Clark smiled silently as he looked at Jenny with a hesitant gaze. Jenny expected an immediate and emphatic "no" to the last question and felt her heart sink as she waited for Dr. Clark's answer. At first he just looked at her thoughtfully, as if he felt that any answer he gave would be incomplete without a long explanation.

"Jenny," he started slowly with measured words, "I teach that the marks were made naturally over time, by progression."

The look on Dr. Clark's face revealed the inner conflict that such an answer, and the underlying reasons, created for him. Without knowing what the underlying reasons might be, it was evident to Jenny that Dr. Clark had wrestled with how best to answer such a question, and, although seemingly at peace with his stance, he clearly struggled with his conscience. The look of confusion and disappointment on Jenny's face prompted him to continue.

"This is a difficult issue, Jenny. Lutrell is a public university, and because of that I am not permitted to teach any theories contrary to progression. I am not even totally free to teach the problems with progression—and there are many. And the university absolutely forbids me from teaching, or even hinting, that the cave painting is of intelligent design."

"What about the evidence?" Jenny asked slowly, feeling suddenly empty and still dumbfounded at what she was hearing. "Is there no place for evidence-based ideas in a university science class?"

Dr. Clark looked around, turned to Dr. Welch and then looked back at Jenny, sighing, "Jenny, I detect in you a real enthusiasm for the truth, and I—we both," he nodded toward

Dr. Welch, "can appreciate that. But there are powerful political forces at work in the public square, including in this university. It is almost impossible to stand against these forces and get, much less keep, a good professorship. In a sense, science has been traded for "scientism," which is a kind of science in which only certain theories are acceptable, regardless of the evidence. On this topic science is being held hostage so that the truth hardly has a chance."

"Scientism?" Jenny asked.

"I hate to admit that I've succumbed, in a sense, to the dictates of scientism myself. I just feel that I can at least do some good by doing things like sponsoring this group and being active in off-campus activities." Dr. Clark said, with a wistful look.

"But ... you *do* believe The Legend," Jenny asked, fearing the worst. "And, you do believe that science can confirm intelligent design, regardless of the designer, don't you?" Again, Jenny feared the worst—that Dr. Clark might actually believe progression to be the true explanation for the cave painting.

"Yes, I do," he answered, smiling. Jenny's tension must have been visible, so he continued. "In fact, I'm quite sure of both. But the time is not right. In the end, however, truth will prevail."

"Are there many like you?" Jenny asked.

Dr. Welch spoke this time. "There are some dissenters among credentialed scientists, but not many, relatively speaking. Progression has made real inroads into the academic consciousness. Like Dr. Clark said, unlike other areas of academia, there is little academic freedom in this area and it can be very risky to oppose its dominance as the only explanation for the cave painting. We are probably the only two here at Lutrell. And we are very junior, and we have to keep a low profile." Dr. Welch ended with a "sorry, but that's the way it is" look and a smile.

Jenny wanted to hear more, but the talk in the room had turned to traditions, and The Legend was brought up often, interrupting her conversation and distracting Drs. Clark and Welch. At one point it appeared someone would speak to the group, and chairs were gathered, facing one side of the room, prompting Alex to dart over to Jenny.

"Jenny, this is weird. Let's go," Alex whispered.

Jenny wanted to stay, but she saw the blaze in Alex's eyes as she mouthed the word "now" and headed for the door. Jenny excused herself, explained that she and Alex had to study for a test, and the two left after seemingly endless handshakes.

Alex didn't speak all the way home. Jenny had tried to make small talk, but to no avail.

☙ ❧

Jenny was surprised to see Alex already seated in Dr. Hawkins' class. She shot Alex a look to say, *"I thought you weren't coming,"* and then smiled. Alex slowly looked away, her face expressionless. Jenny found her seat and looked around. Most of the students seemed disinterested and tired. This course was required, but still the only reason most kids even came to class was because it was considered a "weed out" course. You had to be there to hope for a good grade. *Maybe that's why Alex is here,* Jenny thought.

Dr. Hawkins walked in with a bundle of papers and books under his arm. He always wore the same brown sport coat with patches on the elbows. It appeared old; Jenny figured it must have been a gift at his graduation. The class gradually got quieter as he erased the blackboard behind him. He was usually cheerful, but today Jenny thought he looked as if teaching was a chore.

"Open your books to page 446," his voice carried about the large room. Jenny's book practically opened itself to page 446; the binding was cracked at that spot. "Today we are going to start one of the most important subjects in all of science. Does anyone know what it is?" Usually no hands would go up, but today several hands shot up.

"Progression," one student in the front row stated proudly without waiting to be called on.

"That's right," Hawkins said as he carefully made a mark on a paper in front of him. "Progression. What does that mean?" More hands shot up.

"Change over time!" Another voice, another mark on the paper.

"Yes, but what does it really mean?" Dr. Hawkins asked, pausing as he shifted papers and looked around for another student. Jenny thought the questions to be a bit strange. Dr. Hawkins usually lectured and asked very few questions.

"It means that the earth is always changing. The features we see around us are always changing," came a voice from Jenny's right.

"Not just the earth," Dr. Hawkins reminded, "but all of the cosmos, the known universe, is in a state of constant change." No mark for that student, apparently.

"Progression," continued Dr. Hawkins, "is the unifying principle of the universe, the mainstay of all the disciplines of knowledge. Nothing in science makes sense except in the light of progression."[1] Students scribbled notes mindlessly. "Without the concept of progression, science would be in disarray," Dr. Hawkins intoned.

"Let's look at specific examples of progression," said Dr. Hawkins as he flipped a switch to turn on the overhead projector hanging from the ceiling above. A blue light flooded the screen to his right as he pushed various buttons until side-by-side images of a river came up. "Take this river," he started. "These pictures were taken ten years apart, and notice ..."

Many students were daydreaming now, occasionally taking notes, often checking the clock.

"Now," Dr. Hawkins droned, "let's look at some more remarkable achievements that nature has accomplished through progression." A picture of a quartz crystal shone on the screen.

---

[1] *See*, End Note [52], page 359.

"Here is an example of how nature can produce a very ordered structure, a beautiful structure, actually, on its own." He scrunched his nose under his glasses as he looked behind him, head up, gazing at the quartz formation as if it were a Rembrandt. "A snowflake is an even better example, but it melted before I could get the picture." Those paying attention laughed for him; the others looked up, wondering what they had missed.

"Let's look at our own Lavoux Cavern," Dr. Hawkins said, slightly louder. Jenny straightened up, heart beating faster. Alex put her head down on her desk. "It contains by far the most remarkable natural formations ever seen by man." Dr. Hawkins was winding up.

"This," he scrunched and looked, "is a stalactite. Formed over thousands of years, bit by bit, little by little, crafted by nature. Why?"

Dr. Hawkins stopped abruptly and the sudden silence woke the class, which was also silent as he looked around. No hands. He looked down and studied the paper in front of him. "Kathleen?"

Kathleen was one of the Ancient People Jenny had met the night before. *Did Hawkins know that?* Jenny thought.

Kathleen looked up, startled. "Um, why?"

"That's what I asked *you*," Dr. Hawkins said, to more laughter.

"Well, I guess because ..." Kathleen paused uncomfortably as she struggled for the right answer. "Maybe ... because it *wanted* a stalactite?" The class laughed. Jenny looked sympathetically at Kathleen as if to say "*nice try*."

"Actually, nature made the stalactite for NO reason," Dr. Hawkins boomed out the word no. "Nature doesn't want anything, it doesn't know anything, it doesn't mean anything." Dr. Hawkins paused to let the thought sink in. "That was the basis for Dr. Neihil's work which we will get to shortly, the monumental work that led to the remarkable theory that explains how nature could make this." Dr. Hawkins timed it perfectly. Immediately on the screen before the class was the picture of the hunter and the hunted. Most of the class had already seen the picture; it was right in front of them on page 446, so there was only a mild stir. Jenny's head swirled. Kathleen looked at Jenny hopefully. Alex groaned under her breath.

"Dr. Neihil's theory of progression, sometimes referred to as 'events with modification through natural selective inclusion,' explains how the specialties, the apparent paintings, shown in these figures could come about gradually over many, many years, in perfect agreement with the theory of progression. In fact," Dr. Hawkins was in his element, "the theory of progression by natural selective inclusion explains ..."

A raised hand stopped him mid-sentence.

"Yes?"

Alex wanted to get up and leave when she saw Jenny's hand go up. The rest of the class sensed something interesting and paid attention; no one had ever stopped Dr. Hawkins to ask a question.

"Dr. Hawkins, I've read the material, and I don't see where the book discusses any alternative theories to explain the picture we see here." Jenny thought her voice sounded nervous, and she hoped her pounding heart was not audible to others around her.

"And what theory might that be?" Dr. Hawkins was polite.

"Well, perhaps a person painted the picture, for a reason," Jenny said, in a thoughtful tone as if the idea had just occurred to her. Kathleen shot a disbelieving glance at Jenny. "Yes, come to think of it, it appears to me like it was painted," Jenny finished. She was surprised that there were even a few nodding heads. But Dr. Hawkins had been here before.

"Uh, Jenny, is it?" He pretended to look at the seating chart on the lectern in front of him. "Yes, Jenny, let's explore that a bit, although I don't want to waste too much class time. Believe me, yours is a common misconception, perfectly understandable. But it gives me the opportunity to remind you all about the basic purpose of science. You see, science," he spoke as if science were a person, "attempts to explain how certain things that appear to be designed, even designed for a purpose, actually come about naturally, by accident, in one sense. We could, alternatively, rely on myths, legends, or folk tales to inform us about the world around us, but then we would not be scientists, would we?" Dr. Hawkins was being mildly condescending.

The class was silent. The reference to legends struck home with several in the class. Jenny raised her hand again. Kathleen found some absorbing reading on her desk. Alex wondered what was wrong with the clock.

"Dr. Hawkins, I agree that Neihil's theory is worth studying. I'm just wondering why it is unreasonable to consider other explanations. Hasn't Neihil's theory come under criticism lately for not being an adequate explanation?"

Leaning forward, Dr. Hawkins rested his elbows on the lectern as he thought momentarily. "Jenny, let's be clear. Criticism is part of the healthy debate that distinguishes science from the more dogmatic disciplines, disciplines that cannot tolerate dissent. Furthermore, Neihil recognized that many and grave objections may be advanced against the theory of events with modification through natural selective inclusion; he did not deny this.[2] However, even though Neihil's theory is sometimes criticized, such criticism should not in any way to be considered to be challenging the fact of progression. Progression happened, even here," he said, straightening up and pointing to the screen showing the hunting scene, "and that's a fact. Whether it happened by Neihilian progression, or, as Dr. Silva would venture, by interrupted stasis, remains to be shown conclusively. Regardless of the specific theory, no reputable scientist challenges that progression happened and produced the figure you see reproduced on page 446. *How* progression happened might be in question, and *that* we can discuss; but today Dr.

---

[2] *See*, End Note [53], page 360.

Neihil's theory of gradual progression is accepted as having the most explanatory power. Do you have another theory?"

Jenny's heart beat rapidly as all the heads turned expectantly toward her. She decided to go for it all.

"Dr. Hawkins, with all due respect, a person painted that picture, and it was painted for a purpose—in fact, it is the very image of the painter!" The class hummed as everyone whispered something to a neighbor. The bell rang.

"You are wrong, Ms. ..." Dr. Hawkins searched his seating chart, but to no avail. Immediately, the hubbub of moving chairs and closing books was drowned out by the sound of students bustling in the hallway. "I'll not tolerate such outbursts in this class," he sputtered loudly to no one as he gathered his papers.

Alex rushed out, not even looking at Jenny. Jenny sat stunned, regretting that she had been so bold.

Kathleen also sat, slowly shaking her bowed head in disbelief, and smiling.

Chapter 14

"Alex, what is *this*?" Jenny exclaimed. Jenny stepped over boxes and clothes in the floor as she entered their dorm room. "I know you're not *cleaning*; that's never crossed your mind before." Jenny was trying to be funny, but in truth she hoped beyond hope that Alex was just cleaning. Alex's posters were off the wall; her desk was cleaned off into boxes, and all her clothes were either on her bed or folded on the floor.

"Jenny, you weren't supposed to be back so soon." Alex did not look at Jenny. "I don't know what to tell you."

"Alex, tell me *something*," Jenny pleaded. She walked over and held Alex's shoulders, forcing her to look at her eyes. "Alex, if I didn't know better I would think you were moving out."

Alex's silence brought sudden panic to Jenny's heart. "Are you quitting school?" This was the only possible reason Jenny could conceive that Alex might move out. "Was there a death in the family? What, Alex, *what*?"

"Jenny, I found a place off campus."

"Off campus?" Jenny sat slowly on her bed, trying to understand the words and not wanting to take her stunned eyes off of the scene before her. She watched as Alex, her back turned, slowly put hanging clothes into a large box. The sound of a squeaky cart in the hallway got closer before stopping at the door.

"Cart's here," someone yelled in the door. Jenny felt the world closing in.

"Alex, look at me," Jenny said, crossing the room and again gripping Alex by the shoulders. "This can't be happening." Their eyes met only by accident. "Alex," Jenny tried to force eye contact, "what is this about, tell me. Is this about Chris? Is this about the graffiti on our door? Alex, stop for a moment and talk to me. This can't be happening. I thought we were best friends." Jenny rambled on, confused and hurt. Alex sorted clothes, occasionally finding some garment of Jenny's that she awkwardly set aside with a mumbled, "I think this is yours."

Jenny's thoughts flew as she remembered the day she returned from visiting Mr. and Mrs. Hunn. She recalled considering accepting the status quo so as to avoid driving a wedge that would fracture their relationship. Although knowing that she could not tolerate such a thing, she now found it unbearable to feel the wedge pressing in and to hear the unnerving splintering of their relationship. In desperation, Jenny crossed the room again, and again held Alex's shoulders to stop her from packing and to look her in the eye.

"Alex, stop for a moment. Look at me. I don't know what you see right now, but I'm looking at more than just a roommate. I'm looking at the reason I'm here at Lutrell in the first place. I'm looking at the only girl who invited me to a birthday party in the sixth grade, the girl who forced me to ride a roller coaster in the seventh grade, the girl who taught me to dance in the eighth grade, the girl who taught me to dive, well, pushed me off the high dive, in the ninth grade," Jenny paused, smiling briefly at the memory before continuing. "Alex, you are the girl who taught me to stay up late, to go to movies, to drink cappuccino, to … to *let go and have fun*."

The two stood momentarily as Alex sighed, impatiently avoiding Jenny's intense gaze. Finally, Alex broke loose from Jenny's hold and returned to packing. Jenny again watched Alex as if in slow motion, fighting feelings of despair at the prospect of her leaving. She apparently was only a few minutes from being hopelessly alone.

"Alex," Jenny said softly, "you are the one who finally taught me to laugh again after my mother died. When I moved here I was lost, I was sad, and I was confused. You helped me through that time in a way that no one else could, or would. So I don't know about you, but I'm not just looking at a friend, or even a roommate. Alex, I'm looking at the person who gave my *life* back to me. Please, Alex, say something."

"Jenny, I know," Alex said with some feeling as she started moving faster, putting things in boxes and taking boxes to the cart. "Jenny, I just can't take it any more. I know you've always been intense, and you've always approached your interests with passion. But this has gone too far, this crusade for accuracy ..."

"Truth!" Jenny interrupted with a sharpness she immediately regretted. "Alex," she finished softly, holding back tears, "I only want truth to have a chance."

"Why you? Why *you*, Jenny? Of all the people at this university, why *my* roommate?" Alex was also close to tears as she opened and closed desk drawers. "I can't take it any more, Jenny, all this about cave men, old people ..."

"*Ancient* People," Jenny interjected abruptly, again wishing she had let Alex finish. Alex carried the last of her belongings out into the hallway.

"Whatever. Jenny, I just want to have fun. All this serious talk about Legendism, and Neihili ..."

"Legend-*ism*?" Jenny had never heard the word.

"People talk, Jenny, in case you haven't noticed!" Alex shouted, deliberately looking at their door. Jenny looked as well and was surprised to see that someone had added an "S" on the end of the painted message. "This is not for me, Jenny," Alex finished.

"So that *is* what this is about, right? The writing on the door? Is that what this is about?" Jenny cried. Alex was ignoring Jenny. "The painting on the door? Alex, we'll find out who did it. I'm sure it was just a prank!"

Jenny fumed inside as she contemplated the graffiti on their door. She had requested numerous times to have the door and walls repainted. The picture and letters remained, however, because security personnel said it was the only proof they had of the vandalism. There was no additional evidence to implicate any particular person, so the painting had to remain as the only indication that a crime had been committed. Until some suspects could be apprehended, "the painting remained the only evidence of a painter," according to the university's security office. Seeing the added "S" only infuriated Jenny more as she considered her situation.

Sitting down on her bed, Jenny put her head in her hands. Angry with both herself and the university for letting the graffiti stay for so long, she was also frustrated at the length of time the investigation had taken. Whenever she enquired into the status of the case she was told to be patient and to be assured that such investigations were routine in such matters. The investigators had "several leads" and were evaluating "multiple hypotheses" to reach a best explanation. Jenny always considered the response to be a cop out, an excuse for not simply closing the case as unsolvable. She was lost in contemplation of painting the door and wall herself when her thoughts were interrupted.

"I'll see you around, Jen." Alex began pushing the squeaky cart down the hall.

Jenny stood outside their room, staring in helpless disbelief until the squeaking could be heard no more. After she heard the elevator door shut and she was sure the creaking would not return, Jenny turned and reluctantly entered the unbalanced room, her tear-filled eyes blurring the one picture Alex left, a picture of the two of them standing in front of the cave painting in Lavoux Cavern.

☙ ❧

"Hey, what's up?" Chris said as he came in the open door. "Whoa!" He stopped mid-stride. "What *is* up?"

"Alex found another place to live," Jenny said listlessly from where she sat on the edge of her bed. Chris slowly lowered his raised foot to the ground as he looked around in disbelief.

"Alex? Your best friend Alex? The …"

"I *said*, 'Alex found another place to live,'" Jenny repeated, glaring.

Chris stood awkwardly in silence, finally noticing that Jenny had been crying. "OK, I won't ask any questions. Is there anything I can do? Oops, sorry, that was a question." He hoped there would be a laugh; he didn't even get a smile.

"Chris, there's a lot going on in my life now, and I'm not sure you want to hang around me."

"Try me," Chris said in convincing assurance.

Jenny looked closely at Chris, knowing that she had to get out of the room and be distracted from her suddenly small world. "I want to go to a meeting of the Student Ancient People tonight. Will you go with me?" Jenny asked hopefully.

"No." Chris answered immediately. "See, at least you tried!" Chris said with a big smile. Jenny smiled.

"Seriously, not today. There's a party at my place tonight. You're welcome to stop by later if you like. We'll behave."

Jenny would ordinarily never consider going to a boy's dorm room, and parties interested her very little, but she found herself nodding and thanking Chris.

"See you there," Chris said, as he left.

☙ ❧

At the meeting of the Student Ancient People Jenny found herself talking to a small group that included Kathleen. The topic was how best to counter Dr. Hawkins on the subject of progression.

"The theory has some validity," said Kent, a sophomore who had taken Dr. Hawkins' class the year before.

"But it's not responsible for the cave painting," Kathleen quickly responded.

"I agree," said Kent, "but we need more than The Legend to counter his arguments. Essentially, if I remember correctly, he is arguing that a mechanism that can produce stalactites can also produce cave paintings; it's simply a matter of the degree of complexity, which is really just a question of time."[1]

Jenny was impressed at Kent's memory and grasp of the concept of progression. But she was troubled slightly at his initial approach to countering Dr. Hawkins.

"We don't just need *more* than The Legend," Jenny responded. "We need to convince Dr. Hawkins on the basis of facts alone, regardless of the truth of The Legend."[2] The looks of the others suggested a clarification. "Look," Jenny said, "I believe The Legend. But I believed a person made the painting long before I believed The Legend. We should be able to convince people that the painting was made by a person whether or not The Legend is ultimately true." Some weren't convinced, but all were thinking.

"I agree," Kent finally broke the silence. "As soon as we try to include The Legend into the explanation, we lose. It has been tried for years, without success. Right or wrong, that seems to be a losing argument ... for us. It gives the Doctors the easy occasion to frame the argument as being between science and myth. As the old saw goes, 'Science deals with what is real, The Legend's lot is what is not.'"

"If we stick strictly with science, where does that get us?" Jenny asked.

"The collection of data through observation and experiment, and the formulation and testing of hypotheses," Kent offered. The others looked with disbelief.

"Wow, someone was paying attention!" Kathleen laughed.

---

[1] *See*, End Note [54], page 363.

[2] *See*, End Note [55], page 365.

"How do we do that with the cave painting?" Jenny asked. "We can't really reproduce it. It was a one-time historical event."

"Think of it as a criminal investigation," someone offered. "How would you go about finding the perpetrator?"

"Look for clues, a smoking gun," Kathleen offered, pretending to blow smoke from an imaginary gun. "You get all the leads and make up competing scenarios, see which one makes the most sense. You keep testing and re-testing against competing hypotheses." Kathleen sat back as heads turned with expressions asking where did that come from? "I took a forensics class." Kathleen smirked.

Jenny looked up in sudden interest, her eyes staring through Kathleen. Before she could speak, Brian, a senior and an earth science major, introduced himself to Jenny and joined the group. I've been thinking about this for years," he said, "and I keep coming back to a key question: What is the fundamental difference between, say, stalactites and pictures? If you can pinpoint that, you may be able to make some progress here. But I wouldn't count on it. The topic is too rhetoricized any more. It's hard to communicate."

All the heads turned. "Rhetoricized?" someone asked.

"You know what I mean. The stereotypes, the clichés. It seems just mentioning The Legend conjures up images of naïve simpletons trying to turn back the clock of science.[3] The subject is in danger of becoming too polarizing for any resolution."

Jenny was thinking. "You're on to something, Brian."

"What did Brian the rhetoricizer say?" Kathleen asked.

"The part about the fundamental difference between a stalactite and a painting. What is it? That's a key piece of the puzzle that's missing. Help me here. What can we say about the differences?" For a moment Jenny's heavy heart forgot about Alex and she considered Brian's question, knowing as he did

[3] *See*, End Note [56], page 366.

that there was a difference, but the challenge would be to articulate it.

"Well, there is a difference in complexity, you know, the level of detail and the way all the parts are combined," Kathleen offered. "One is certainly more complex than the other."

"Yeah, but like Kent was saying, Dr. Hawkins will just say that is a difference in degree, not in kind," Jenny said. "But that's still good; keep thinking."

"But a stalactite is not that complex," Kent said. "I mean, I guess it depends on how you define complexity. But a stalactite is really nothing more than the same thing over and over again. All it requires is mineral water, gravity and time."

"All a painting requires is pigment and a wall," Kathleen said, playing the adversary. "And time."

"Keep going," Jenny said. "We're close. Is the painting really just pigment and stone?"

"I said *a wall*," Kathleen smiled. "But, seriously, it seems that it is not just the level of complexity that's different. There is a difference in organization. A stalactite can form by just allowing enough time for natural laws to organize the outcome. But it doesn't seem that natural laws could organize a picture—at least not a picture that isn't just a simple repeating pattern."

"I think you're right, Kathleen," Brian added. "Just because something is complex does not necessarily rule out natural causes. The way the parts are organized, however, could have some relevance to determining its origin." The group continued discussing the topic until Jenny realized it was getting late, and she did want to see Chris.

"I've got to go, but let's meet again and consider this some more," Jenny said. "Do you mind if I join your meetings more often?" Jenny asked, knowing the answer but being polite.

"Of course not," Kathleen said as she stood in the doorway, "at least not as long as you believe The Legend."

"The *truth*," Jenny smiled. "It's the truth that I believe."

Brian, standing behind Kathleen in the doorway, gave Jenny a knowing wink and a smile.

Chapter 15

My watch must be wrong, Jenny thought. Every light in the dorm was on, and music seemed to be coming out of every room. She hesitated before Chris's door, not really wanting to go in. But the prospect of returning to her half-empty room was too much to even think about. She reached up to knock when suddenly the door swung open.

"Don't hit me!" someone yelled.

"Sorry," Jenny laughed. "I'm looking for Chris."

"Are you going to hit him?"

Jenny laughed again.

"If not, then *my* name's Chris."

Again, Jenny laughed, and it felt good.

"Actually, my name's Ben, but you can call me Ben." He opened the door wide, and looked back. "Hey, loser, there's a serious babe here to see you."

Jenny stepped in and looked around. She wasn't sure if she was hearing two stereos from Chris's room, or if one was coming from another room. Chris jumped up and gave Jenny a hug. "Thought you might need that," he said.

"Four computers?" Jenny asked as she looked around.

"Six, actually; we both have laptops."

"What possible use can you have for six computers?" Jenny asked in amazement.

"Three. I only have three," Chris said, plopping down before a large monitor. "Watch this." Jenny watched as Chris brought the keyboard over his lap. "I'll show you what this computer's good for." Chris started clicking away, his back turned to Jenny, who was still standing.

Jenny looked around, thinking, *this is a party?* Still, the noise and distractions held back the persistent heavy air of Alex's absence.

Ben looked over his shoulder, "You on?"

"Yeah," Chris said, leaning over to look at another monitor. Soon Chris was dividing his time between the two monitors, periodically leaning over the second, typing a few words, and then returning to the first.

"What are you doing?" Jenny asked, partly to remind Chris that she was there.

"Talking to Ben," said Chris, typing non-stop. Jenny looked across the room where Ben was typing at his own computer.

"Excuse me?" Jenny asked.

"We're I.M.'ing," Chris said. Jenny looked again at Ben, and again at Chris.

"You're doing *what*?" Jenny asked.

"I.M.'ing, *instant messaging* each other," Chris said. "You really got to get with it, Jen." Alex was the only one who called her Jen. She wanted to sit down. *Why is it so hot*, she wondered.

"I use this old computer for I.M.'ing, bulletin boards, certain e-mail accounts, etc." Chris said, leaning over and typing. "That way I don't expose my new baby to so many virus threats," Chris continued, patting his computer like it was a favorite pet.

"Can I sit?" Jenny asked, wondering where she might sit if the answer were to be yes. Without looking away, Chris pulled a plastic crate from under the computer table. "Here," he said, "make yourself comfortable."

Jenny sat, trying to remember why she was there with two guys, both ignoring her and each other, except for "I.M.'ing" and an occasional whistle. Jenny watched Chris's screen for a few minutes as he typed in a cryptic code back to Ben.

"Why don't you two just ... oh, never mind."

"Because we're talking about you," Chris laughed. Jenny searched the screen.

Chris and Ben clicked away. Music flooded the room. But Jenny sat in virtual silence. *Multiple hypotheses, competing scenarios, smoking guns*. The ideas were persistent but ephemeral as Jenny sensed their importance but unsuccessfully sought a connection. Thoughts of Alex cast over Jenny's mind like a great shadow, dimming all other thoughts, pushing them to obscurity. Over and over she recalled the determination in Alex's voice as she packed to leave.

"Oh, crap!" yelled Chris. "Look at this!"

Jenny snapped back to the party and looked at Chris's screen.

"Looks pretty to me," Jenny said, "what's wrong with it?" The screen was filled with random colors in a somewhat zigzag pattern similar to flame-stitched fabric.

"That," said Chris, matter-of-factly, "is a picture of my sister and her new dog. Crap."

"You're right," Jenny said. "They look like crap." She surprised herself with the comment. "Looks like your new baby caught a cold. Why don't you e-mail Ben for advice?" Jenny offered sarcastically.

"It's not e-mail, Jenny, get with it," Chris said.

Jenny was glad he didn't say Jen, but even thinking that triggered a tumble of thoughts of Alex.

"Try again. You probably just lost the four-one-one," Ben said as he shot down something that looked like a space ship.

"Yeah, you lost the four-one, one," Jenny said in mock agreement.

"It's not the four-one, one; it's four-one-one, as in the telephone number for information. Get with it, Jenny," Chris joked.

"OK, I don't really care, but because this party is winding down, tell me what four-one-one is," Jenny said with a tired voice.

"Information," Chris said, hitting the keys harder in an effort, apparently, to make the computer do what he wanted. "I lost some information in the transmission."

"Oh, no," Jenny said, sarcastically. "You want me to go out and look for it?"

"If you don't get the four-one-one, you might get an interesting picture, but it's not what you want. I mean, if you want something meaningful." Chris was distracted now, checking cable connections. Jenny was suddenly tired and would have lain down except for the fact that the beds looked like they had not been made since the start of school and were piled high with clothes, books and sports equipment.

"You got to tell the bits where to go," Chris said, partially submerged under his desk. "You do know what bits are, don't you?" He looked up at Jenny. "You don't *live* in that cave, do you?"

The cave. Jenny didn't want to even think about the cave right now. Thinking of the cave made her think about Alex, and thinking about Alex made her want to re-wind time and start over.

"I could put your cave picture on here, if you could find a way to send it to me." Chris said, back at his seat and pointing to his monitor.

"I know, but only if you have the four-one-one," Jenny said with a sneer as she shifted uncomfortably and looked around for a pillow.

"You got it!" Chris and Ben said together.

"Well, anyway, the bits have to come through the way the program says, or else you just get noise. It may be neat noise, but it's just noise."

Ben and Chris were agreeing with each other. Jenny sat with her elbows on her knees and her chin in her hands. Ben continued typing as he talked, occasionally pausing so as to avoid distraction from his screen.

"Noise and neat noise can be interesting, but if you have the four-one-one you get a picture."

Suddenly Jenny bolted upright. "Chris!" she practically yelled, "Say that again!"

Chris jumped. "What? Say what again? I wasn't even listening to myself. Ben, were you listening to me?"

"Nope," Ben replied, "not unless you I.M.'ed me. But I'm sure whatever it was, it was *brilliant*!"

"You said something about noise, neat noise, and pictures." Jenny pulled her plastic crate up closer. "Tell me about that."

"Of all the things you could talk about, you want to talk about neat noise?" Chris asked. "Jenny, you need to ..."

"I know, I need to get with it. Just tell me, quickly."

"It's easy," started Chris, "and it's all about the four-one-one: information. You see, this computer is really dumb; it uses bits, but it can't just use random bits, the bits have to be in a certain order. Only when the bits come into the computer in the right order can the computer do something useful—like make the picture of my sister and her dog." He tried a few more clicks. "Something, or really *someone*, someone smart like me," he leaned back, pushing his shoulders back in mock pride, "has to tell the bits where to go—that's the four-one-one, to know what to do with the bits."

Jenny interrupted, "So, if the informa— I mean the four-one-one gets lost, or scrambled ..."

"Or is never generated in the first place," Chris interjected.

"Yeah ... yeah, that's right, I suppose," Jenny said, thinking before she continued, "then the computer doesn't know what to do. It tries, but with no information content the completely random bits make a random picture, and that's called noise."

Chris stopped and turned to look at Jenny, surprised both that she understood and that she cared.

"It's not just for pictures; if you want this computer to do anything—word processing, download songs, *whatever*, you've got to tell it what to do with the bits."

"Keep going," Jenny moved closer. "Tell me more about noise—and pictures."

Chris looked at her curiously, shrugged and said, "You can send a lot of bits over randomly in some complex pattern and you will get some kind of image, maybe even something you might describe as a picture. The picture itself will be complex, hard to describe precisely because there is no order to it, no specificity, you could say, just a bunch of spots of color. That's noise."

"Can I write?" Jenny suddenly interrupted, looking around for paper.

"Sure," Chris said, still looking strangely at Jenny. "On paper?"

Jenny stared at Chris.

"Well you could use my old computer," Chris said, motioning to another keyboard.

"Can I just have a piece of paper?" Jenny asked impatiently, looking around the desk. Chris pulled a few sheets out of the printer and gave them to Jenny. Jenny scribbled to get the pen writing and then jotted down some notes before looking up at Chris expectantly.

Chris started, hesitantly, "So ... you got noise, right?" Jenny nodded as Chris continued, "Or you can send the bits over in a simple repeating pattern, much less complex, but much more specified. You will get another picture, a little more interesting. It might be a recognizable pattern, lots of specificity, but not very complex. Just the same thing over and over again. That's what I call 'neat noise.' Or," he turned again to face Jenny, who was already practically in his face, "you can put in the ..." He stopped while Jenny said loudly, "The four-one-one, and you get a picture!"

"She's got it," Chris yelled to Ben. The two of them stood up as Jenny finished scribbling feverishly.

"Four-one-one, four-one-one, four-one-one," Chris and Ben started in a slow chant as they high-fived each other in the center of the room. Jenny shook her head, laughing.

"You guys are nuts!" Jenny laughed. "But good nuts; you may have even helped me crack a tough one!"

"Want to know more?" Chris asked Jenny as he sat back down, remembering that he was trying to get some four-one-one to his screen.

"I think I get the idea. To make something like a picture, an image, let's say, that is itself complex and specified, it takes information. The information, in the case of a computer image, is in the form of a string of bits having a highly complex, but specified pattern or sequence. Right?"

"Right!" Chris said. "Actually, strictly speaking this is *syntactic* information—information that is expressed in complex strings of smaller elements, like bits. This is information theory stuff. Are you sure you're interested?" Chris asked with a contorted face of unbelief, still surprised that Jenny cared. But Jenny's bright eyes were eager for more.

Chris continued, "Well, more importantly for us is another type of information called *semantic* information, that is, information in which the order is so well specified that, in addition to being syntactic, it is *meaningful* to us. Granted, there's some subjectivity in determining meaning, but for some purposes meaning is clear and undeniable." Chris stopped while Jenny wrote. "Still with me?"

Jenny nodded without looking up.

"Chance alone could produce only disordered, complex sequences that show no consistent patterns—no *specificity*. Some kind of mathematical algorithm could produce relatively simple repetitive sequences that could produce a picture like you saw earlier on my screen. Somehow the ordered data got scrambled and a portion of the sequence just repeated itself all across my monitor. But to produce a meaningful picture you have to have some way of getting semantic information, or *complex, specified information*.[1] And it has to be perfect. If there is one bit out of order, you're not likely to get anything useful. That happens

[1] *See*, End Note [57], page 366.

sometimes when you copy data, for example. Very, very rarely, but it happens. If there's a copying mistake, your information can be seriously degraded, if not destroyed."

Jenny was listening intently now as she continued to write. "So you can't get a good copying error?" Jenny asked, being facetious.

"Well, I don't know if you *can't*. I guess anything's possible, but such a thing is called a permutation, and it's almost always fatal to information," Chris said, in all seriousness. "And the results, like this picture of my sister and her dog," Chris again struck the keys, looking sideways at the screen as an hour-glass symbol went around on the screen, "are called permutants. I've never seen a good permutant."[2]

Chris was lost in his monitor again. Jenny finished her notes and now wanted to go back to her room and think about this some more. As she got up to leave, she said, "Chris, I've got to go now, but before I go, tell me, just so I know for sure, where are all the places the four-one-one can come from?"

"Only one place to get the four-one-one," Chris said, tapping his head. Jenny looked at Chris with a wide grin.

"I thought so. Chris, you are a *genius*!" Jenny said, giving him a hug.

"With a capital J!" Chris chimed.

"What about me?" Ben said, getting up for a hug as well. Jenny hugged him and went for the door.

"Hey, I'm going to leave you party animals," Jenny said. "I'll see you later."

"Wait. Here they are!" yelled Chris, looking at his screen.

"Which one's the dog?" Ben asked, dodging a fake punch from Chris.

Jenny practically sprinted down the hallway. Ben and Chris stood looking at the computer, smiling.

---

[2] *See*, End Note [58], page 371.

Chapter 16

Jenny stared at the ceiling, wide-awake. She found that as long as she didn't look at the other side of the room she could forget for a moment that Alex wasn't there. But no matter what she tried, her eyes were drawn like a magnet to the bare walls with their patterns of holes where thumbtacks used to be.

I wonder what Alex is doing now? Jenny kept thinking.

Jenny got up and sat at her desk. The only distraction effective in displacing the thought of Alex was the notion of pictures, noise, and neat noise. Jenny began to write, making charts and diagrams, attempting to put her thoughts on paper.

Information. That's the fundamental difference Brian was talking about. To make the picture on the wall of Lavoux Cavern requires information that the stalactite doesn't. Does it?[1]

Jenny was getting tired thinking, but each time she lay down she thought of Alex, so she forced herself to stay up.

Noise corresponds to random, complex patterns, like wind-blown leaves. Chance can do that. Neat noise is like crystals, highly specified, simple, repeating patterns. Natural law "algorithms" can do that. But a picture has specified complexity; the parts are placed in a complex manner with specificity. Semantic information. Can nature alone do that?[2]

She continued writing far into the night, eventually falling asleep on her desk, her head resting on her notes.

---

[1] *See*, End Note [59], page 381.

[2] *See*, End Note [60], page 393.

Jenny entered Dr. Hawkins' classroom like most of the other kids, with the energy of someone who had stayed up most of the night. She looked over at Alex's seat—empty. Her heart ached and her brain was tired, but her spirits were high as she reviewed the notes she had made the night before.

"Kathleen," Jenny whispered across the room. She motioned for Kathleen to come over, which she did. "Hey," Jenny whispered, "I think I might have the answer."

"The answer?" Kathleen asked excitedly. "No way!"

"When can we get together again?" Jenny asked.

"Well, tonight I am meeting with another group, not just students; maybe we can meet then." Dr. Hawkins entered the room. "Talk to you later," Kathleen said as she hurried back to her seat. Still no Alex.

"Yesterday we left with a very bold statement from one of your classmates." Dr. Hawkins was in good spirits also as he got right to the point. "Maybe now Jenny would like to explain some of her reasoning, even though I'm reluctant to be overly indulgent of re-hashing settled controversies. Jenny?"

"Yes, I'm here."

The class was full today, except for Alex. Jenny's mind was racing. She hadn't planned on engaging Dr. Hawkins today. *What was I thinking?* She thought.

"Jenny, you are of the *opinion*," he stressed opinion, "that the cave markings in Lavoux Cavern were made by a person. Right?"

"I think that is abundantly evident," Jenny said, unable to suppress taking the bold approach again.

"'Abundantly evident,' you say. Maybe you can just give us your most convincing argument, keeping it to, say, five minutes?"

"Dr. Hawkins," Jenny started, her courage waning, "actually, if I could, I would like to have some time to prepare.

I've come across some new," she paused, "information on the subject."

"Jenny, we don't have a lot of time in this class. We are learning science, and there is a lot to learn. Perhaps we can debate this outside of class."

A debate! Jenny had never thought about that, and, frankly, the thought was somewhat frightening.

"A debate?" Jenny asked cautiously, suddenly feeling very conspicuous.

"A debate. You know, you and someone else each presenting opposing positions … before your peers." Dr. Hawkins made it sound like a dare.

"A debate might be very fruitful." Jenny found herself sounding a little arrogant, as if the voice were coming from someone else.

"I think a debate can be arranged. And just whom would you like to debate?" Dr. Hawkins asked casually, getting ready to jot down a note.[3]

"I would like to debate Dr. Seeting," Jenny said loudly, now feeling very, very overconfident. She shot a desperate glance to Alex, forgetting that Alex was gone. Looking to Kathleen, Jenny found her staring back with a wide-eyed look of shock. Again the class buzzed, finding the prospect of a student-professor debate enlivening.

"Dr. Seeting it is," Dr. Hawkins said slowly as he wrote, clicked his pen and returned it to his pocket.

"Now, class, if you will open up your books, back to page 446, let's pick up with Neihil's mechanism of …"

Jenny was there in person only; her mind was a bog of thoughts, her stomach becoming a tightening knot of fear. She only vaguely heard the bell and Kathleen speaking to her.

"Jenny, let's go." Kathleen was leaning into Jenny's face. Jenny got up, realized where she was, and hurried out with Kathleen, who slipped her a piece of paper with an address on it.

---

[3] *See*, End Note [61], page 394.

"Be there, if you want. There will be others, teachers and people like that."

"Teachers?" Jenny asked, "Any science teachers?" Recalling Drs. Clark and Welch from her first meeting of the Student Ancient People, Jenny hoped to discuss the origin of the cave painting with them.

"Dr. Clark? Dr. Welch? Jenny asked, staring intensely at Kathleen.

"Probably. And at least one more, Kozak," Kathleen stated, looking in the air to think of more. Jenny was also thinking. She knew the name Kozak.

"Old Man Kozak!" she suddenly remembered. "I think I know him! He's a science teacher? Didn't he used to be a principal?"

"I think so," Kathleen replied. "But he lost that job, as well as a wife, and a lot of weight somewhere along the way."

"He's of the Ancient People?" Jenny was awake again. "And a *science teacher?*" Her mind was in high gear once more.

Jenny drove to the address Kathleen had given her. She was early—she didn't like getting to social gatherings late as a visitor. She hoped Kathleen had already arrived. As she walked up to the door it opened and an older man in a suit greeted Jenny.

"Hello, my name is Rex, Dr. Rex Bender. You must be Jenny; Kathleen told me you might be here. Come on in, and welcome."

Another doctor! Jenny thought. *A medical doctor, perhaps? How many doctors do the Ancient People have?* Somehow the thought of educated Ancient People still struck her as odd. Progressionist literature always insisted that no scientists question progression as the cause of the cave paintings. Dr Hawkins once referred in class to those scientists who believe the painting was made by a person as "those who only pretend to scientific credentials."

"You're the first one here. Come in and have a seat. I'll be right back in."

Rex led Jenny into a large living room and then excused himself into another room. Jenny sat at the end of a long sofa facing a large wall of bookshelves decorated with art and souvenirs from around the world. Soft music emanated from hidden speakers. A large decorative mahogany Chinese checkers board was on the end table next to her, with brightly colored marbles perfectly placed in a star pattern. *That's odd,* she thought; *I haven't seen Chinese checkers in years.*

The entire house was neat and uncluttered. Pictures of children and grandchildren adorned the walls. Jenny could see through one door into an office that had at least three diplomas on the wall. She was tempted to go in and take a closer look when the doorbell rang. Fortunately it was Kathleen; she arrived with Brian. Jenny was glad to see them both. Soon the room was filled with people, old and young.

"Kathleen," Jenny whispered, "which one is Kozak?"

"He's not here yet, but he will be." Kathleen winked. "Here, talk to Rex."

Rex? Seems informal, Jenny thought. She found Rex easy to talk to, however, and soon she was lost in conversation as Kathleen went off to mingle with the other guests. As they talked, Jenny was pleasantly surprised to see Dr. Clark coming in the door. Rex noticed Jenny's look of surprised.

"Have you met Dr. Clark?" he asked.

"Yes, we met once earlier this year."

Jenny stepped over to get Dr. Clark's attention and shake his hand. "Dr. Clark!"

Dr. Clark smiled broadly and gave a questioning look to say that he recalled seeing Jenny somewhere, but could not remember her name.

"Jenny. Remember we met at a meeting of the Student Ancient People?" Jenny said. "You were with Dr. Welch; you were 'earth' and she was 'wind.'"

"Yes, I do remember," Dr. Clark said, laughing at the recollection. "Wow, it's good to see you again Jenny. I thought that after that night we would never see you again. You can call me Jim."

Jenny, Rex, and Dr. Clark were enjoying a thoughtful conversation when Kathleen returned and hit Jenny's elbow, nodding toward the door. In came a man Jenny would never have recognized; he appeared remarkably young for someone known as "Old Man Kozak" ten years ago.

Rex looked toward the door as well, reaching out his arm, "Bernie!"

Bernie? Jenny thought. *Old Man Kozak is a 'Bernie'?* She smiled at the thought. Kozak entered, shaking Rex's hand without taking his eyes off of Jenny. He and Jenny stood looking at one another for a few seconds, each mentally fitting the old face with the new. Kozak then reached out to Jenny.

"Jenny, it's been awhile."

"You remember me?" Jenny was aghast.

"Yes, I do. We should talk."

Kozak floated over to sit down. Jenny looked at Kathleen to ask what to do. Kathleen leaned her head in the direction of Kozak, mouthing, "*Go sit.*"

Jenny nodded to Dr. Clark and Rex before following Kozak, eventually finding herself squeezed onto the same sofa from which she started. Looking closely at Kozak's face, Jenny's thoughts went back to the days of seeing him troll the hallways. *He always seemed so mean*, Jenny thought.

"Mr. ..."

"Bernie, call me Bernie," Kozak said kindly as he looked intently at Jenny. "It's been a long time, Jenny, but I am very glad to see you again. You've become a very beautiful woman."

Jenny blushed. Kozak looked away.

"How in the world do you remember me?" Jenny was still amazed. "You must have seen thousands of kids go through your school."

"Because, Jenny," Kozak's eyes looked like they were watering, "you are the only student to whom I ever owed an apology." Kozak looked away again.

"An apology!" Jenny exclaimed, "Whatever in the world for?" Jenny felt genuine tenderness in his voice. She sensed he had known more than his share of pain since she saw him last.

"Do you remember your report in the fifth grade on the cave figures of Lavoux Cavern? And Ms. Hart?" Kozak asked as if Jenny might have difficulty recalling the day. Jenny nodded, still questioning with her eyes. Kozak continued, "I was the one who told Ms. Hart you could not report that a person painted the picture in Lavoux Cavern."

"I know that," Jenny said, softly. "Ms. Hart said it was because of science guidelines, or something like that."

Kozak looked down. "I was, I *am* of the Ancient People, Jenny; of course, I should have known better anyway. It was hard, Jenny. I had a career. I had a family. I'm not making sense, am I?"

Kozak looked up at Jenny with an expression between laughing and crying. She wanted to put her arm around him.

"I finally let my views be known. I'm teaching science—but not Earth Science. I'm not an administrator anymore."

"Mr. Koz … I mean, Bernie," Jenny said, putting a hand on his shoulder, "I think I understand. Although not at all necessary, I accept your apology. In fact, one of the reasons I'm here is to help try to change the way things are, to change the rules, so to speak, for people like you, and ultimately for everyone."

☙ ❧

Kathleen and Rex had begun to get everyone's attention. The room was full, but eventually everyone was quiet and seated, some on the floor, some on the stairs leading to the second floor. Jenny was surprised to find that they expected her to speak. She looked around from where she sat; all eyes were on her. Jenny noticed that Dr. Welch had joined the group, and gave her a nod and a smile.

"I'm not sure what you all expect to hear from me. I hadn't actually anticipated being the featured speaker." Jenny looked around nervously.

"Jenny," Kathleen's voice was kind, "I've told a few of the people here all about Hawkins' class and the stand you took. I've also told them that you are going to debate Dr. Seeting, and that you had some new information that might help prove that the painting in Lavoux Cavern was painted by a person."

The thought of debating Seeting brought a brief sick feeling, and Jenny actually shuddered. She found a smile, however, and decided to stand.

"Well, that was certainly nice of you, Kathleen," Jenny said with friendly sarcasm, smiling toward Kathleen. "Actually," she addressed the group, "I sometimes think the world must be crazy to even have to debate whether the painting in Lavoux Cavern, or any painting, was painted by a person." Everyone agreed. "But be that as it may, I think we can put forth a strong scientific argument that at least should be worthy of consideration." More nods and expectant faces.

"I said *scientific*, and," she hesitated, "what I mean by that is an argument that does not rely on, or even refer to, The Legend."

She waited, seeing only a few knowing nods. *Don't lose them*, she thought.

"I'm not saying The Legend is not important," she tried again. "In fact, I believe The Legend. But, we must, or *I think* that we must, find a way to communicate to scientists in their own language." *More nods than before*, she thought.

"Maybe you should tell us what you have in mind," Rex said in a friendly, interested voice.

"Well, it can sound complicated at first, but let me try. Maybe this group can help me refine my arguments for the debate."

Jenny fished in her pocket for the paper with her notes. Unfolding it, she looked around her.

"Brian started me thinking when he asked, 'What is the fundamental difference between a stalactite and the wall painting?' Both are found in Lavoux Cavern, but only one can reasonably be explained to be a result of progression by natural selective inclusion, while the other can't." Jenny glanced at Brian, who indicated his agreement.

"The question, though, in general is, how did any of the things we see around us get here?[4] In other words, what does it take to make, or create, the things we see or discover in nature? To answer that question, we must consider how complex the thing is, and how neat it is, so to speak. In other words, is it only complex, like ..."

Jenny was struggling with her own notes. She paused for a moment and then, looking about the room again, had an idea. Pointing to the Chinese checkers board, she asked, "Can I use this?"

"Sure," Rex answered. The others looked around in anticipation as Jenny moved over beside the board.

"OK," Jenny replied slowly. "Be patient, I'm making this up as I go."

Everyone laughed. Jenny began gathering up all the colored marbles in her hands.

"If I were to pick up all these marbles," she said, holding her hands in the air, "and drop them on the floor, the pattern they would make after coming to a rest would be very complex and not very neat, not specified in any way, but randomly distributed like blowing leaves or falling raindrops, or," she paused, searching for another example, "like a rockslide!" Jenny said, suddenly recalling the cave. "I won't do it," she laughed as

---

[4] *See*, End Note [62], page 396.

she continued to hold the marbles high. Others laughed and shifted to more comfortable positions. "But you see how such a pattern would be complex in the sense that it is random; there is no particular design to it. In other words, it would have very little specificity, that is, there's no particular significance to where any particular marble ends up."

"Now," Jenny continued, "suppose that instead of dropping the marbles on the floor, I dropped them on the Chinese checker board." She looked around. "What would happen?"

"They would probably bounce and roll around until they came to a stop, and most would probably stop when they dropped into one of the little depressions," someone said.

"And together the marbles would make a star pattern," said another.

"That's right!" Jenny was seeing that her illustration might work. "The marbles would make a pattern that is not random, that is, its design is less complex, and it has quite a bit of specificity determined by the constraints of the board, these little holes." She leaned over the checkerboard, pointing out the holes. "The marbles would end up in specific places to make an identifiable shape, like atoms of crystals, or the minerals of stalactites. And, like the atoms in a crystal formation, the marbles would exhibit the same pattern over and over again." Jenny was thinking out loud; the thought connected with Rex and Kozak.

"Now, consider this," Jenny slowed down. "What if," she paused, "I was lost in the desert, and I wanted to use these marbles to spell the word "help"? Could I drop them, or throw them, or let the wind blow them, and expect for them to form the letters H, E, L, or P, much less in the right sequence?"

The looks all around the room said "of course not."

"Even if I had a million years and a million tries a year?"

Again, looks on the faces indicated "no, not even then."

"Why not?" Jenny was gaining momentum, but she waited for a response as she struggled to quickly think through one herself.

"It's because," Jenny finally said impatiently, "the marbles don't know where to go!" She looked around. "Get it?" Jenny looked about the room, eyes dancing expectantly from face to face.

Rex said, "I think so; keep going."

"In the first two examples, the marbles didn't know where to go either, they were just going where they had to go, you know, just going where nature determined."

"To the lowest available energy state, within the constraints of the system," Rex stated.[5]

"Right, second law of thermodynamics," added Dr. Welch.

"The downhill slide to maximum entropy," Dr. Clark added knowingly, with a hand motion adding emphasis. The three continued talking among themselves for a moment, eventually ending with nods of agreement.

Jenny wasn't sure what they were talking about, but she nodded as well while they spoke. She was glad for the side conversation; it gave her time to think.

"That's right," she assumed as she continued. "In the first case the marbles would make a complex but unspecified pattern, not very repeatable. In the second case, the pattern would be very repeatable, specified by the little indentations in the board. But it would be relatively simple. Nature can do both of these by itself. Chance can make complex patterns; natural laws can make simple, repeating patterns."

"Given the marbles and natural laws to start with," Kozak clarified.

"But in the third case," Jenny continued, still trying to keep her thoughts ahead of her mouth, "there is both specificity

---

[5] *See*, End Note [63], page 397.

and complexity. But for this nature alone is of no help. Something is missing. What is it?"

"Intelligence!" someone shouted.

"That's part of it," Jenny said quickly, looking toward the voice with a smile. "But there's more. What is it that intelligence can do that nature alone cannot do?" She couldn't wait for more guesses, "The marbles need direction—they need *information*, you know, *four-one-one*."

Nobody got the four-one-one part, but some were beginning to see where she was going.

"Something, or, I should say, some*one*, needs to tell the marbles where to go, where to be placed, so a meaningful message can be formed."

Rex and Kozak were nodding to each other. "I think you're right," Rex said. "In the third case, something needs to make choices, to make particular selections out of an infinite number of possible selections. That is the essence of information theory."

"That's right," Jenny continued. "And where is the only place such information can come from?" Jenny asked as she waited again, looking closely at each face. "That's where intelligence comes in. Information can come only from the mind of an intelligent being, that is, a *person*! Only a person can make specific choices among multiple possibilities and place the marbles in a pattern that is both specified and complex!"

She finished with a flourish, and almost forgot to connect the final dot.

"The painting in Lavoux Cavern is like the word 'help' in the desert. It is ordered, it is complex, and the placement of each bit, each line, each color, each *marble*, so to speak, is specified! Each component of the painting had to be directed as to its placement, its size, its color, its *whatever*, and all this direction is *information* that must be supplied from the mind of a person."

The room was now buzzing as each person began adding to the theory with various thoughts and comments. Jenny found herself suddenly tired, but the conversations went on late into the night, and they tended to have an uplifting effect on her. She was beginning to believe that a debate just might be effective.

Finally only Jenny, Rex, Dr. Clark and Dr. Welch were left, each reluctant to leave. Jenny's enthusiasm increased as they talked, and all four sensed a breakthrough in the debate. In addition, Jenny learned more about the pressures facing scientists who believe the cave painting to be the work of an artist. She learned of those like Kozak who had lost jobs, or had been silenced with the threat of litigation for suggesting human origin of the painting. But she also discovered that there was a small but growing number of scientists who questioned the naturalistic explanation of progression. These scientists, doctors, and other academic and professional people were quietly working to change the institutional bias that insisted on and enforced the progression-only teaching of science at universities like Lutrell. The realization that she was not alone in her convictions and her scientific understanding gave Jenny new-found strength to continue to question, and new energy to face her fears in the upcoming debate.

After several attempts to leave, they all finally stood, apologized to Rex for overstaying their welcome, and headed to the door. Standing at the door, Rex shook Jenny's hand hard and long. "Good luck with the debate. Let me know when it's scheduled; I would like to come see it myself." he said warmly.

Drs. Clark and Welch followed closely behind.

As they left, Dr. Welch leaned over to Dr. Clark, and said with a grin, "I think we may have found fire." They both smiled at the thought.

Chapter 17

The stares of those sitting at nearby tables finally registered with Jenny as she suddenly became aware she was tapping loudly on the counter. Forcing her hand down flat, she turned away self-consciously, feeling the annoyance linger until each stare slowly went back to reading. Behind the counter where she stood was the most prized collection in Lutrell University's library—the Benton Collection, otherwise known as the rare book room. Jenny tried to match memory with reality as she scanned the fine woodwork with its intricate details and polished finish. The dark wood wainscoting below transitioned into long, narrow windows that extended to crown molding around a high ceiling that was also made of dark wood panels.

While Jenny was looking up a woman appeared from behind a dividing wall that separated the counter area from the protected shelves. Jenny thought she might be a librarian, but according to her nametag she was "Sarah, Assistant Historian, The Benton Collection."

"May I help you?" asked Sarah as she lifted a stack of papers up and onto the counter. Sarah and Jenny looked hard at each other, each sensing that they had met before.

"Have we met?" both said, almost simultaneously before starting to laugh. After a few minutes of comparing classes and teachers, they decided that possibly they had seen each other in the library.

"Well, anyway, can I help you?" Sarah asked again, still smiling and studying Jenny's face.

"Yes," Jenny said in a loud whisper, trying to be quiet as she glanced inconspicuously toward the readers. "I understand you have some materials here about the Ancient People."

"Oh, not another one!" Sarah almost laughed. "What is so popular about the Ancient People these days? I haven't seen so much interest since our original manuscripts were removed."

"Were removed?" Jenny asked with a look of surprised disappointment.

"Well, we only had two of those books; those Ancient People don't seem to write much down," Sarah said with a questioning look.

"What happened to the two you had?" Jenny asked, showing her disappointment and looking around the counter as if perhaps the books were simply misplaced.

"We haven't had those in years, probably ten years," Sarah said, thinking as she shook her head. "They've been out of this collection since before I started here. Those books had apparently sat undisturbed for years, and suddenly there was intense interest in them."

Jenny's heart sank as she realized that she may not be able to compare The Legend she heard with the writings she remembered reading as a child. She looked Sarah in the eye hoping there might be some kind of mistake.

"Are you sure?" Jenny asked. "Could you just check?"

"I'm sure. Look, come back here," said Sarah with a smile and a quick glance around. She motioned Jenny to a side gate actuated by a control at the counter. The loud click opened the gate and again turned heads of some who watched Jenny as she carefully closed the gate while closing her eyes in a vain effort to suppress the sound of another noisy click. Once back with Sarah, the two went into the repository that consisted of polished wood shelving holding not only books but also paintings, sculptures and other works of art. Jenny was impressed with the carefully arranged glass-fronted cabinets of books. Most of the cabinets were locked. Several had large books laid open with elaborate stands and special lighting to show hand-written lettering. The entire room looked like a cross between a fine library and a museum. The two walked into a small room furnished with leather chairs and a small table.

"Right here," Sarah pointed to an empty cabinet. "Here's where our books about the Ancient People were. You can see there are a few related books, but the main manuscripts have been gone for a while. They were removed to be placed in an offsite collection of historical documents."

"Offsite?" Jenny asked as she walked over and ran her finger over the empty cabinet. Two small cards lay on the cleanly dusted shelf marking the spot as placeholders for the missing manuscripts. Jenny leaned near the open cabinet and then looked around with a strange look.

"Are you permitted to smoke in here?" Jenny asked, still looking around with a questioning look.

"No, but you have a good nose!" replied Sarah. "These books are often studied by Professor Manning. He's known around here as Manny, or, if you're one of his faithful, as 'The Man,'" Sarah said as she looked around in affected reticence. "What a character," Sarah laughed. "He's been around from the time when you *could* smoke in here, and he still insists on smoking at home with these materials. They always come back with the smell of stale cigar smoke. That old man has given us unending grief," Sarah said, shaking her head. "But you can't do anything about it; he's the leading authority around here on the Ancient People; he is The Man."

"What does he do with the books?" asked Jenny, looking at the two small cards and still concerned as to why the books were gone. "I mean, why does he have to take them out of here?"

"Do you know what the books are?" asked Sarah.

"Yes, sort of," answered Jenny. "I actually got to look at them and read parts of them some years ago, when I was a kid."

"Really! Here?" Sarah was surprised. "That must have been ten or more years ago, right? You must have been a *little* kid!"

"Yeah, I was here with my aunt, doing some research," Jenny said.

"As I understand it, that was about the time when the importance of the books was first recognized. At least that's when suddenly people like Professor Manning became interested. He's studied them thoroughly since, including studying the original languages, comparing translations, writing commentaries, that sort of thing. Much of his work has been published and he's the leading authority on interpretation of ancient writings."

"What does he write about? Do you know?" asked Jenny.

"Mostly boring stuff," laughed Sarah, "even for an assistant historian like me!" Sarah laughed again with a glance down at her name badge. "Seriously, his last book was subtitled 'Systems of Interpretation,' and he bored me with the details all afternoon one day. His big thing is textual criticism. He takes the original documents and by his method of what he calls 'higher criticism' he's able to strip away mythical elements of the writings. It seems he is very effective—he's shown how little of the existing writings of the Ancient People is actually of any use as history. Most of it can be attributed to progressionary revisionism. Although he's a historian, he's had a huge impact on other disciplines, including science and philosophy."

"You seem to know a lot; I'm impressed," Jenny said.

"Well, I've worked here for the last couple of years as a grad student, and I *am* an assistant historian," Sarah said, acting important with her nose in the air. "Also, I see Manny a lot. He's one of the few people with permission to take materials out of this room on a limited basis. So I see him quite frequently. He's not very sociable to most people, but he has quite a following, and he always seems to want to tell me all about his work. I listen patiently; I guess he thinks that means I'm interested!" laughed Sarah. "So he goes on and on."

"Well, I was sure hoping to take a look at those books myself," Jenny said with a sigh.

"I wish I could help you," Sarah said. "Most of these books," she motioned around her, "rarely get out of their protective glass. But those Ancient People books—it seems they became real important recently because of a debate or something."

"A debate about the Ancient People?" Jenny asked, trying not to look startled.

"Yes, a student is going to debate a professor," Sarah said as she shook her head. "I think it's even a girl! I feel sorry for her," Sarah added.

Jenny wanted to smile and had to turn away.

"A debate, huh?" Jenny forced a serious look as she turned back to Sarah. "And what does 'The Man' have to do with that?"

"He's prepping Dr. Seeting—he's the professor in the debate—on the latest interpretations of the traditions and legends of the Ancient People. I suppose that's why Manny came and got these books," Sarah pointed to the empty place on the shelves.

"I thought you said they were moved offsite years ago," Jenny quickly answered, looking intently at Sarah.

"Well …" Sarah looked slightly uncomfortable, "they were offsite. Actually, Manny, Professor Manning, keeps them at home most of the time; that's what I mean by offsite. They were returned here recently as part of an inventory audit." Sarah looked confused. "But now they are offsite permanently. Manny picked them up and told me that he had arranged to have them transferred to a museum collection. After the debate, that is." Sarah looked satisfied as she nodded slowly.

"After the debate. So he and … Dr. Seeting, did you say? They believe that understanding The Legend will help them in the debate?"

"I guess the student is going to try to defend the legends. I feel sorry for that student."

"Really? Why?" Jenny asked, partly amused but growing concerned. "She must have agreed to do it. Maybe she knows something about The Legend herself."

"I know Professor Manning," Sarah said with a solemn look. "There's a reason his students call him The Man. He knows the traditions and legends of the Ancient People backward and forward. When it comes to showing that the legends are actually myths, there is none better. He knows the original writings, he knows the history of their development, and he knows how to present the evidence. No one wants to be on the other side of a disagreement with him. It's said that when he gets you in his scopes he'll make a monkey out of you," Sarah said. "You'll think you were on trial or something."

"Really!" Jenny said, surprised at Sarah's seriousness. "Has he ever actually talked to any of the Ancient People?" she asked.

"I don't know. Are there really such people, I mean around here?" Sarah asked, surprised. "Actually, for the longest time I thought it was all just an elaborate fable. It still seems that if it were true there would be a lot more writings around. These two or three manuscripts we have here—had here—are very old and in some ways incomplete. I don't even think they were written by Ancient People, but by others who had studied the traditions from the Ancient People."

"I think there might be Ancient People around, even here at Lutrell," Jenny smiled, and then quickly changed the subject. "Do you really think Professor Manning is helping Dr. Seeting prepare?"

"Yes, I'm sure of it. Manny sees this as a unique opportunity to prove the 'death of the king,' as he says. Of course, Dr. Seeting is much more tactful than Manny, but as I understand it, he's also taking the challenge seriously. If you want my opinion, I think that's one reason these materials were removed, so that student couldn't get to them as well. That poor girl," Sarah said, chuckling and shaking her head. "She doesn't know what she's getting into."

Jenny sensed the gravity in Sarah's voice, and began to feel a little queasy. Although she didn't want to base her entire case on The Legend, she felt that reviewing the original writings would help her to at least have some basis for countering any misinformation that might surface in the debate.

"Listen," Jenny said as she turned back toward the counter, "if I leave you my phone number, will you call me if for some reason the books are returned? It's really important."

The two started back toward the counter, and Jenny walked over to the gate as Sarah pushed a button that unlocked it, again with a loud click. Jenny held the gate open with her foot as she turned and wrote her name and phone number on a slip of paper.

"Like I said, it's important. I need those books to prepare for an upcoming debate."

With that, Jenny let the gate swing shut with a loud clap, and turned to glare at the hard faces that quickly turned her way. She barely held back a smirk as she quickly walked away.

Sarah stood looking after Jenny. Gradually a grin came across her face. Soon the grin was replaced by a broad smile as she slowly turned back to her work on her stack of papers.

Chapter 18

Jenny and Alex were each shocked to see the other—Jenny because she would never have expected to see Alex, Alex because she was not paying attention.

"Alex!" Jenny exclaimed in astonishment. She resisted an urge to hug her, as she looked on surprised and delighted.

"Hey, Jenny," Alex said politely. She appeared uneasy with the encounter, standing awkwardly with nothing to say.

Jenny was thrilled with the opportunity to speak to Alex. Except for occasional eye contact in Dr. Hawkins' class, Alex had avoided Jenny. A flood of questions surged inside Jenny, but she constrained herself, studying Alex's hesitant expression. A few tense seconds of silence prompted Jenny to try to save the moment.

"It's good to see you, Alex," Jenny said, stepping forward slightly.

Although it had been only four weeks since Alex moved out, Jenny thought she appeared thinner, and her eyes looked tired. But for an instant she imagined that things were as they had been, and she and Alex were again hanging out together. The thought was fleeting, however, as Alex's distant look and reluctance to talk confirmed the continued rift between them.

"What are you doing here?" Jenny asked, trying not to sound incredulous. Over the last weeks she had rehearsed many times how she would act and what she would say to Alex, given the chance. She determined the best approach was to reach out and assure Alex that there were no hard feelings.

"I'm here with a friend," Alex said, glancing back at a young man arranging books on a book table.

The "here" each referred to was a conference sponsored by Lutrell University in conjunction with the group Scientific Hegemony Assures Materialism, called "Progression, Triumph of a Theory." Exhibitions by various scientific and research organizations devoted to the teaching of progression filled the University Conference Center. Students could even get extra credit by attending one of the mini-seminars offered, which was the main reason Jenny had come. In fact, she had just attended a presentation by a representative of the SITI project: the Search for Intraterrestrial Intelligence. SITI sought to confirm the existence of intelligent life underground. For although scientists were sure that the Ancient People's warrior king did not make the cave painting, they did not rule out the possibility of some other kind of intelligence living underground that may have been responsible for the pictures.

Alex and Jenny were standing uncomfortably before a booth selling a book entitled *How Do We Know What We Know?* Ordinarily such a topic would attract Jenny's immediate intense interest. But now the whole conference seemed inconsequential. She longed to once again be back in the Student Center sharing a double latte, or even to be at an amusement park on a roller coaster, anything to once again be with Alex, even to endure her incessant prodding to let go and have fun.

"A friend?" Jenny asked, looking back at the person arranging books. Jenny could not help but wonder if he was the one she had seen leaving their dorm room. For now, however, it hardly mattered.

"You wouldn't like him," Alex said quickly.

"Alex, any friend of yours is a ..."

"You wouldn't like him, Jenny. Trust me," Alex said sharply with a curt look indicating Alex's suggestion that Jenny move on to another booth. But by now her friend had noticed the conversation and had walked over to them.

"Hi! Malcolm. Malcolm Lloyd," the friend said with an extended hand and a big smile.

"This is Jenny," Alex said, clearly bothered at the prospect of having to explain who Jenny was. "You know, the one I told you about," Alex said, looking directly at Malcolm's face.

"Oh, *Jenny*!" Malcolm said knowingly, still smiling. "Yes, I remember. You're Alex's old roommate!"

The words were friendly, but they nevertheless stung and Jenny wondered if he might be a future old roommate of Alex's. But the thought distressed Jenny and she decided to quickly change the subject.

"Tell me, Malcolm, are you the author?" Jenny asked, looking over the neatly fanned-out collection of books. Alex, seeing that Jenny was not going to leave, went behind the table and sat down, pretending to attend to paperwork.

"Oh, no," laughed Malcolm, flattered that Jenny would even think such a thing. "This book was written by a former professor of mine. But I did a lot of the research." He flipped through the book as if to impress Jenny by the number of pages.

Intent on being pleasant and hoping to talk more with Alex whom she kept in the corner of her eye, Jenny took the book and also flipped through the pages. "So, how *do* we know what we know?" Jenny asked cheerfully.[1]

"Do you really want to know?" Malcolm asked, surprised. Apparently his carefully arranged books had remained undisturbed so far. The convention buzzed about them, but Malcolm's booth was just one of many set up around the outside wall of the large room adjoining the auditorium.

"Yes, really!" Jenny said enthusiastically. "I mean, can we really know anything for certain? Can we know truth?" Jenny asked.

---

[1] *See*, End Note [64], page 408.

Alex shifted positions and tried to become more distracted, putting her elbows on the table and her head in her hands and sighing. Malcolm was thrilled.

"You know, hardly anyone ever asks that question!" Malcolm enthused. "But that's sort of what this book is about!" Malcolm said as he hurriedly tried to find a particular page.

While he was looking, Jenny looked at Alex who returned her stare with a forced, but polite smile.

"Look, here's a chapter called *Truth or Consequences*. Get it? If there is truth then there should be consequences." Malcolm held the book out, evidently taken at the profundity of his insight. Jenny took it, dutifully examining the chapter title, nodding in serious approval.

"So what you are saying, then," Jenny started loudly, trying to get Alex's attention, "is that *if* there is truth, unchanging truth, then there are *consequences* associated with ignoring this truth."

Malcolm nodded vigorously.

"And does it work just like natural laws, so if you plant wheat, you will get wheat, and if you sow weeds, you will reap weeds?"

"Yes, yes," Malcolm nodded. "Of course, that is only if there is absolute truth. In this book we argue that there is no such thing as absolute truth—quite convincingly, I might add."

"Can we say absolutely that there is no absolute truth?" Jenny asked, pretending to be playfully rhetorical.[2]

Malcolm had to think about that, but he was enjoying the exchange. Jenny was still focusing on Alex; she knew Alex was listening.

"I think so," Malcolm said tentatively, "but I don't remember if that is addressed that way in the book. I can look." He started thumbing pages.

"That's OK," Jenny said. "I'm actually more concerned about something else. What about contrivances in nature? Do

---

2 *See*, End Note [65], page 408.

you know what I mean by contrivances? I mean things that have all the appearances of having been designed."

"Then you mean *apparent* contrivances," Malcolm said, pleased that he could make such a contribution.

Normally Jenny would have been quick to counter such a statement with an extended monologue. But with a constant awareness of Alex she pretended to browse the book, saying casually, "Yeah, maybe that's all we can say. Some things in nature only appear to be contrived."

"Well, that *is* the question, isn't it? Can we ever know? Take the cave figures of Lavoux Cavern, for example. They appear to be contrived. Are you familiar with the figures I'm talking about?"

Malcolm was serious. Jenny wondered what Alex had told him about her that he would ask such a question, but Jenny acted indifferent.

"Yes, I've seen the pictures. There are a lot of people here with theories of how they were formed naturally."

Jenny and Malcolm both looked around, scanning all the booths and exhibits, many of which featured various depictions of the cave art of Lavoux Cavern. Jenny was glad that, for the moment, there seemed to be little interest in Malcolm's booth; it gave her more time to think of a way to speak to Alex alone. Most convention-goers were at the main booth sponsored by the National Institute of Science.

"Tell me, Malcolm, what can we know about the cave markings of Lavoux Cavern?" Jenny asked.

"Well, most people agree that there are only two ways to account for those markings. One, a person, or people, made them; and, two, that they are the product of blind, natural causes, you know, progression."

"Do you think it matters which option is true?"[3] Jenny asked. "They are mutually exclusive, you know."

---

[3] *See*, End Note [66], page 409.

Malcolm had obviously never considered that question. He thought for a moment with his hand to his chin. "No …" he started. "Well …" he thought some more. Finally he was sure. "No, really it's just an academic exercise. Actually, in the end people just believe what they want to believe. I suppose it doesn't hurt to believe that people made the figures. It's kind of a lowland idea," Malcolm said with a slightly arrogant smirk, "but people seem to really believe it. But science assures us that the cave markings are the result of natural causes. I've seen the cave figures myself. They look to me like they were formed naturally, by progression."

Jenny let Malcolm ramble, not really expecting him to have a well thought out answer. Her intent was not to convince Malcolm, but to buy time until she could speak with Alex before she left.

"And by *science* you mean …" Jenny asked, looking hopeful as she put the book down on the table, nudging it against Alex's elbow. Alex looked up with a pained look and moved her elbow over slightly.

"Science? That's easy. Science is concerned solely with what can be explained in natural terms—based on the evidence, of course.[4] Observation and experimentation—that's science." Malcolm sat on the edge of the book table.

Jenny smiled slightly. "So, science that, by definition, can concern itself only with natural causes has determined that the Lavoux Cavern cave art must be the result of natural causes?"[5]

"Exactly!" Malcolm said, standing and stretching. "That's the beauty of science. It focuses the inquiry to things we can know, things we can see and touch, none of this legendary stuff."

"Doesn't that beg the question?" Jenny asked, now finding herself intent on helping Malcolm to understand. "Think about that sometime."

---

[4] *See*, End Note [67], page 410.

[5] *See*, End Note [68], page 411.

A group of people had moved in their direction and were now milling around near Malcolm's booth, and Jenny could tell Malcolm wanted to get their attention.

"I'll buy the book," Jenny said quickly, picking up the book she had laid by Alex's elbow. "Does Alex make the sale?"

Alex looked up with another pained smile and took the book from Jenny. Malcolm was busy flipping through the pages of another book for the new visitors.

"Jenny, you don't have to buy a book," Alex said dryly, holding the book out to make sure Jenny really wanted it. Jenny noticed that Malcolm was lost in conversation with someone else and took advantage of the opening to talk to Alex.

"It's really good to see you, Alex," Jenny said quietly, wanting to say so much more as she bent down beside the table so their faces were level.

"Same here, Jenny," Alex nodded, almost reluctantly. Alex turned away to start writing a sales slip. Jenny waited quietly, hoping Malcolm would stay distracted for a few more minutes.

Jenny wrote a check, circling her phone number. "This is my phone number, in case you need it, and here's my address," she said as she slid the check over to Alex. Alex sighed and took the check, giving Jenny her "Oh, please" look. Jenny's heart beat faster as Alex put the book in a bag. Malcolm's distractions had left.

"Alex … if you ever wanted to … I wouldn't ask any questions …" Jenny stopped abruptly, grabbed the book and left quickly, not looking back.

Alex stared straight ahead.

Malcolm smiled as he adjusted the remaining books on the table.

Chapter 19

Jenny could see the note taped to her door from the end of the hallway. The pink paper stood out against the new white paint. *Probably Chris,* thought Jenny. Taking it down she read: *Jenny, someone called for you, something about Ann DeHon, wants to meet you for dinner tonight. Something about Ancient People. Please call her back.*

Ann DeHon? Jenny thought. *Who in the world?* She looked up and down the hallway to see if there might be someone who could tell her who put the note on the door. Seeing no one, she folded the note as she opened her door and entered the room. *Ann DeHon, Ann DeHon,* she tried to recall any connection with that name as she put Malcolm's book away. Looking again at the note she decided to call the phone number scribbled at the bottom.

"Yes, hello," Jenny spoke to a woman's voice, "May I speak to Ann DeHon?"

"May I ask who's calling?" the nice voice replied.

Jenny was leery of giving out her name, "Well, someone put a phone message on my door ... something about dinner tonight."

"Oh, you must be Jenny."

Jenny was surprised. "Yes, that's right."

"Hello, Jenny. My name is Bryn Scott. I'm one of the organizers of a conference going on at the University Convention Center called 'Progression, Triumph of a Theory.' I believe you are aware of the conference. Ann DeHon is of the

Ancient People and asked that I call you to invite you to a dinner tonight honoring the contributions of Ancient People to understanding the origin of the cave figures of Lavoux Cavern. I know it's short notice, but I had trouble contacting you."

Now Jenny was dumbfounded. She suspected this was some kind of a joke. Her phone number was listed. Imagining Alex behind this, she almost laughed.

"Who is this Ann DeHon?" Jenny asked, suspiciously.

"Apparently the two of you have met before, but you may not remember. I can try to get more information if you like; I was just asked to try to contact you because of your interest in the Ancient People, as well as the cave figures, and because you might want to meet other Ancient People interested in the origin of the cave figures."

Jenny thought for a few seconds. Something didn't seem right. She had walked by most of the exhibits at the conference earlier that day and did not see any sponsored by the Ancient People. It was difficult for her to even imagine what contribution the Ancient People could even make to a conference called "Progression, Triumph of a Theory." Why didn't Katherine or Brian know about this? Surely they would have informed her of any involvement in the conference by the Ancient People. And who is Ann DeHon? And why didn't Ann, whoever she is, call her herself?

"Are you there?" asked Bryn.

"Yes, sorry," Jenny said, still thinking. "OK, I'll be there. What time should I arrive?"

ଊ ଋ

Returning to the convention center, Jenny approached tentatively, still suspecting that someone was playing a practical joke. It didn't seem like the kind of thing Alex would do; she loved to joke, but she was never malicious. Jenny entered cautiously, looking around before giving her name to one of the hosts at a table in the lobby. Surprised that the host found her name on a pre-printed nametag, Jenny relaxed a little as she took the nametag and a table number and walked toward a large room set up as a dining room. She wished she were not alone.

"May I help you?" asked a waiter dressed in black and white. Jenny was amazed as she gazed about the lavishly decorated room. She handed her table number to the waiter and followed him around one round table after another, each set with white linen and crystal. Chandeliers hung from a high ceiling, lighting the room and reflecting in large floor-to-ceiling windows along one wall of the room.

"Here you are," said the waiter as he pulled out a seat for Jenny.

"Thank you," Jenny said as she sat, looking around. There were no others at her table yet, but the room was slowly filling as the waiter began filling drink glasses by each place setting. Each time a woman neared her table Jenny would search her face to see if she might be someone Jenny knew as Ann DeHon. She waited, feeling conspicuous, and wondered if she should walk around a little until some others came to her table. Before she could move, however, she was startled to see a man walking briskly toward her. She recognized the man but could not immediately recall where she had seen him before.

"Jenny! I'm so glad you came."

"Andrew!" Jenny said, astonished to see the "other" Mr. Hunn, and glad she suddenly remembered his name.

"Oh, call me Andy. Only my father calls me Andrew."

"Andy ... *Andy Hunn!*" Jenny said as she stood up to shake his hand. "So *you're* the one who invited me to this dinner!" Jenny laughed as she realized the mix up.

"Yes, that I am," Andy said as he moved over to sit beside Jenny. "Sorry I didn't give you a little more advance notice; I wasn't even sure I would be here until the last minute. Anyone sitting here?" he asked, pulling a chair back.

"No, I'm the only one here so far," Jenny said, still smiling and glad that she had decided to come. Although her first impression of Andy had not been favorable, seeing him now brought a sense of relief. She recalled her first impression of him—that he did not share his father's interest in the traditions of the Ancient People. Apparently she was wrong. Here he was, part of a larger group of Ancient People involved in the struggle for understanding of the origin of the cave painting. She sensed that his father would be proud of his being at the banquet.

"Tell me, how are your parents?"

"They're fine, Jenny. Doing just great."

The two made small talk as Jenny studied Andy. In his suit he looked very different from the man she saw guarding the front porch in overalls—and uncomfortable, as if the suit and tie were seldom worn, and then only reluctantly. His face showed signs of hardship, and his rough hands seemed out of place before such a fine place setting. Even his hair appeared unaccustomed to being forced into place as it was.

Soon their table had filled, with introductions going around as each person joined them. Andy explained to the other guests at the table how he met Jenny, how she had investigated the cave painting, The Legend, and even about the upcoming debate with Dr. Seeting. Jenny sensed that Andy was trying to convey to the table that she was not of the Ancient People, but that she was at least sympathetic to their plight.

The talk was pleasant as the dinner progressed. Jenny enjoyed talking to the other guests, but she noticed that whenever The Legend was brought up the discussion was less than enthusiastic. There were even some uneasy looks her way when she asked about others' thoughts on how to prove that their warrior king made the cave painting. Finally, Andy spoke up with some clarification.

"You see, Jenny believes The Legend to be literally true," Andy said in a friendly tone, as if this would be unexpected news to the table. Indeed it was unexpected news as the expressions ranged from surprised but friendly to surprised and confused. Jenny herself was in the latter group. She sat stunned, not sure if she heard right, or what Andy meant.

"Oh, that's nice," one of the friendly surprised faces responded. "I used to believe it was literally true as well." Others around the table nodded.

"You see, Jenny," explained Andy, "although we are Ancient People, we believe the cave figures were formed naturally over time, by progression." The words rolled out and around the table prompting more nods and smiles before coming back to Andy, who continued, "It's an idea called *themeistic progression.*[1] It recognizes that science has shown the cave figures to have been formed naturally, while also accommodating our beliefs in a warrior king."

Suddenly Jenny felt very uncomfortable. She gazed about the large room. *All these people are Ancient People? And they don't believe The Legend to be true?* Everything looked different now as she slowly sized up the situation, feeling trapped between rage and politeness. She looked down at her lap, pretending to have trouble straightening her napkin and wishing someone would speak and change the subject.

"Themeistic progression avoids conflict with science by recognizing that progression occurred, but that the warrior king may have played a role in getting the process started," came a voice from across the table.[2]

"That's right, Jenny," added Andy. "You see, The Legend does not have to be literally true to be meaningful. We all believe there really was a historical warrior king, and we believe we all have a common ancestry. But the reason for continuing our traditions and The Legend is to promulgate certain *themes* related to purpose and morals. As long as the themes are preserved, the literal truth is not important; The Legend is still figuratively true. Thus the term *themeistic* progression. We recognize that certain themes of the Ancient People are not threatened by advances in science, including progression."

---

[1] *See*, End Note [69], page 412.

[2] *See*, End Note [70], page 413.

Jenny's uneasiness must have been visible as many around the table offered various explanations to try to help her understand.

"The Legend was never meant to be literally interpreted," one person explained.

"What's important is that we revere the teachings of our traditions; not argue over whether certain things actually happened," another helpful voice chimed in.

"After all, our traditions and The Legend were never intended to be a science book," someone laughed as all the faces agreed.

Having had some time to sort through her thoughts, Jenny decided to be polite as she asked, "So, you don't believe your warrior king made the painting in Lavoux Cavern?"

Each person around the table hesitated as each gave the others an opportunity to speak. Finally Andy clarified, "It's not that we don't believe he made them. We just don't believe that he literally went down in the cavern and drew on the wall."

Jenny laughed with a half smile and hoped it didn't look like she thought the idea was silly, although she did.

"You don't understand. That doesn't mean that the warrior king didn't have anything to do with the cave paintings," someone interjected. Jenny's puzzled look prompted further elucidation.

"One theory is that the warrior king supplied the starting materials, and nature did the rest. For example, maybe on that dark night while he was out with his council the warrior king accidentally dropped his war paint down the opening to the cavern, and it was subsequently washed up on the wall by an event such as water in the cavern—in full agreement with the theory of Neihilian progression."

Serious agreement was reflected around the table as Jenny continued to be shocked at what she heard. She noted the conspicuous use of the article *the* instead of *our* before any reference to the warrior king.

"Or maybe the warrior king really did fall into the cavern, and the war paint he carried fell from him before he could get out. That could explain his three-day absence. Maybe he was unconscious at the bottom of the opening, or maybe it just took that long to find a way back up out of the opening. But the cave painting is pretty far from the original opening, so it's unlikely he went that far to make a painting."

Andy was agreeing as he added, "Although most people today don't really believe he was actually in the ground for three days."

"In any event," started an old man that reminded Jenny of "old" Mr. Hunn, "the important thing to remember is that science and the traditions of the Ancient People can be reconciled—they can even complement one another."

"Why?" Jenny asked, looking around the table. "What purpose is served by this so-called reconciliation?"

"The purpose is to accord our beliefs with overwhelmingly mainstream scientific thinking," Andy started. "I mean, if Neihil is right, as he seems to be, our beliefs should not contradict the evidence of progression."

Dessert was being served as some movement at the head table indicated someone was going to speak. As she scanned the head table Jenny noticed Dr. Silva and Dr. Seeting seated with several other men and women she didn't recognize. Most were smiling and laughing. Jenny wasn't sure if she had the energy or the desire to continue the table conversation, so she changed the subject.

"Who are the speakers?" she asked, looking toward the head table.

"The main speaker, in the blue suit to the right of the lectern, is Dr. Otis Sutton. He's president of Clarent College, an Ancient People college on the other side of the mountains. Clarent has recently begun advocating themeistic progression as the best explanation for the origin of the specialties in Lavoux Cavern," Andy said.

Jenny had recently learned of Clarent College and recalled hearing it was connected with the Ancient People. She had even considered making a trip there to speak with science professors to prepare for her debate with Dr. Seeting. To learn that a college purported to be an "Ancient People" school actually denies the literal truth of The Legend struck Jenny as a profound contradiction. Again the questions piled up in her mind, but she resisted the urge to ask them now.

"Did you go to Clarent?" Jenny asked, trying to make polite conversation with Andy.

"No. I thought about it, but I had to work on the farm. I always thought I would go to college someday, but I could never get away from the duties of our farm, especially after Dad got older."

Andy sounded almost embarrassed. Jenny looked at Andy, smiling kindly.

"How did you find out about this convention? Why are you here?" Jenny asked.

"I'm just here because Professor Manning, a professor here at Lutrell—he's up there," Andy pointed to the head table, "called me a few weeks ago and invited me. Actually, he's the one who told me about themeistic progression and introduced me to these people." Andy motioned around the table. The others were involved in their own conversations, which Jenny was glad for.

Jenny was taken back at the mention of Professor Manning, suddenly both suspicious and curious as to the connection.

"Professor Manning called you?" Jenny asked in disbelief. "Out of the blue, he just called you? How did he know about you?"

"He said he got my name from a student in a discussion group he sponsors. At first I thought it might be you, but it was some guy named Alex."

Jenny tried to look steadily at Andy as she replied, "And why did he ask this guy Alex for names of Ancient People?"

"Apparently Manny—that's what he asked to be called—is working on a project about Ancient People and Alex knew about me. Maybe he's a friend of Chris's or something. Manny was looking for Ancient People to help him do some research. After talking to me he invited me for an interview at Lavoux Cavern. I needed the work, so I looked into it, and here I am."

"So here you are," Jenny said knowingly. "And let me guess, you're working on some higher criticism of some Ancient People writings, right?"

Andy stopped and stared in stunned disbelief. "How did you know? That's exactly right. You would be surprised at how many mistakes and inconsistencies are in some of these writings. I'm helping Professor Manning catalog all the errors."

"And let me guess again, when he found out that you know me, he asked you to invite me here tonight," Jenny said with a suspicious look.

"Well, not exactly. He knows Dr. Seeting, who mentioned your name in a meeting with Professor Manning. I was in another room—I couldn't take the cigar smoke." Andy made a look of disgust as he waved the air before his face before continuing, "As they were discussing how to 'cement the case' as they said, for an upcoming proposal or something, your name was mentioned."

Jenny was suddenly both frightened and interested.

Andy laughed as he continued, "I didn't know what their conversation was all about, but Professor Manning was assuring Dr. Seeting that there was no way for you to see the evidence, because you would never find it. They were at odds with one another; I guess Dr. Seeting wasn't pleased that you couldn't find the evidence. I assumed that you were helping Professor Manning with his research as well, so I went in to tell them I knew you; at least I had met you."

"And what did they say?" Jenny asked, becoming more uncomfortable.

"They looked surprised that I was there. Dr. Seeting said he knew you as well and then suggested that I invite you here. This all happened in the last couple of days."

Jenny finally looked away, put her napkin on the table and sighed. Looking around the room again, she noticed that other than Andy and some people at the head table there was no one else she recognized. No students from the Student Ancient People, no people from Rex's house.

"Andy, I need to go now," Jenny said, turning to Andy. "Do you mind?"

"Are you sure?" Andy asked, surprised. "The main speaker will start in a few minutes." Others around the table expressed disappointment that Jenny might leave.

Jenny smiled and looked around the table and then around the room before looking back to Andy. "Yes, I really have to go, Andy. But I do appreciate being invited tonight, and it is good to see you again."

Andy looked at her with a look of polite sophistication; Jenny looked at him with eyes of sorrow and sympathy.

"I'll walk you out," Andy offered as Jenny pushed her chair back. After saying good-byes to the others at the table, the two made their way to the lobby. Jenny was glad to have some time to speak to Andy alone. As they passed through the doors of the dining room Jenny turned abruptly to speak.

"You're being used, Andy," she blurted, barely restraining herself.

"What?" Andy asked, wide-eyed.

"Do you understand the theory of progression?" Jenny asked. "Because if you did, you would realize that it simply does not provide a scientific basis for the origin of the cave paintings in Lavoux Cavern. The Legend offers a much more reasonable explanation—and I'm not talking about any written legend."

Andy appeared to be offended at the suggestion that he didn't understand progression—or The Legend. "I've known about progression for years, and I always wondered if it might be true. Since coming here I've learned not only the scientific basis for progression, but also that The Legend can be interpreted different ways. And it can be interpreted in ways consistent with progression."

"But *why*?" Jenny implored, gesturing in frustration. "The Legend explains the cave paintings perfectly. Why 'reinterpret' it—only to remove what it explains so well? Why give that up?"

"*You* are going to tell *me* what The Legend explains?" Andy asked, becoming defensive.

"Look, I learned The Legend from the same place you learned it, Andy—from your father! And you know as well as I do that The Legend is true, so why give it up for something as tenuous as progression?"

"I still believe The Legend is true. I just don't believe it's literally true; I don't believe it has to be interpreted literally!" Andy said confidently, as if he had carefully reasoned his way to such a position.

"What's the difference between something being 'not literally true' and being not true at all? Don't you see, Andy, this so-called 'reconciliation' of two incompatible theories is really just a capitulation of one in favor of the other. That would be OK if the theory of progression was clearly the more reasonable of the two. But that's not the case here. I'm still wondering what it is that you feel you gain by giving up your beliefs."

Jenny looked kindly at Andy as she waited for a reply.

"I'm not giving up my beliefs," Andy protested weakly.

Jenny could see through the open doorway that Otis Sutton was beginning his talk. Every once in a while loud laughter would spill out into the lobby area, and Jenny could tell Andy was anxious to go back into the dining room to hear him. Andy looked back also before continuing, "I just believe that by partnering with scientists, the Ancient People can end a long history of being embarrassed, to some extent, by scientific advances. That's all, it's just a partnering, recognizing that both camps have something to offer."

Jenny was not sure what embarrassing episodes of history Andy was referring to, but as she studied Andy's face she could see Mr. Hunn's kind yet intense eyes. She could only imagine what Mr. and Mrs. Hunn would think about the whole idea, and she didn't want to ask if they knew about Andy's newfound "partnering" concept.

"Andy, you need to understand what is happening here, and I'll try to make it quick. I know Dr. Seeting and the other professors and scientists here at Lutrell. I even have some respect for them, but they have no respect for the Ancient People or The Legend. Some of them—Dr. Hawkins, for example—despise the Ancient People with a passion. And you play right into their hand by relegating your beliefs to the irrelevant sidelines of inquiry. While you see yourself as 'accommodating' they see you as surrendering to their ... their *dogma*—some kind of *rule* that only natural, non-human causes can be considered in explaining the origin of the cave painting. The bias is inflexibly applied regardless of—*in spite of*—the evidence of human causes; it prejudges the question of origins before it is asked. That's what makes it dogmatic!"

Jenny paused for a moment as she tried to gauge Andy's interest and understanding. Andy surprised her with his reply.

"There is a rule," Andy replied slowly, almost reluctantly. "I overheard Professor Manning talking about it. But it's OK; it is simply a working rule of science to keep science objective. It's really very simple—where there's no evidence of intelligent, that is, human design of the painting, only natural, non-human explanations for the cave painting can be considered. That way beliefs like The Legend are not permitted to unduly influence science. Science must remain objective."

A Rule? Objective? A flurry of thoughts filled Jenny's mind as she tried to reconcile how such a reasonable-sounding position could reach such a skewed conclusion. How can objectivity be achieved by a rule that biases the question to admit only certain answers prior to any inquiry? She strained to make sense of the idea, knowing that Andy's attention was getting shorter by the second.

"Andy," Jenny started. "Think of it this way. It seems that this Rule, as you say, has the opposite of the desired effect. Once certain explanations for the origin of the cave painting are ruled out by definition—by application of The Rule—it seems that objectivity has been discarded in favor of a ... an *assumption,* that turns the inquiry into a *presumption*. The bottom line is that The Rule has the effect of rendering the beliefs of the Ancient People irrelevant—if your warrior king did not actually and literally make that cave painting, then what good is the rest of your Legend and traditions? They should be discarded as irrelevant. Who cares if you have beliefs as long as they are not permitted to have any impact on anything that matters?"

Andy stood for a few moments thinking. He looked back in the dining room where the speaker was now well into his talk. Jenny continued, softly, "Andy, I'm still waiting to understand what it is that you feel you gain by denying the literal truth of The Legend."

"I don't look at it as gaining anything, Jenny. It just seems that if scientists have proven that all the various cave specialties were formed naturally over time, that ... well, our traditions and The Legend should be adjusted accordingly. What's the point of believing a lie?"

Jenny sensed she was finally getting an honest response. She paused as she could tell Andy was considering his last words.

"Andy, that is exactly the right question. What is the point of believing a lie?" Jenny waited while Andy shifted uncomfortably. "Andy, you can go back in and hear Dr. Sutton. But if you want, we can go over the Student Center and talk some more about these things. You can probably be back before he's finished."

For a few seconds Andy looked undecided as he glanced back at the dining room. "How will you feel if I do go back in?" he asked.

"Like a father whose son just sold his birthright—and that for a pittance!" Jenny said, looking intently at Andy.

Andy hesitated a second more and then turned and asked, "Where's the Student Center?"

"This way," Jenny said, taking Andy's arm and leading him toward the door.

"Can I take off this tie?" Andy asked, laughing.

Andy took one last look back, seeing Dr. Sutton still holding forth, with arms spread in a welcoming gesture, smiling around the room.

Chapter 20

The argument was becoming heated as Jenny laughed. She and some friends from the Student Ancient People had just returned from a quick one-day late season ski trip and were comparing notes in the Student Center before returning to their rooms. Kathleen and Jenny had skied together, as did their friends Peter and Eric. However, they had become separated at the top of the first run and didn't get back together until the end of the day. All the way home the girls argued with the boys as they compared their experiences. It was the first time any of them had skied at nearby Assumption Peak Resort and although the girls wanted to go back, the boys vowed never to return.

"You two are deluded," Peter almost shouted as the four huddled around a table with hot chocolate and dessert.

"It's almost as if we were on two different mountains," added Eric. "There's no way you can describe those slopes as gentle or fun."

"I thought I was going to die at almost every moment!" Eric added, only half laughing. "They shouldn't even let skiers down that mountain. I don't know how that place ever got to be a resort."

The girls looked at each other and shook their heads, laughing some more.

"I'm telling you," Jenny said, "You are the deluded ones. It was great skiing. The weather was perfect; the blue sky, the sun ..."

"The sun?" Peter interrupted. "It was shady, even dark in most places!"

A busboy was working a nearby table and overheard the discussion as it became more and more impassioned. He rolled his cart over and listened some more as the emotionally charged conversation continued before finally he started laughing. The four turned to look as the busboy's laughter continued.

"Sorry, I hope you don't think I was eavesdropping, but I couldn't help but overhear you all," he said as he continued to laugh.

Jenny and the others looked at each other as they realized that they had been loud enough to be heard and boisterous enough to attract attention.

"Let me guess, you all skied at Assumption Peak, right?"

Jenny and the others looked at each other nodding as they turned back to the busboy with expressions of surprise and wonder. He laughed more as he pushed his cart aside and pulled up a chair near their table.

"And let me guess again, it was your first time there, right?" asked the busboy. More nods generated more laughter.

"Sorry," the busboy repeated. "But this happens all the time. I used to work the ski patrol at Assumption, and I saw this all the time." He looked at his watch and then around the room as he continued, "I have a couple of minutes. Mind if I sit?"

"Not at all! Please!" Kathleen said.

The busboy sat down and looked around again, before saying, "Assumption Peak is unusual in its layout. Do you remember how the lift was situated? The lift runs up a sharp ridge oriented west to east. At the top of the lift you ski down the ridge along a gentle path called Assumption Run. Remember?"

All the heads were in agreement as they listened intently.

"Yes, that's where we got separated," said Peter. "You girls wanted to stop and look at a silly map or something. We wanted to go ahead and you challenged us to go on saying you could catch up to us. Remember?"

Jenny and Kathleen nodded.

The busboy continued, "Now, not too far down Assumption Run you come to Commitment Point and you start down the mountain. Which way did you go?"

"Right."

"Left."

The four started laughing as they looked at each other in disbelief. "You went left?" Jenny asked wide-eyed.

"You went *left*?" the busboy echoed. "Well, that explains it."

"What? Explains what?" Eric and Peter asked together as the girls started laughing.

"You went down the dark side!" said the busboy. "Didn't you see all the warning signs?"

"Of course we did. But for us 'danger ahead' means 'fun ahead.' We weren't about to let that stop us." They all laughed as they began to realize what had happened. "Besides, we didn't know there was another way."

"Look, here's the deal," said the busboy. "That mountain has two sides and you went down the north side. Did you notice that you could have gone to the right?"

"That was a ski trail?" Peter asked. "I remember what you're talking about, but that trail didn't look like it was for fun skiing; that path looked too straight and narrow to be enjoyable. And besides, the sun would have constantly been in our eyes."

"That's the path we took, Mike," said Jenny, noticing the busboy's name tag and looking back at Eric and Peter with a smirk.

"The north slope," Mike shook his head solemnly, "they should have closed it years ago. But the authorities decided that everyone should have a free choice in the matter, and all they can do is try to inform people of the alternatives."

"What are you talking about?" Eric asked, looking at Mike intently.

"The north slope gets very little sun, sometimes none," Mike continued. "That's why it usually has more snow on it, which attracts a lot of skiers. However, the extra snow hides rocks, tree trunks, and all kinds of other obstacles, including sharp drop offs and dangerous precipices. The lack of sun removes the contrast necessary to detect changes in terrain. The shady darkness makes the slope even more hazardous."

"Yeah, you're telling us!" Eric and Peter agreed, looking at each other. "That's why we're not going back. People should call that hill 'Death Hill.'"

Mike looked at Eric and Peter without smiling. "We did. Hardly a day went by that the ski patrol didn't have to go down that mountain with someone hurt, or worse." Mike paused. "On the other hand," he stated cheerfully, "the south side, which is the sunny side, is much safer. There are still many obstacles, and people still have problems, but everyone makes it down safely. It's all about the position of the sun. It's either for you or against you."

"Now you tell us," Peter laughed. "Why didn't someone tell us beforehand?"

"They did, you idiots," laughed Jenny. "That's what Kathleen and I were doing at the top of the first run. We had a guidebook with a map. It was all spelled out very clearly. It told us the benefits of choosing that 'straight and narrow' path," Jenny said sarcastically, continuing to poke fun. "The book made it clear what dangers lay on the dark side and the benefits of staying on the side with the sun. It even said that we would see lots of people taking the broad way to the left."

"That's right," said Mike. "And so you see, your experience on the mountain depends on which way you go at Commitment Point. Sometimes we called that the 'leap of ultimate assumption.' Once you start, you are committed—and the end point is predictable. It's very difficult to change course after that."

The four looked at each other as they considered Mike's words and realized why they had the earlier misunderstanding. Suddenly they appreciated that, although they had all been on the same mountain, their perception of it was completely different due to their decision at the point of ultimate assumption.

"Hey, thanks, Mike," Peter said. The others agreed as Mike looked around, got up and began pushing his cart again.

"No problem. I actually used to see this kind of thing a lot. We used to lose a lot of business from people who came to our slopes for the first time, didn't read the book, took the wide path that looked like the most fun, and then didn't return. It's a real problem. And now look," he said as he motioned around the room, "I'm still cleaning up after people who refuse to heed the posted signs—to return their trays!"

As they watched Mike go off to the next table, Kathleen said, "So, we were on the same mountain after all."

"You're right," Peter said. "It's amazing how differently the same world can look based on the simple decision to go away from or toward the sun."[1]

Jenny looked outside. "Hey, the sun's down, and I've got studying to do before tomorrow. I'll see you ski bums later."

☙ ❧

The ringing phone irritated Jenny who reluctantly picked it up.

"Jenny?"

"Yes, this is Jenny," Jenny said into the phone as she continued studying.

"Jenny, this is Sarah. Do you remember me, from The Benton Collection?"

The Benton Collection ... it sounds very familiar, Jenny thought.

"The rare book room," Sarah clarified. "A couple of weeks ago you were looking ..."

---

[1] *See*, End Note [71], page 414.

"Yes, Sarah!" Jenny said, suddenly listening closely as she pushed her book and papers away. "I remember you. Did you get the books I wanted?"

"No, sorry," Sarah answered. "But after you left I remembered another book that might interest you. And I remembered where we first met each other!" Sarah said excitedly.

"Really?" Jenny asked.

"Yes, I'm sure of it. Can we meet somewhere now? There's something interesting that I want to show you in a book I found. We've had copies of this book sitting around here for years. They're not part of the rare book collection, but for some reason they've been stored here for safekeeping."

Jenny hesitated as she considered the books and papers on her desk. "When can we meet?" asked Jenny.

"Well, if you don't mind my bringing some friends, we can meet now. How about meeting off campus somewhere?"

Meeting off campus struck Jenny as odd, and she still had lots of studying to do, but she also wanted to meet as soon as possible. "That's fine, where?"

"Some place noisy where we can get a private booth—how about Pies 'R Round?"

Jenny laughed. Pies 'R Round, often referred to simply as "The Round," was a popular pizza spot, known for being a late night hangout. It was loud with plenty of distractions, including a game room and even a small bowling alley. Jenny loved pizza and often went there with friends.

"Sounds great. See you in a few," Jenny said as she quickly hung up the phone.

☙ ❧

The Round was crowded as Jenny entered the front door looking for Sarah. Along the way there she had continued to try to mentally place Sarah's face, finally deciding it was of no use. She would just wait and let Sarah tell her where they had met.

Looking around the room, she didn't see Sarah, so she made her way to a booth where she could wait while watching the door.

Soon Sarah came in with two guys. For a fleeting moment Jenny considered the possibility that Alex was again trying to set her up with a date. She even secretly wished that it were Alex coming in with a surprise date. Jenny shook the thought and stood up to get Sarah's attention, and then sat again as the three joined her at the booth. Sarah was carrying a large brown bag that she set beside her on the bench.

"Jenny, this is Stew and Tal, short for Talbot," Sarah said as they each shook Jenny's hand. "And this is Jenny. I told you guys about her."

Jenny smiled as the three made themselves comfortable. Tal sat by Jenny, with Sarah and Stew across the table. Soon a server was standing by to take orders. Anxious to see what Sarah brought, Jenny patiently waited as each ordered. Jenny ordered a slice of pizza, although she wasn't hungry at all.

"I love this place," Sarah said. "I come here about three times a week."

Jenny smiled, nodding. "Yeah, this is a pretty nice place," she said, looking around. "It's kind of noisy, but nice," Jenny said loudly. "I like the pizza here, too. But I just ate not long ago, so I'm not real hungry."

"I wanted noisy," Sarah started, "because I didn't want anyone overhearing us." Sarah looked around furtively and then turned toward Jenny smiling. "Guess where we've seen each other!"

Jenny was getting impatient. She liked Sarah and was also anxious to know where they had first met. But she was more anxious to see the book Sarah brought and was not really in the mood for guessing games. Nevertheless, she looked at Sarah and pretended to think hard.

"I don't know. Give me a clue," Jenny said.

Sarah reached into the brown bag beside her and pulled out a quilt, which she quickly unfolded and put over her head.

Jenny immediately recognized her as she pulled the quilt down around her head and under her chin.

"Yes! You're right! Now I remember," Jenny said excitedly.

Stew and Tal gave Sarah a funny look as Jenny got up to give Sarah a hug.

"This is unbelievable. How in the world did you remember?"

Sarah removed the quilt and sat back down laughing. "Well, it bugged me to no end that I couldn't remember," said Sarah. "Then, just as you left the rare book room that day I realized that you were the one to debate Dr. Seeting. I was flabbergasted! Then, as I pictured you debating, it occurred to me that you were out there on that cold day in the Open Forum. It just came to me."

Sarah turned to Stew and Tal. "This is the person I told you about that day at the Open Forum."

Stew and Tal both gave Jenny nods of approval.

"According to Sarah, you have quite a gift for piercing questions," Tal said as the four leaned back to let the server place drinks before them. "We are also Pro-Wife and it seems you gave that Pro-Voice guy a run for his money," Tal continued.

"And when I realized who you were and that you were going to be debating Dr. Seeting, I was thrilled," said Sarah. "In a way I feel indebted to you, after how you stood up for the Pro-Wife group like you did. I wanted to find a way to help you. I even tried to convince Professor Manning to bring the Ancient People books back for a day or two," continued Sarah.

At the mention of Professor Manning's name, both Stew and Tal gave Sarah a funny look. "You tried to convince Manny to let the Ancient People books out of his sight?" Tal asked. "He's been so funny about those books lately I'm surprised they haven't simply disappeared completely."

The conversation continued as Jenny learned that Tal and Stew both worked with Sarah staffing The Benton Collection. As they talked, it became clear that for some reason Professor Manning had taken a greater interest in the Ancient People manuscripts this school year than he ever had in past years. Sarah confessed that even though the books really did stay with Professor Manning at his home much of the time, it was only recently that he had gained approval to have them removed to a permanent offsite location and made inaccessible in a museum display.

"I'm sorry, Jenny," Sarah said, looking down. "But I was given instructions to simply tell people the books had been removed permanently. Manny has really been putting the pressure on lately."

"Sarah," Jenny started, "I understand. Thanks for being honest. I don't know this Professor Manning, and I'm not going to ask you to say anything you aren't comfortable saying, but there's something odd going on here. Is this Professor Manning guy on the level?"

Sarah, Tal, and Stew looked at one another as their body language indicated that they had their suspicions. Finally Stew spoke up, apparently for the group as they each nodded in approval, "Manny is a strange person. I don't know if he's on the level or not. He seems to play things close to the vest, if you know what I mean. He is very protective of the writings of the Ancient People, but it's not clear if it's because he wants to be the only authority on their traditions and legends, or if he's trying to keep important information from the public. But," Stew paused as he thought about what to say, "he carries a lot of weight around Lutrell; I wouldn't want to cross him."

For a few minutes Jenny forgot that Sarah had brought a book as she told them about her interest in the Ancient People. The noise of video games and bowling pins beeped and clattered around them as she told them about her efforts to show the cave paintings were painted and expressed her frustration at the dogmatic, unyielding resistance to the idea she found when trying to express her views at Lutrell. "It seems that the idea of a human painter is dismissed out of hand, as if it was simply an impossibility," stated Jenny. "It's as if we are not talking about the same cave painting."

After a few moments, Sarah spoke up. "I can't say how the cave paintings originated. I must admit it seems hard to believe that they are an accident of nature. But who knows? It seems that all the scientists agree that Neihil's theory explains them pretty well."

Jenny rolled her eyes as she sighed. "Sarah, you're not helping me here. Remember, I'm the one getting ready to debate Dr. Seeting."

"Oh yeah! That reminds me. Here, take a look at this." Sarah reached in the bag and pulled out a book that looked more like a loosely bound stack of papers. "This is a book that Stew reminded me about," Sarah said. "I had forgotten about it. A pile of these has been sitting around the rare book storage room for years and years. We often thought about throwing them away, but for some reason they stayed there. We weren't allowed to put them into general circulation, so they stayed in the rare book room."

Jenny looked with interest as Sarah handed her the large flat book. The stiff cardboard cover was printed in silver on blue with the title: *Reports to the General Board on the Exploration of Lavoux Cavern*.

"Wow," said Jenny as she gently ran her hand over the cover, "sounds very official." The book was dated the year before her fifth grade field trip. "Wow," Jenny repeated, turning the book over as she inspected its cover. "This is unbelievable. What is it?" Jenny laughed. "I mean, what was this for? Who published it?"

"Look inside," said Stew. "It's written by your friend Dr. Seeting." Jenny's raised eyebrows indicated her increased interest.

"Apparently for the first couple of years of exploration he was required to report to the university's Board of Governors what he found in his explorations, due to 'the sensitive nature of the discoveries,'" Stew said, with a suspicious look. "I've always wondered about that. What sensitive discoveries are they referring to? Except, of course, the cave painting."

"And, look here," Sarah said as she took the book from Jenny to look for a certain page. "I know a little about the legends of the Ancient People …"

"Actually, there is only one legend, *The Legend,*" Jenny corrected her with a friendly smile. "Many traditions, but only one Legend."

Sarah looked at Jenny with mock annoyance. "OK, smarty. Let me start over. I know a little about THE legend," laughed Sarah, "and I know there is something about a guy who went underground and left his mark."

"Hey, that's good. I'm impressed," Jenny said. "You're right. Their warrior king apparently fell into the cavern and was missing for three days. It is entirely reasonable that his mark is the cave painting."

"Well, wait until you see this," Sarah said, finally finding her place and turning the book back around to Jenny. "You are going to flip."

As Jenny read, her face turned ashen and then began to turn red as her heart beat faster. With the noise of The Round fading in the background of her mind, her neck felt warm as her eyes followed the entry on a page having the heading "Daily Log." The date of entry was apparently shortly after the purchase of Lavoux Cavern, and the written account chronicled the first complete inventory of all the lines and colors of the cave painting.

Sarah, Stew and Tal ate and waited patiently as Jenny continued reading, visibly shaken at what she read. At one point Jenny looked up and asked, "You said this has not ever been made public?"

"That's right. It's part of the university's private collection. We always had strict instructions to keep it stored in a private location," Sarah said. "I had forgotten about it, until, like I said, Stew reminded me of it. Is it helpful?"

Jenny blinked as she looked at Sarah. "Yes, very. Let me read a little more." Jenny's eyes read the following hand-written entry:

> Today we once again made visual and analytical tests on the apparent painting. Without question the marks appear to be made by a person. Even though the marks scream design I must keep in mind that what I see was not designed, but rather formed naturally to have the illusion of design. However, to support this position without a plausible alternative theory will be difficult. (Even though there seems to be no trace of a human painter otherwise). There is definite appearance of order, definite composition. To suppress teleological explanations as a competing hypothesis will be most difficult. (Although to prove such explanations will likely be just as difficult). Until a theory of natural origins can be developed, suggest that any theory of origins implicating design be scrupulously avoided.

Jenny was stunned. *A theory of "natural" origins? Suppress explanations? Competing hypothesis? Design?* The group sat silently for a few minutes as Jenny absorbed the information. Jenny recalled her conversation with Andy. *This is The Rule! Explanations of design are simply not permitted—they are to be scrupulously avoided. But why?* Looking at Sarah and Stew, she turned a few more pages as she asked, "Are there any more revelations like this one?"

"You mean basically admitting the inference of design? I don't think so," said Sarah, looking at Stew. "No, I think that's it," she said as Stew shook his head in agreement.

"The rest of the book is pretty boring—just natural formations measurements, data, et cetera, et cetera. And, of course, lots of detail about the apparent painting—the specialties," Stew said.

"Why do you think Dr. Seeting would write something like this?" Jenny asked, not really expecting an answer. "How could someone as smart as Dr. Seeting fail to consider his own instincts as at least debatable evidence of human existence, or at least presence, in the cavern? Why simply rule out an entire line of evidence?"

Stew and Tal were finished eating. Jenny had not touched her pizza and now had even less of an appetite. She fanned through the pages again, occasionally stopping to read certain entries.

"Well," Sarah said, "I was pretty sure you would like to see this. I don't know if I can be much more help, but I'm willing to do what I can."

"Sarah, this helps me a great deal," Jenny said thoughtfully. "For one thing it confirms for me something that I've long suspected—that this controversy about the cave painting is about more than a search for the truth. Something very powerful is at work here to skew the professional thinking of someone like Dr. Seeting. And it tells me that any facts I can marshal about the verity of The Legend are going to be of little use in debating the likes of Dr. Seeting. This is not about the objective evidence. This is not about seeking truth. As hard as it is to believe, it is clear to me now that this is about construing evidence according to notions determined by a preordained starting point. It's almost like no matter what the evidence, it is going to be inflexibly interpreted in the light of some kind of *a priori* philosophical assumptions."

The four sat silently for a few minutes as Jenny continued to turn pages, scanning various entries. The server came with the check. Sarah quickly finished her pizza. Stew and Tal were both eyeing Jenny's untouched pizza.

"Sarah, thanks a lot. Really. You have helped me a great deal. If you don't mind, I'd like to go now and consider this," she said as she handed the book back to Sarah and picked up the check. "And, guys, would you like to split some pizza?" Jenny laughed as she pushed her pizza over to Stew and Tal.

Stew and Tal both grabbed for the pizza as Jenny got up to leave. She said good-bye to all three as she hurried out. As Sarah watched her leave, Stew and Tal pretended to fight over Jenny's slice of pizza. They were smiling as they each grabbed a side and yelled, "One, two, three, pull!"

Chapter 21

"Is this a note from the principal?" asked Jenny's father as he put the paper down and returned his glasses to their place beside the bed.

Finally a date had been set, and the yellow flier announced "The Great Debate," as it was billed. There was even a picture of Jenny, along with Dr. Seeting.

"Sort of," Jenny said, nervously excited. "I'm not sure how it all happened, really, but I'm going to debate Dr. Seeting, *the* Dr. Seeting!"

Aunt Sema barged in carrying sandwiches on a platter and fussing under her breath about "leaving well enough alone." She didn't look at Jenny as she quickly turned toward the door. Jenny watched her leave and then looked back at her father.

"Dad, I wish you could be there," she said excitedly.

"Jenny, I wish I could be lots of places. But more than anything I wish I could be with you somewhere other than here in this room." He looked around him, and then back at Jenny, studying her face. "You know, more and more I see your mother in you. She had some fire ..." he paused and coughed, "*whoo,* she had some fire."

He stared with a smile in the distance. Jenny thought he was looking much older lately, and his voice was interrupted by the cough more and more.

"One time," he was remembering, "she didn't like the way the town council was planning on re-routing some major roadways. Supposedly required for safety purposes, the proposal would have put a highway right across our property and that of our neighbors. Their reasons for insisting on implementing their plans didn't make sense to her, but the town leaders insisted it had to be just the way they said. Your mother," he chuckled, and coughed, "she dug in; she uncovered what she called a hidden agenda." He laughed again as he recalled more. "She found out that the real reason had nothing to do with safety—it had to do with a plan to lower our property values to create an industrial park. You should have been at the next town meeting. Even *I* went," he looked at Jenny thoughtfully, "and I was too busy to care about any of that stuff."

Her father paused for a long time, thinking. "As I listened to her, I realized what a very brave thing she was doing, standing up to that town council. If she hadn't stood for the truth, we would have lost our land. After she spoke," he said softly with a slight smile and a far away look, "I stood up and clapped." Jenny's father's eyes were moist as he turned to look at Jenny. "I don't know why. I just ..."

Aunt Sema bustled in, "Are you going to eat those sandwiches, or just look at them?" She glared at Jenny, "Aren't you supposed to be in school, girl?"

☙ ❧

The debate had been moved from one of the lecture rooms to the main auditorium. Jenny wasn't sure if that was good or bad. But as she entered, she decided it was the latter. A large banner, large enough to span the entire stage, hung over the stage curtain. On it was screen-printed, in letters that must have been as tall as a man, the words "Science, Myth or Fact?" She was early, as usual, and sat in the back of the room, trying to quell her stomach, repeating her now-memorized theory over and over.

She watched as men with walkie-talkies and lots of keys jingling from their belts moved two large lecterns onto the stage, one on either side. "Testing, testing …" one of them said, motioning to someone above Jenny's head. More jingling sounded as a large table was moved to the middle of the stage; it was covered with another banner that read "Lutrell University, Academic Freedom."

Jenny re-read the rules of the debate. Thirty minutes for each side, ten minutes of rebuttal, and then open for questions. She had rehearsed her thirty minutes, but she really had no idea what to do with her ten minutes of rebuttal.

As Jenny pondered, she slowly breathed in the fragrance of a single red rose given to her by her father earlier that day when she had stopped by for a visit. He said it was the way he and her mother remembered each other when they couldn't be together, particularly if they had to be apart in a difficult time for either one. Jenny smiled as she enjoyed the smell of the rose, turning it in her fingers and feeling the petals gently brush her nose. She was sad that her father could not be there, but comforted by knowing he was at least with her in his thoughts. She closed her eyes as she imagined him standing behind her with his hand on her shoulder.

Although it was still early, the sound of students wandering in through the doors behind her brought Jenny back to reality. Apparently a group of Dr. Seeting's students had decided to make the debate an event, sitting together on one side of the auditorium in the front, before one of the lecterns. *That must be his,* Jenny thought, *I wonder how they know?*

Soon more and more students were streaming in. Jenny did not want to see anyone she knew, so she decided to go backstage, where she had been told she would meet the moderator and wait to be introduced. She gathered up her papers and the rose and headed down the side aisle to the front.

Standing in the back she could peek through the curtain at the auditorium. *Mistake,* she thought as her shoulders quivered and she quickly drew back. *No, I may as well see it now.*

She peeked again. The sea of seats was filling quickly. She noticed there was even a balcony, with rows of seats blending into shadows. A large video camera was pulling a long tail of wires as several men maneuvered it to the front of the auditorium. A row of spotlights above pressed a brassy sheen onto the wooden floor of the stage.

"Jenny!"

She jumped and turned. There was Dr. Seeting, smiling, hand outstretched.

"You scared me, I mean, even more than I already was, or am," Jenny stammered, shaking Dr. Seeting's hand.

"You're a brave girl, Jenny," he said with a calm smile. "But, remember, just have fun. I'll go easy on you." He smiled and walked over to his side of the stage behind the curtain.

Just have fun? Jenny thought. It sounded like something Alex would say, and Jenny wished for all the world that Alex was backstage with her. "Just have fun—sure," she muttered as her stomach fluttered.

Jenny heard announcements from the other side of the curtain, and her mind went blank. She desperately looked at her notes that now looked like gibberish to her. Soon, without even having heard her own introduction, she was being ushered in front of the curtain, and found herself on stage, the row of spotlights in her eyes. She wondered if people could tell she was shaking. She tried to smile. There was, surprisingly, applause, some of it enthusiastic.

"And on my left," the moderator boomed, "representing the opposing viewpoint in The Great Debate, professor emeritus, president of the National Institute of Sciences, published author and noted speaker, our very own, Dr. Seeting!"

It sounded like a boxing match. Wild cheers went up, mainly from the section in front of Dr. Seeting. Dr. Seeting came out and started walking over to Jenny. *What's this?* she thought.

Instinctively, she walked over toward him, and the two met in the middle and shook hands.

"Just have fun," he whispered again with a broad smile.

"Going first, Dr. Seeting will have thirty minutes," the moderator said as he started an oversized timer.

Something you might see on a game show, Jenny thought. Jenny looked out over the audience. Every seat was filled. There were people standing around the walls, and students sitting on the steps in the aisles. The video camera was pointed toward Dr. Seeting. Jenny was glad she didn't have to go first as Dr. Seeting stepped up to his microphone, turning all the faces his way.

"Thank you, students, faculty, parents and friends."

He was a polished speaker, clearly enjoying the chance to address such a large crowd. His voice was sure and confident; words flowed out naturally, not as a memorized speech.

"Let me begin by stating my purpose in engaging in this debate. My views are not only my own, but represent the views of the National Institute of Sciences, this university, and virtually every scientist knowledgeable on the subject of progression. I studied personally with Dr. Neihil, and I can vouch for his integrity and his love for the scientific method."

He paused for a few moments and cleared his throat before continuing.

"And it is for the integrity of science, and the scientific method that I stand here before you tonight. I stand for the proposition that we can know all we need to know based on the evidence we see before us, and only the evidence we see before us, through observation and experimentation. There is no need to rely on myths or legends."

He paused once more. Jenny winced inside. A murmur went through the crowd. His section whistled.

"Now, let's get right to the point, because our time is limited," he said, "and consider how nature, and nature alone, in the dark and over long periods of time, could have made the markings on the wall—the apparent paintings—of Lavoux Cavern. Come," he continued a little louder with emphasis, "let's reason together as we consider how Neihil's theory of events with modification by natural selective inclusion not only perfectly explains all the variety of specialties of Lavoux Cavern, but it does so without invoking any mysterious, unseen entities—entities born only in the minds of men. Traditions that become an opium for the people, a sort of mystical booze …"

Loud laughter interrupted Dr. Seeting. He waited momentarily before continuing, "Careful research has shown the ancient writings of the traditions to be mostly fabricated from unreliable sources …"

His voice trailed in and out of Jenny's mind. He heaped praise on Professor Manning and his "overseas" colleagues for their untiring work in the area of writing interpretation, particularly as it applies to legends and other myths. Jenny pictured the empty shelves in the rare book room, imagining their empty case being cemented permanently shut.

Dr. Seeting kept the attention of the audience as he continued, "After careful reflection one finds that there is grandeur in Neihil's view of nature, with its several powers, having been originally breathed into a few forms or into one; and that, while this planet has gone cycling on according to the fixed law of gravity, from so simple a beginning endless forms most beautiful and most wonderful have been, and are being, progressed."[1]

Jenny felt weak and tried to concentrate, but she found herself only half listening to Dr. Seeting. He seemed to say Neihil in each sentence, Jenny thought. His statements were often punctuated by a gesture to the great banner hanging above them and the question, "Myth or Fact?" Such dramatics always got enthusiastic applause from his cheering section. Jenny's mind tried to concentrate as Dr. Seeting's words seemed to ebb and flow into her consciousness.

"… like a blind artist, nature can …" Dr. Seeting was expounding on the ease with which natural selective inclusion

---

[1] *See*, End Note [72], page 420.

could make complex designs when some commotion in the back caught Jenny's attention. *Is that Aunt Sema?* Jenny's eyes squinted in the lights to see Aunt Sema holding open one of the double doors in the back. *Unbelievable, why would Aunt Sema be here?*

"… climbing a seemingly insurmountable mountain peak by way of a long, gradual slope …" Dr. Seeting continued to expound, mentioning Dr. Hawkins often.[2]

More commotion distracted Jenny as people standing in the back spread to make room for a wheel chair. *Dad?* Jenny's heart raced. She had never seen her father in a wheel chair unless he was making a required trip to the doctor's office, and that happened only occasionally.

"Progression, therefore, is the best scientific …"

Jenny heard Dr. Seeting's words from afar as she thought about her father. She tried to make eye contact and thought she saw her father and Aunt Sema smile at her with a nod.

"It is clear as the sun and evident as the day that there is no painter; and still more, that there can be no painter," Dr. Seeting made exaggerated motions of a painter with a brush.

Jenny thought she caught her father's attention momentarily and lifted the rose just enough for him to see it. His smile calmed her ever so slightly as she breathed deeply in anticipation of her time to speak.

"… is safe to say that if you meet somebody who claims not to believe in progression, that person is either ignorant, stupid, insane, or wicked …"[3]

Dr. Seeting's bombast caught Jenny's attention briefly. It seemed he was coming to his conclusion and fear once again swept over her as she realized her time to speak was approaching. She again looked back at Aunt Sema and her father, wishing she could go back and hug them both.

---

[2] *See,* End Note [73], page 422.

[3] *See,* End Note [74], page 425.

"… is not just some sort of *ruse*! No, progression is a fact, *fact*, FACT!"[4]

A loud "ding" startled Jenny's mind back to the stage. After what seemed like hours of applause, the moderator indicated to Jenny that her time had started. *This is it*, she thought. *Just have fun, just have fun*.

Jenny leaned forward into the microphone, "Thank …" feedback squealed, she jumped back and then leaned forward again slightly, "um, th- …" more feedback. Seeting's section giggled. She heard keys as one of the men with a walkie-talkie came out and slid her podium back toward the curtain. He stood there while Jenny tried again.

"Thank you," she said tentatively into the microphone, looking at it as she spoke. She continued, trying to speak firmly, although her voice was quivering. "My name is Jenny …" Her mind was mushy, *why did I say that? I was just introduced*, she thought. *Concentrate*!

"Let me also begin by stating my purpose in coming to this debate." It felt good to get a whole sentence out. "I represent only those interested in truth. My interest is in knowing truth, because," she looked at her father, "truth is not changed by what I, or," she looked at Dr. Seeting, "anyone else thinks about it." She tried to say that with authority, but she was sure it sounded weak. The audience was quiet.

"Truth stands like a beacon," she tried louder, trying hard to remember her memorized opening, "its light being seen in incontrovertible facts. Those who follow the light will find the truth." Jenny was still nervous; the words came out as someone else's. The audience was politely silent. *I'm not connecting*, Jenny panicked. *Just have fun*. Jenny looked at the crowd. She had never seen so many faces at one time. Overly conscious about looking at her notes, she lost track of her note cards and her hands would not stop shaking.

---

[4] *See*, End Note [75], page 426.

"Look, I'm not a scientist. I'm not even a senior," she laughed and was surprised that the audience did as well. "But when I was a little girl I followed Dr. Seeting into a hole in the ground, and what I saw in the darkness of that empty cavern changed my life. I saw one of the most beautiful pictures I have ever seen—a painting, the work of an artist. It was not a Van Gogh, of course, but it had color, it had a message, it had *life*."

The audience was still silent, but the faces asked for more.

"I knew then that a person painted that picture. And I know now that a person painted that picture for a purpose."

Seeting's kids hooted.

Dr. Seeting leaned over to his microphone, "Like The Legend says," he quipped before stepping back. Jenny shot a look at the moderator.

"You'll have ten minutes, Dr. Seeting," the moderator intoned.

"But you do believe The Legend, don't you?" Dr. Seeting stepped back quickly as the moderator glared and motioned for silence.

Jenny was flustered. She imagined the large timer about to ding. *Don't take the bait.*

"I am not here to defend The Legend," she said slowly but firmly, "and I *will not* attempt to defend it in this debate," she continued, finding her voice. "I am here to speak for objectivity in science—for truth. Others are apparently here to suppress the truth. How do you know? Well, when the origin of the cave painting is the question, you look for someone who takes the position that any hypothesis inferring design should be scrupulously avoided, regardless of the evidence."

Jenny thought she saw Dr. Seeting flinch slightly. In the past few weeks she had come to fully appreciate that Dr. Seeting's flawed approach to science was due to the assumption of the premise of no human design. This premise was inflexibly applied as an unwritten rule of science—*The* Rule. Only after discovering The Rule did it make sense to Jenny why origins science was limited to "progression-only" explanations—any theory of human design was simply deemed unscientific. Knowing that science practiced objectively cannot rule out an inference of human design, The Rule was nevertheless ruthlessly enforced to limit possible explanations of the facts to exclude any that might infer human design.[5] Although not fully realizing why Dr. Seeting would wish to suppress such evidence, Jenny came to realize that The Rule imposed an artificial requirement on science—a surreptitiously adopted essential premise of "no-design" that was enforced as dogma.[6]

Further, Jenny understood that just because a past, historical event cannot be tested by experimental methods, this does not mean that hypotheses to explain such events can't be scientifically tested. Just as forensic methods are used in other fields, good historical researchers focus on formulating multiple competing (versus single) hypotheses to explain unrepeatable happenings.[7] Recalling the graffiti on her dorm room door, she realized that proper investigation requires that all available clues are considered in an effort to converge upon the best causal explanation for any event. With respect to the cave painting, without a competing "design" hypothesis, the "no-design" hypothesis of Dr. Seeting in effect goes unchallenged. Scientific inquiry is ended.

Jenny continued, "You just heard Dr. Seeting offer his theory of the origin of the cave painting, a theory of 'no design.' But you did not hear one word of how the evidence might support a competing theory suggesting intelligent, human design. Why is this? Because there is no evidence for such a competing theory? You know this is not true. In fact, there is abundant evidence of human design of the cave painting. You could say the very image of the painting itself *screams* design,

---

[5] *See*, End Note [76], page 428.

[6] *See*, End Note [77], page 433.

[7] *See*, End Note [78], page 434.

but Dr. Seeting seems to be constrained by a Rule that permits, actually *demands*, only certain explanations."

Dr. Seeting again looked uncomfortable and annoyed, wishing to speak but impatiently held back by the glare of the moderator.

"The Rule is so inflexibly applied that it clouds the judgment of even the brightest minds in science so that they think they are doing science while preventing competing hypotheses that might implicate human origin of the painting. The Rule is a method of doing science, the intention and effect of which is to convince the public to believe its central tenet, that the cave painting is not painted by a human artist."

Now Jenny paused in spite of the pressure of the ticking time clock. Realizing that fully explaining The Rule would probably prevent her from getting to her theory, she mentally considered her remaining topics, straining to think how best to continue.

"Maybe Dr. Seeting can speak to that later," Jenny said with a look to the other side of the stage. "Maybe Dr. Seeting can explain why he is permitted to apply The Rule to exclude certain lines of inquiry, regardless of the evidence. Maybe he can explain why he is not required to show how he tested his own theory against competing theories. In short, maybe he can answer the question: Why are theories implicating human design ruled unscientific, regardless of the evidence? Should it matter that a design inference also supports The Legend? If so, why?"

At the mention of The Legend, many in the audience applauded. Once again, Jenny resisted the urge to make The Legend the focus of debate.

She continued, "I think that if you will hear me out, you will agree with me that this question is not one of 'myth or fact.'" Jenny turned to gesture to the banner behind her. "No, this is a question of science itself—what qualifies as a scientific explanation for observable, empirical data? And what hypothesis can be reasonably formulated to explain the historical, unrepeatable event of the cave painting? Not only can my theory be tested against the competing theory of Neihilian progression, I submit that it can be supported *scientifically*," she emphasized the word, looking at Dr. Seeting, "that the cave painting of Lavoux Cavern is exactly that—"

"Go back in your cave!" came a sudden shout from laughter rippling out of Seeting's section. Jenny pretended she didn't hear the comment.

"—a painting, painted by a painter for a purpose."

Dr. Seeting smiled with body language that said, "Go ahead, try!" He was having fun. His students were alive.

"Progression can, indeed, maybe even by natural selective inclusion, explain many interesting structures. Take stalactites, for example; these structures are clearly made by nature according to natural processes, natural laws."

The looks on the faces were back to polite.

"But progression cannot explain the cave painting of Lavoux Cavern," Jenny said forcefully, trying to regain momentum. "Take, for example, the various theories of progression presented tonight, including those of our own Doctors Silva and Hawkins. These theories are far from consistent and fall short of being scientific explanations. For example, although Dr. Hawkins simply finds it easy to believe that the cave painting is a result of progressionary gradualism, Dr. Silva finds it difficult to explain how the cave painting could have come about in a reasonable sequence of intermediate forms. At least Dr. Silva is honest with the facts. Nevertheless, he asks us to *suppose* many scenarios involving abrupt changes." Jenny looked at the crowd and said forcefully, "Silva supposes what he has to suppose, and Hawkins finds it easy to believe what he wants to believe, but supposing and believing are not enough to make a scientific explanation."[8]

---

[8] *See*, End Note [79], page 436.

The last statement brought some more applause, but Jenny was beginning to panic, realizing that she was still not connecting with the crowd.

"As I mentioned, I intend to give a reasoned explanation for the painting having been painted by a person." Jenny had completely lost her place in her notes and was becoming more flustered as she imagined a long gap of silence while she searched the lectern. She could see restless movement in the audience.

C'mon, Jen, let go and have fun! The voice inside was Alex's this time. Jenny looked down at her notes once more for what seemed like an eternity. The video camera hummed. Finally, she turned her notes over, folded her hands, and looked out over the audience.

"Imagine I have a big Chinese checkers board here," she said, smiling as the audience also smiled at the thought. "And here are the marbles," she continued, hands cupped high in the air. She noticed renewed interest from most of the faces.

Dr. Seeting stopped making notes and was listening. The crowd was engaged as Jenny continued with the illustration just as she had at Rex's house.

"Complex but not ordered—rockslide ... ordered but not complex—stalactites ..." she continued on, smiling, getting steady eye contact now from the carpet of faces. She was back in Rex's living room. With each concept there were nods and smiles coming back to Jenny.

"Now, stick with me here," she said, looking around, becoming more comfortable with the crowd. "What if I were lost in a desert ... no, in a *cave*," she changed the illustration, her mind reeling to keep up, "and I wanted to use these marbles to spell 'help'?"

The audience was with her now. Steady gazes nodded with inquisitive faces. She even thought she saw Aunt Sema nodding.

"Could I just throw these marbles up in the air and expect them to land on the ground and spell 'help'?"

"Nooo," came the loud, unexpected response from the now-lively audience. Dr. Seeting's students looked at him; he was looking at the audience in surprise.

"What if I threw a few marbles in the air today, a few more next year, a few more the following— "

Jenny was cut off by more loud "no's" from the audience, together with applause and laughter.

"That's right," Jenny said more loudly, leaning into the microphone. "No matter how many times I tried, once the marbles left my hand, nature alone would be unable to make an intelligible message—no matter how much time nature had, the marbles would never spell the word 'help.' And why not?"

Jenny held the pause; the audience waited with anticipation. She asked again, glancing at the speeding timer, "Why could I not just leave these marbles to natural processes and expect them to spell the word help, or any other words, for that matter?"

The audience was startled when Kathleen yelled out from her seat, "Because they wouldn't know where to go!"

Jenny laughed; the audience roared.

"Thank you!" Jenny said, still laughing. "That's right, they wouldn't know where to go!" She held her arms out in a "*Who knows?*" position and waited while the audience got quiet again. She again looked at the large timer—*looks like about seven minutes*, she thought.

"That's the basis, in fact, for my theory about the cave painting in Lavoux Cavern. I'll explain it to you now."

Dr. Seeting looked on impatiently, projecting "it's about time."

"I call it the 'four-one-one theory,'" she said, hoping that Chris and Ben were in the crowd. "Four-one-one is the phone number for information, right?"

A tremor of nods went through the audience.

"The only way the marbles can make a pattern of semantic information, that is, a meaningful message by spelling out the letters H-E-L-P, is if each marble knows where to go.

And, because the marble itself can't know, it must be directed, even placed by something, or someone that can make choices among possible alternatives, someone who can supply the necessary *information*, the *four-one-one*!"

Jenny's eye caught Dr. Seeting with a concerned frown that quickly migrated to his students, who were quieter now.

"If the information—the four-one-one—is missing, the marbles go wherever nature dictates based on natural laws and chance. Once the marbles leave my hand, the law of gravity would make them fall to the ground, and chance would dictate where they end up. But to make a meaningful pattern of marbles, the required information must come from somewhere to permit something or someone to make informed choices of placement among an infinite variety of possibilities. Information content is required for meaningful, purposeful design, and it is the key to detecting design. Purposeful design can be detected by ruling out law and chance as adequate explanations for any thing or event. And it is this third factor—information—that is the difference between stalactites and paintings; between dictate and design; between a mess and a message."

Jenny was again thinking out loud, but her voice was sure and the audience was energized and understanding her theory. Jenny thought about her last sentence and continued, "And it doesn't matter how long you have or how many chances you get—time is not a key factor here. The difference between a *mess* and a *message* is more than *age*!"

Dr. Seeting rolled his eyes as Jenny paused, but the audience was with her, laughing and applauding. Seeting's kids kept looking at him to gauge his reactions.

"Specified complexity," Jenny continued softly but deliberately, "is the hallmark of design, whether it is a written message or a painted message. A book cannot be written by chance; a painting cannot be painted by the dictates of natural laws. Such things can be the product of only an information-rich source—where choice displaces chance and logic channels laws," she stated, now her voice rising. "And now, I ask you, where is the only place such information can possibly come from?" Sensing that time was running out, she didn't wait for an answer but pointed to her head and leaned close to the microphone, "From an intelligent source, from the mind of a person!"

There was more loud applause. Jenny glanced to see Dr. Seeting looking serious as he continually checked the timer, anticipating his chance to stop Jenny's momentum.

"And it's not just for spelling 'help,'" Jenny continued forcefully, also eyeing the timer. "Imagine if, instead of 'help,' you wanted to spell something much more complex, such as 'mighty warrior king kills seven bison single-handedly ...'"

The audience erupted in loud applause at the thought, the sustained clapping forcing Jenny to wait for the noise to subside before continuing.

"*That* would take a lot of FOUR-ONE-ONE!" she shouted. Applause continued for several minutes, drowning out the ding of the timer. Several more minutes passed before the moderator could again introduce Dr. Seeting. He approached his lectern and stood for a few seconds studying some notes. After making a written notation he looked up, straightened his tie and smiled.

"Very interesting theory," Dr. Seeting started quietly and calmly. "Very clever. However, speaking as a scientist, I can inform you that you've only succeeded in falling into a very common, but comfortable, trap—belief in a theory that cannot be falsified. You see, *if,* and I can't stress 'if' enough, a person painted something in Lavoux Cavern, that *person,*" he said with derision, "didn't leave a trace of evidence of his or her existence. There is simply no way of ever knowing about his existence—of proving his existence. Of course, there is no way of *dis*-proving his existence either, although if he was ever there he sure did a good job of covering his tracks!"

Much of the audience laughed and clapped. Dr. Seeting continued, "It seems the deciding factor in this debate is simply one's personal belief. And one's personal belief has proven to be a shaky foundation on which to build scientific theories."

Jenny was still shaking slightly as she came back to reality and realized she needed to pay attention—this was the rebuttal.

"If I understand your theory, you are proposing that the specialties in Lavoux Cavern could be made only by a person, because only a person could supply the necessary information, the four-one-one, to choose among possible marks and figures to make the apparent painting we see." Dr. Seeting spoke the last words sardonically. "Is this right?"

Jenny nodded.

Dr. Seeting calmly took a drink, smiled, and said, "Look, I'll prove to you how convincing your arguments are to a scientific mind. I contend that the real basis of your conclusion is *belief* in myths and legends," Dr. Seeting said, "wrapped, ingeniously I might say, in pseudo-scientific language. Let me illustrate." He looked out over the audience. "How many of you are convinced Jenny is right, *absolutely* right? Stand up."

A relatively large number stood up. Jenny was surprised and thought, *nice illustration, professor.*

"Now," Dr. Seeting said confidently, "I want all those standing who are of the Ancient People to sit down."

A slight stirring began, growing louder and then quieter as, one by one, every person standing sat down. Jenny was stunned, not so much by the general outcome, but by its totality. Across the auditorium, *not one person was left standing*. Even the people standing along the walls looked like they were trying to squat down. Dr. Seeting even seemed surprised. Jenny felt ill and wanted to leave. *Be strong, it's only a few more minutes.* Seeting's students were noisy again.

Dr. Seeting paused for effect before continuing. "It seems to me, Jenny, that what we have here is not a scientific debate, but, rather, what we have is a great ideological divide, split between those disposed to believe myths, and those who insist on rigorous evaluation of independently confirmable facts. As this banner behind me says, science can be based on either myth or ..."

A rush went through the crowd. Someone had stood to his or her feet. Jenny saw the figure on the far side near the back. *Alex?* She strained her eyes against the floodlights. *It is Alex. Her hair is different, but it's Alex for sure. Maybe she's just getting up to leave,* Jenny thought. But Alex stood, motionless, eyes staring straight ahead.

Dr. Seeting stopped, amused at the distraction, but not sure how to go on. The pause drew two more students to their feet. Another. Then another. Two more. *Ben and Chris,* Jenny thought.

"That's good, you've made your point," Dr. Seeting tried, motioning for people to be seated. "If I could continue ..."

More people stood. Jenny could not believe what she was seeing. She fought to keep her mind clear. The steadily louder rustle of people standing spread all over the auditorium.

"I think we can all take our seats now," Dr. Seeting said, speaking loudly into his microphone while forcing a smile. He motioned to the moderator to stop the timer. "Please, can everyone now sit ..."

Clap ... clap ... clap.

Someone started clapping. Jenny's eyes searched for the sound, finding it in the back where Aunt Sema had convinced two big students to hold up Jenny's father, one student under each arm. Jenny watched as he hung there, suspended at his shoulders, clapping—slowly, and at first alone, then faster as the clapping spread through the audience.

Clap, clap, clap, smiles could be seen as people looked around, clapping in steady unison.

Now almost everyone was standing, except for a conspicuous hole of somber, seated bodies in front of Dr. Seeting. The clapping continued in a constant rhythm. Dr. Seeting moved closer to the microphone, looking at the moderator, who sat helplessly amused.

After seconds that seemed like minutes, from the clapping crowd came some voices in unison: "four-one-one, four-one-one, four-one-one." Louder it got as the wave of noise spread.

"Four-one-one, FOUR-ONE-ONE, FOUR-ONE-ONE."

Ben and Chris were turned around, leading their section like orchestra conductors. Jenny noticed Alex had her face in her hands.

Gradually Jenny realized what was happening.

FOUR-ONE-ONE, FOUR-ONE-ONE.

The chant filled the room, seeming to grow stronger with every effort of Dr. Seeting's to stop it.

FOUR-ONE-ONE, FOUR-ONE-ONE, FOUR-ONE-ONE …

Between each "four-one-one" chant, Dr. Seeting's students attempted to squeeze in a loud "NO," a shortened form of what they had started shouting. A livid Dr. Seeting glared at the moderator, who sat shrugging and laughing.

"This is not the way science is done," Dr. Seeting yelled in the microphone.

"FOUR-ONE-ONE, FOUR-ONE-ONE," the chant continued, louder now, with synchronized clapping. The video camera was turned around, frantically trying to capture the room.

"This theory has not been peer reviewed!" Dr. Seeting shouted as he made a last, futile attempt to regain control.

At this time a large man smelling of cigar smoke appeared from behind the curtain behind Jenny and walked heavily over to Dr. Seeting. The two deliberated for a few seconds before going over to confer with the moderator. Jenny watched with interest as the moderator stopped the timer, the smoker walked gravely off stage, and Dr. Seeting returned to his lectern, leaned into his microphone and shouted, "Thank you. That's it. We are *finished!*"

The noise in the auditorium continued as Dr. Seeting quickly walked off behind the curtain. Jenny remained at her lectern, looking out over the activity in the room. Alex was trying to make her way to the stage. Jenny's father was sitting again, coughing but still clapping. Aunt Sema was mouthing the words "four-one-one" while standing guard over the wheel chair. Dr. Seeting's students left empty seats.

Jenny considered the spectacle in front of her, then looked up and smiled.

THE END

Epilogue

Jenny looked on in trepidation as she stood before the opening in the ground before her. Her father had requested to be buried beside her mother and her brother on the property their family had owned prior to her mother's death. The current owners had obliged his request, and Jenny had made the long trek over the mountains to the lowlands of her childhood. The open field where Jenny stood was distinguishable from the surrounding area only by the circle of friends gathered in the afternoon sun, and the deep, dark hole into which her father's coffin was being lowered.

Beside Jenny stood Alex, one arm around Jenny's shoulder, the other holding a child. Aunt Sema held Jenny's other hand as they waited for the large casket to come to rest. Brian, Kent, Rex, Kathleen, Kozak and many of the Student Ancient People made the journey to be with Jenny on this day. Chris had moved away and could not be reached. Some of the Hunn clan and other former neighbors and friends came to honor her father as well. To one side a young mother was trying to keep her child quiet as they sat on a colorful quilt on the grass. Unknown to Jenny, even Dr. Seeting stood back listening and watching.

In Jenny's last two years at Lutrell University she and Dr. Seeting had become friends of sorts, each admiring the other's intellectual tenacity. He had not seen Jenny since she graduated with honors. With her double major in earth science and information technology he had hoped she would pursue advanced degrees at Lutrell. He had even urged her to apply to his post-graduate program. Caring for her father, however, had consumed all Jenny's free time as a senior, and almost all her time and effort in the year since graduation.

The Great Debate had far-reaching results. Jenny agreed to be a founding member and advisor to a new student group called the Discovery Academy that formed as a result of the debate. The Discovery Academy sought to bring a different perspective to the study of culture, particularly in the arts and sciences. Soon similar student groups and clubs were being formed at other colleges, and even high schools throughout the land.

The Discovery Academy soon grew into a separate organization off campus and Jenny was active in the Academy's Center for Science and Culture. Even more significant, however, Dr. Seeting had bowed to pressure and permitted subsequent editions of his earth science book to include a small section on the "difficulties" with the theory of progression. Also, Mr. Kozak had persuaded his administration to permit him to "teach the controversy" as it was called. It seemed to Jenny that there might be hope for truth after all.

But cave paintings, debates, and even truth itself now seemed insignificant as Jenny gazed through brimming eyes at the single red rose she had placed on top of the dark casket. In detached solitude she barely heard the words of comfort offered by those about her. Very few people knew her father, and almost none knew him well. Only two people knew him as a friend in his last days. As his health failed, his eyes remained bright, and for Jenny and Aunt Sema, even though not unexpected, his death seemed, as death does, too sudden.

After all was said, Jenny whispered to Aunt Sema and then slipped away with Alex as the first scoops of dirt fell over the rose. They walked slowly away from the gravesite, arm in arm, Alex shifting the baby on her hip as they talked softly. Alex had married while carrying her first child and divorced in less than a year while carrying her second. It was a very difficult and lonely time for Alex as she considered her options. With Jenny's encouragement she decided to keep both children, a decision she never regretted.

"Things have changed a lot, huh?" Alex said, trying to make conversation. Jenny just sighed in agreement.

"We've changed a lot, huh?" Alex said, pulling Jenny's elbow close.

"*We've* changed?" Jenny asked with a grin and a sniff as she wiped her eyes.

"Yes, Jenny, *we've* changed. You and I. Look at us. We're older. We're smarter. One of us is wiser."

Jenny laughed softly. Even after Alex returned to the dorm, and later after the two found their own apartment off campus, Alex continued to see Malcolm and his friends. Jenny tried to avoid them; she preferred the company of Chris and the Student Ancient People. Eventually Alex also distanced herself from her circle of friends, but not before she found herself pregnant for graduation. Now as a single parent of two Alex appreciated Jenny's counsel as much as her friendship—a friendship that extended with open arms to help whenever needed.

"Why is it that way, Alex?" Jenny asked, looking at Alex.

"*Hey, cutie!*" Jenny said before Alex could answer, bending down and squinting as the baby in Alex's arms smiled and reached for Jenny.

"Why is *what* that way?" Alex asked as she handed the little girl to Jenny.

"Why is life the way it is? Why is it that answers are so hard to come by sometimes? Why do lessons often have to be learned the hard way? Why can't there be a handbook, or something like that, for living?"

Jenny bounced the baby girl as she talked. Alex didn't answer and the two continued to the edge of the clearing and sat on a bench facing the small group still gathered around the gravesite.

"That's what parents are for," Alex said before she could stop herself. "Sorry."

"That's OK," Jenny said softly, gently bouncing the girl on her knee. "You're right."

"Your father did a pretty good job," Alex said quietly.

Both girls cried for a few moments as Alex held Jenny's quivering shoulders. Jenny held Alex's baby tight. The graveside group was slowly dispersing.

"Alex," Jenny sniffed as she looked up and smiled, "look at this little cutie. How can they be so perfect when they're so small? Can you believe we used to be this small?"

Jenny and Alex welcomed the diversion to play with the baby, making her laugh.

"I often wonder how we can start out so innocent and small and then grow so big, and then ..." Jenny looked up at the mound of dirt in the distance. "How does it all work? And why?"

"You wonder too much, Jen," Alex said. "You've always wondered too much."

"Well, how does it happen?"

"I don't know, Jen, I guess it's just the way we're designed," Alex said.

Jenny froze momentarily, looking at the baby that now sat still on her motionless knee. Alex noticed, looking concerned as Jenny slowly looked up at Alex's face.

"*What*?" Alex said, giving Jenny a strange look.

Jenny continued with a distant stare, looking from Alex to the baby and back. Alex knew that look and tried to imagine what Jenny might be thinking.

"What are you thinking? Tell me!"

It slowly dawned on Alex what Jenny was thinking, as Jenny grinned.

"I think you're right, Alex. I think you are *exactly* right!" Jenny was standing now.

"Oh, Jenny, *no*! Not *again*," Alex groaned in mock despair as she took the baby and started walking away. Jenny followed right behind her.

"Do you know what that means?" Jenny asked excitedly.

"I'm not listening. I'm leaving you, Jenny," Alex laughed, picking up her pace.

"Just think about it ..." Jenny called from behind as she caught Alex's arm.

As they rejoined the small group remaining at the graveside, the evening sky was bright behind them, the unseen smile of the Sun silhouetting Jenny talking as Alex shook her head.

Notes and Quotes

Much of the allegorical content of *The Cave Painting* is based on published positions, views and quotes of leading Darwinists and intelligent design theorists. The following notes provide documentary support for selected portions of *The Cave Painting*, as well as additional insight into the underlying themes of the novel. Further, by examining specific arguments in a non-fictional context the reader can better understand the distinctions between intelligent design theory and naturalistic evolutionary theories, particularly Darwinism. With reference to primary source materials where possible, these notes seek to emphasize the logical and factual differences between the competing theories.

<u>Quick Reference to Major Topics</u>

These Notes reflect the author's opinion and current understanding with respect to people and ideas represented. Notes and Quotes will be updated for future editions/versions of *The Cave Painting*. Comments, corrections, and criticism are welcome and can be sent to the author at roddybullock@idnetohio.com. Please reference the relevant Note number in all communications.

[1] *"After all, the search for intelligence is routinely conducted in other contexts such as archeology and the Search for Extraterrestrial Intelligence (SETI)."*

Archeology and the SETI project, each discussed below, are just two of many examples of scientific endeavors that rely on the ability to recognize intelligent design as a basis for making scientific judgments. Without even making a conscious decision to do so, scientists in these disciplines use intelligent design principles every day—with no thought as to justifying the scientific basis for doing so. Intelligent design is simply taken for granted as being recognizable, and as being a reliable source from which to draw inferences of unseen and otherwise undetectable persons or entities.

A good example of design detection being utilized to resolve a scientific dispute is the debate over whether a rock found in 50,000-year-old strata in North Carolina "is of human manufacture, or the result of natural processes."[1] The finding scientist, advocating a 50,000-year-old status for the rock contends it has been shaped and made for a purpose. However, another scientist, holding 13,000 years ago as being the earliest period of human habitation in North America, claims the shape is just the result of natural processes. To show design as the best explanation, the scientist making the 50,000-year-old claim will have to show both a clear specification and why it is unlikely that natural processes produced the shape. In this instance, to resolve the dispute an inference of design must be "tested."

Searching for intelligence behind apparent design is not limited to objects found on Earth. Scientists tested the inference of design when an apparent sculpture of a human face appeared in a photo of the landscape on Mars. The photo, taken by NASA's Viking 1 spacecraft nearly 30 years ago, revealed a landscape pattern bearing a striking resemblance to a human face.[2] NASA, considering it to be a Martian mesa having unusual shadows, released the photo to the public, reasoning it would be a good way to engage the public and attract attention to Mars.

Attention indeed! Many people saw the photos as evidence of alien intelligence on Mars. The public attention focused on the "face on Mars" prompted NASA to make additional photographs a priority when the Mars Global Surveyor (MGS) was sent for more photos and mapping in September, 1997. Additional photos, at a better angle and ten times sharper, revealed the "face" to be what NASA thought it was in the first place—the Martian equivalent of a butte or mesa. Rather than being the remnants of an alien race, the new photos supported a different inference; the mesa was simply a random formation on the Martian landscape

---

[1] *See*, http://www.archaeology.org/online/news/topper.html; Archaeology, a publication of the Archaeological Institute of America, On Line News, November 17, 2004.

[2] *See*, http://science.nasa.gov/headlines/y2001/ast24may_1.htm; *Unmasking the Face on Mars*, May 24, 2001.

Despite lingering doubts among skeptics and conspiracy theorists, extremely detailed photos and measurements using laser altimeters have conclusively shown the "face" to be just another mesa, like the others that "litter" the region of Mars on which its found.[3] Nevertheless, what prompted the thorough investigation by NASA was the *inference of design*. Based on the first photo, there was a clear, unmistakable appearance of design upon which a reasonable person, even a reasonable scientist, could have inferred, i.e., *hypothesized*, was a result of intelligent agency. Not until additional data was received could the inference of design be ruled out as being truly apparent design. In like manner, intelligent design theorists simply make the assertion, the *hypothesis*, if you will, that life on earth supports an *inference of design*. Until the inference can be shown to be wholly unwarranted, the *scientific* activity of *hypothesizing* design is not only reasonable and proper, it should be readily endorsed by true scientists interested in objectively seeking the truth of life's origin and development.

[2] *"An unearthed artifact that appears to have been designed and made for a purpose is unquestionably and immediately considered to have been the product of intelligence."*

A news report entitled "World's Oldest Art Uncovered in Germany"[4] told of "the incredible discovery" in a cave in Southern Germany of three small figurines carved from mammoth bone. So "intricate in their design" were they, archeologists believe they could change our understanding of the imaginative power of early man's mind. Such a find was not the first in this region of Germany, in the Danube valley near the modern-day city of Ulm. The author of the news article recounted the history of past discoveries as well as the events surrounding the discovery of the three new figurines, and speculated on their use. Most significant, however, was one of the author's conclusions:

> Whatever the meaning behind the little carvings, there's no doubt that they are too intricate to fit with existing theories about the gradual evolution of art, which suggest it began bluntly and has become honed to sophistication over the course of thousands and thousands of years.[5]

The article continues with speculations about the human origins of the art and the purpose for the figurines. Interestingly,

---

[3] Ibid.

[4] *See*, www.dw-world-de, linked from http://209.157.64.200/focus/f-news/k-archeology/browse, (December 20, 2003).

[5] Ibid.

among those who found and are studying the carvings, there was apparently no discussion—no question—of *whether* the figurines were of human origin. Why was there no discussion of alternate theories of origin for these figurines? The question seems silly, but is it? Why simply jump to the conclusion that merely because the objects are "intricate in their design" that they are the product of intelligent design? Have we any evidence otherwise of a human sculptor? Have we seen such a person? How do we *know* (as opposed to theorize) that a human being made the "art" objects? A little reflection will reveal that such questions are indeed silly, and that is the point. If archeologists took the same approach to science that origins researchers do, they would not naively rely on evidence of clear, intricate design to attribute such design to a human designer. However, precisely because design can be detected and attributed to an intelligent designer the field of archeology can yield useful information and increase our knowledge of the world around us.

[3] *"Specified, ordered data received by the radio telescopes of the SETI project even slightly resembling that of the simplest strand of DNA would no doubt be hailed as a breakthrough discovery of intelligent life beyond planet earth."*

SETI is an acronym for Search for Extraterrestrial Intelligence.[6] The SETI Institute is a non-profit corporation headquartered in Mountain View, California, that serves as an institutional home for research and educational projects relating to the study of life in the universe. SETI is an effort to detect evidence of technological civilizations that may exist elsewhere in the universe, particularly in our galaxy. With a budget of millions of dollars per year and a staff of scientists, engineers, educators and administrators, the SETI Institute is a serious, committed attempt at finding life "out there."[7]

Project Phoenix was the name of a SETI Institute research project to search for extraterrestrial intelligence. Project Phoenix used the world's largest telescopes (radio telescopes, not optical telescopes) to scrutinize the vicinities of nearby, sun-like stars for radio signals in the range of 1000 MHz to 3000 MHz. Why radio signals in this frequency? Because, according to the SETI website, the microwave band contains a naturally produced emission line, a narrow-band "broadcast," at 1,420 MHz due to interstellar hydrogen. Every radio astronomer (including extraterrestrial ones, we are assured), will know about this hydrogen emission. Therefore, it serves as a universal "marker" on the radio dial. Specifically, the SETI scientists are looking for narrow-band (less than 300 Hz wide) artificially produced signals. In addition, according to their website, "Other tell-tale characteristics

[6] *See*, http://www.seti.org/site/pp.asp?c=ktJ2J9MMIsE&b=178905, (January, 2006).
[7] Ibid.

include a signal that is completely polarized or *the existence of coded information on the signal.*"[8] Thus, the SETI scientists agree that if they were to find "coded information" on a signal it would be evidence of intelligence. In fact, even though the current search is for "carrier" signals that in themselves do not convey information, the scientists know that a message would be recognizable, even if not decipherable. According to their website:

> Signals sent by a civilization for its own purposes may be impossible for us to unravel. But one thing *we would know irrespective of content* is that *another intelligent civilization* is out there.[9]

SETI is a well-organized, well-funded organization. Its approach to the search for, as well as their protocol for confirming, extraterrestrial signals is impressively well thought out. However, to intelligent design theorists shut out of the "scientific" arena, it is ironic that such expense, effort, and scientific resources are being expended in an attempt to find an unseen, undetected, unknown "intelligence" by looking for "coded information," that "irrespective of content" would confirm "another intelligent civilization" out there. In other words, SETI scientists have a plan for detecting an unseen *intelligent* being based on detectable *design*. How will they do so? Easy—by detecting the "tell-tale" characteristic of "*coded information*" on radio signals.

The correct and scientifically sound recognition by SETI that intelligence can be detected by the presence of coded information is an unwitting admission of the implications of design in detecting the presence of intelligence. Only an intelligent being able to make choices among possible alternatives can organize, i.e., "code," information. One of the most remarkable discoveries in the twentieth century was the discovery of coded information in the DNA of living beings. The genetic code itself is a linguistic system which functions to allow specified genetic symbols to carry biological information. A recent study concludes that the code is the most optimal of a million other randomly selected possibilities.[10] Information organized per the DNA code is much more complex than any signal SETI researchers would attribute to intelligence, and yet, the coded information of DNA is attributed to an accident of nature. [11]

---

[8] Ibid., emphasis added.

[9] Ibid., emphasis added.

[10] S.J. Freeland and L.D. Hurst, *The Genetic Code is One in Million,* Journal of Molecular Evolution, vol. 47, pp 238-248 (1998).

[11] Interestingly, theoretical physicist and cosmologist (as well as evolutionist) Paul Davies recently commented in an interview that the SETI program might be more likely to find an alien intelligence if they were to look in DNA rather than radio waves. *See,*

Unlike their biological counterparts, not only are the scientists at SETI sure coded information is evidence of intelligence, but according to their website:

> Failure to find a signal wouldn't prove that we're the only thinking beings in the Galaxy. After all, absence of evidence is not evidence of absence.[12]

To an objective onlooker such a dichotomy of thought in science between the search for elusive extraterrestrial "intelligent" design to detect an unknown "thinking being" somewhere in the universe, and the denial of confirmed coded information in DNA as strong evidence of intelligent design, is difficult to reconcile. To reject evidence that life on earth may be a product of an intelligent being while simultaneously (and in the absence of any evidence) entertaining thoughts of other intelligent beings elsewhere in the universe is at minimum a profound contradiction in reasoning.[13]

[4] *"... unless, of course, the maker had the foresight to sign, and possibly date, the object. Even then the signature would be suspect. After all, a signature simply has the appearance of design – a trap for the unwary naturalist."*

As discussed more fully below, in an amusing and telling line of reasoning, Darwinists[14] go to great lengths to assure us, if not

---

Miguel Artime, Interview with Paul Davies, Australian Center for Astrobiology, (ASTROSETIEl universo a tu alcance, November 12, 2004), http://www.astroseti.org/davieseng.php).

[12] *See*, http://www.seti.org/site/pp.asp?c=ktJ2J9MMIsE&b=178905.

[13] Note that due to the clear parallels in scientific endeavor, a SETI Senior Astronomer recently published an article seeking to deny that the SETI approach to detecting intelligence supports the intelligent design movement. Seth Shostak, *SETI and Intelligent Design* http://www.seti.org/site/apps/nl/content2.asp?c=ktJ2J9MMIsE&b=194993&ct=1638783, (December, 1, 2005). Shostak is hardly convincing in asserting that "champions of Intelligent Design" are wrong to maintain that the SETI enterprise is logically similar to their own because SETI is only "on the lookout for simple signals" (Ibid.). Further, trying to draw a line between *intelligence* and *artificiality* in an attempt to find a distinction with a difference strikes the intelligent reader as artificial. In the end, SETI can deny the logical connection all they want, but until they change their name to Search for Extraterrestrial Unintelligence, they are only protesting too much.

[14] The term "Darwinist" is used to describe one who holds to Darwin's theory of evolution, that is, the theory that all living things are modified descendents of a common ancestor by means of purposeless, unguided, biological processes, of which Darwinian mutation/natural selection is the leading example. Darwinists, therefore, are those who believe that the theory of evolution explains the origin of all living species, including *Homo sapiens*, based on purely physical process not instigated or guided by an intelligent

"prove" to us, that anything of non-human origin that has all the clear, unambiguous hallmarks of design is not designed at all, but merely has the *appearance* of design. That is, the design is not objectively real; it is "just an illusion, like the rising and setting of the sun."[15] Evolution as understood by Darwinists, therefore, can be characterized as the theory of "no design" in nature. Importantly, intelligent design is simply the scientific challenge to the "no design" hypothesis of Darwinism.

Leading evolutionist Richard Dawkins[16] made the most famous of statements denying true design on the first page of his book *The Blind Watchmaker*.[17] Dawkins, clearly aware that a theory of unguided, purposeless, accidental[18] natural processes creating a complex organism such as humans is incredible at best, and unsupportable at worst, stated confidently that, "biology is the study of complicated

---

being. Most evolutionists today are Darwinists; that is, they subscribe to some form of Darwin's theory of "descent with modification." So-called Neo-Darwinists understand the necessary modifications to be the result of genetic mutations. Consistent use of terms is mandatory for understanding anything, including the issues in the origin-of-life controversy. The term "evolution" is no exception. Whenever the term is used in a propositional statement, the validity of the statement is dependent upon the meaning ascribed to the term. For this reason, Darwinists seemingly intentionally use the term "evolution" without a definite, meaningful definition. When challenged, many Darwinists will claim the term simply means "change over time"—who could deny that there has been change over time? But everyone knows that the term is also loaded and sagging with the semantic baggage of everything from observable change within species to unobservable "atoms-to-Adam" creation of living organisms. Such a chimerical definition is part of the problem in the origins debate.

[15] James Barham, "The Emergence of Biological Value", from Chapter 11 of *Debating Design*, Ed. Dembski & Ruse, (Cambridge University Press 2004).

[16] Richard Dawkins is one of the world's leading Darwinists and a prolific writer of books on the subject for the lay reader. He holds the Charles Simonyi chair of Public Understanding of Science at Oxford University.

[17] Richard Dawkins, *The Blind Watchmaker* (New York: W. W. Norton & Company, 1996), p. 1. The subtitle of this book is, *Why the Evidence of Evolution Reveals a Universe Without Design*. In the introduction to the 1996 edition Dawkins stated that the underlying theme of this book is to explain "Darwinism and design."

[18] The term "accidental" may strike some as odd and others as a mischaracterization. However, in the absence of intelligence to instigate or guide a process of change, only physics and chemistry (which is really only physics) are available to do the necessary "creation." Whatever is created by a process of physical law operating on matter, like a chance pattern of spilled beans on the ground, can only be considered as being accidental. Honest Darwinists do not object to the "accidental" label, they simply believe some accidents are more fortuitous than others, especially if there is a "sorting" process among the accidents to keep the "beneficial" ones. The idea that natural selection can actually blindly "select" anything in the sense of making unguided choices among "good" and "bad" accidents is not supported by any evidence. In any event, there is no denying that natural selection, whatever its merits, can "create" nothing; it can only "select" death for some created things in favor of other created things. Nothing can be selected for death unless first created—survival of the fittest presupposes the "arrival of the fittest. All creation for Darwinism, therefore, is relegated to chance, random, *accidental* variations.

things that give the appearance of having been designed for a purpose."[19]

If biology were merely the *study of* complicated things (irrespective of their "appearance of having been designed for a purpose"), evolution, particularly Darwinism, would likely play only a minor role, if any role at all. Is Darwinism necessary to *study* photosynthesis? To *study* the digestive process? To *study* reproduction? To *study* pollination? To *study* any of the life systems? The answer, not surprisingly, is no. In fact, as shown below, Darwinism is not even necessary to study *biochemistry*.

In the kind of data-driven analysis one would expect from a scientist in search of knowledge, biochemistry professor Michael Behe[20] examined college biochemistry texts for evidence of the importance of evolution to the subject.[21] His interest was piqued by statements in a leading biochemistry text written by a professor of biophysics at Johns Hopkins University. On the first page of the first chapter, the author stated that to understand why biomolecules "appear to be extraordinarily well fitted to their tasks ... biomolecules should be studied from two points of view."[22] The two points of view include, "by the principles and approaches used in classical chemistry" and "in the light of the hypothesis that biomolecules are the products of evolutionary selection ..."[23]

But is evolution really important to a study of biochemistry?[24] As Behe puts it:

> Although a callow student might take [the text author's] word for it, a dispassionate observer would look for evidence of evolution's importance to the study of biochemistry. An excellent place to start is the book's index.[25]

The very detailed index of this text included only two citations to evolution out of 6000. According to Behe, the index that "included

[19] Dawkins, *Blind Watchmaker*, p. 1. (Note, again, that Darwinists must believe that all of creation has no purpose; there is only the appearance of having been designed for a purpose.)

[20] Michael Behe is a molecular biologist, Professor of Biochemistry at Lehigh University, and author of the book *Darwin's Black Box*.

[21] Michael Behe, *Darwin's Black Box* (New York: Touchstone, 1996), p. 180.

[22] Ibid.

[23] Ibid.

[24] By *biochemistry*, Behe includes all sciences that investigate life at the molecular level, even if the science is done in a department with another name, such as molecular biology, genetics, or embryology. *See* Behe, *Darwin's Black Box*, p. 277.

[25] Ibid., p. 180.

virtually everything of relevance to biochemistry," belied that evolution apparently was rarely a relevant topic. A later edition of the same text included only two entries to evolution in an index of 7000 entries.[26] Not until an even later edition (1982) did new authors include more references to evolution—22 out of 8000—but most of these, according to Behe, amount to little more than "a wand to wave over mysteries."[27] This textbook is apparently not an exception. In his book, Behe provides a table of 30 biochemistry textbooks used at major universities dating from 1970 to 1992. Of these, 13 have *zero* entries for evolution in their respective indices. Fully two-thirds have five or fewer such entries.[28]

Contrary to the pretentious claims of today's grandees of evolution, Darwinism is not necessary to *study*, or even to understand, any other scientific subject, including molecular biology. Darwinism becomes necessary in biology only as a naturalistic attempt at explaining the *origin* of living things. There is a big difference between studying things and theorizing about their origins.

[5] *"... but such can be the case only when scientific inquiry is limited* by definition *to exploring only "natural" causes for all effects."*

As with any subject of debate, meaningful progress can be made only if both sides use terms consistently. For example, in the "evolution debate" the term "evolution" is so malleable as to be hammered out to mean practically anything from general "change over time" to "molecules-to-man" human origins. Precise definitions and consistent use of terms becomes essential when discussing *biological* evolution. If biological "evolution" is used to describe change within populations of a species (commonly referred to as "micro-evolution"), such as the change in the beak sizes of "Darwin's finches,"[29] then practically everyone "believes in evolution."[30] Documented evidence

---

[26] Ibid., p. 181.

[27] Ibid.

[28] Ibid., p. 182.

[29] Darwin studied several species of finches on the Galapagos Islands that have subsequently come to be known as "Darwin's finches." The finches play an important role in the teaching of contemporary evolutionary theory, as the finch's characteristics, including beak size, have been carefully studied and monitored for decades. A heritable change in the average beak size for the population from one generation to the next is often cited as "evidence" for the "fact" of evolution.

[30] Sometimes Darwinists object to the terminology "believing in evolution." Apparently evolution isn't something to "believe in" like religion; it is simply accepted, like gravity. One doesn't "believe in" gravity. However, Darwinists themselves use the terminology of belief. For example, in his book *The Blind Watchmaker*, Richard Dawkins stated with respect to Darwin's famous challenge ("If it could be demonstrated that any complex organ existed which could not possibly have been formed by numerous, successive, slight modifications, then my theory would absolutely break down."), that if such a case is ever found "I shall cease to believe in Darwinism" (Dawkins, *Blind Watchmaker*, p. 91).

supports such a theory of evolution. However, if "evolution" is intended to mean the unguided origin of new morphology, new DNA information, or new species (commonly referred to as "macro-evolution"), then the theory of evolution remains highly speculative and not supported by the evidence.[31]

Likewise, Dawkins criticizes those who "desperately want not to have to believe in Darwinism" (Ibid., p. 250). Dawkins has even gone as far as saying that anyone who claims *not* to "believe in evolution" is "ignorant, stupid, or insane (or wicked ...)" (Richard Dawkins, review of *Blueprints* by Donald Johanson and Maitland Edey, *New York Times*, April 9, 1989, sec. 7, p. 34). Dawkins later defended this statement against charges of arrogance. He not only denied that the statement was arrogant, he claimed that it is self-evident. He said, "Examine the statement carefully and it turns out to be moderate, almost self-evidently true" (Richard Dawkins, "Ignorance is No Crime," *Free Inquiry Magazine*, Vol. 21, No. 3, (found at www.secularhumanism.org/library/if/dawkins_21_3.html), April 15, 2004). Dawkins even spoke of *disbelieving* in evolution when he said, "You cannot be both *sane* and *well educated* and disbelieve in evolution. The evidence is so strong that any sane, educated person has *got* to believe in evolution" (Quote found at <www.positiveatheism.org/hist/quotes/dawkins.htm>, April 15, 2004, emphasis in original, quoting Lanny Swerdlow, "My Sort [*sic*] Interview with Richard Dawkins" (Portland Oregon, 1996)). Thus, like any good religious movement, evolution is something to "believe" and, in fact, it has its true believers that carry Darwin's flame with religious zeal.

[31] Darwinists deny that "macro-evolution" is not supported by the evidence. But as discussed elsewhere herein, a search for documented evidence of even one case of speciation is lacking in the literature. What is offered up as "proof" is often ambiguous at best, such as recently discovered plants that appear to have been speciated in recent times. (Of course, even this inquiry is only as valid as the definition of the term "species," which also can get quite distorted in an attempt to verify speciation.) One prominent example of the lack of documentation for speciation will suffice here. In his Pulitzer-Prize winning book *The Beak of the Finch*, Jonathan Weiner documents the extensive work of researchers on the Galapagos Islands as they meticulously record and compile evidence with respect to "Darwin's finches." (Jonathan Weiner, *The Beak of the Finch* (New York: Viking Books, 1994). The subtitle of Weiner's book is *A Story of Evolution in our Time*, and it is widely hailed as "a landmark in evolutionary studies" as quoted from the back cover of the book. It is written to lead the reader to believe that although Darwin never documented the origin of a single species (Ibid., p. 6) finally there is convincing evidence of speciation. Not only can "evolution in action" be observed, but "what has emerged from these studies is how fast divergence [i.e., divergence to speciation] can happen, how much we can actually see in real time. We are not talking about the lapse of ages" (Ibid., p. 178). In a section entitled "New Beings on this Earth," Weiner writes, "The apparent fixity of the species once seemed the greatest argument against evolution, just as the apparent fixity of the earth once seemed a commonsense argument against Copernicanism" (Ibid., p. 202). Therefore, presumably Weiner is going to document the evidence for speciation. However, such a build up is predictably disappointing at the point of providing any evidence for anything other than so-called "micro-evolution"—change within a species. In the following chapter entitled "New Beings" Weiner asks the question, "Why are there so many kinds [*i.e.*, species] of animals?" Right question; here is Weiner's reply: "Adaptive radiations like Darwin's finches are the essence of the answer" (Ibid., p. 207). Thus, beak size variation (fancifully referred to as "adaptive radiation") is the "essence" of why there are so many kinds of animals. The entire book, then, is no

Consider, therefore, the term "science" and its use in the evolution debate. Evolutionists often accuse intelligent design theorists of seeking to "re-define" science." Intelligent design theorists reply that they are simply insisting on consistent application of an "objective" definition of science. Is this all an ivory tower exercise in semantics? No. Far from being an academic question, pinning down the bounds of "science" has far-reaching, everyday importance as placing bounds on what public school teachers, students, and parents can expect to be taught as fact, and ultimately as truth. Consider carefully, therefore, the following discussion.

In its simplest form, science is simply the human activity of seeking explanations for natural phenomena, the explanations being based on observable data. Keeping this idea in mind, consider the following two simple descriptions of science (which are not presented here as comprehensive, but as sufficiently simple and complete to illustrate the issue at hand):

1) Science is the activity of seeking explanations for natural phenomena.

2) Science is the activity of seeking only natural (i.e., unintelligent) causes as explanations for natural phenomena.

What is the difference between the two definitions?

Definition 1 is *objective*.

Definition 2 is *not objective*.

Definition 1 is consistent with the definition of science provided by the National Academy of Sciences (NAS, a pro-evolution organization),[32] state department of education definitions of science and the scientific method,[33] and dictionary definitions.[34]

---

more than a recapitulation of the well-worn Darwinian argument for speciation—minor variations can produce so-called "radiations," which, given enough time make new species. This may or may not be true, but it is only *speculation* at this time. Finch beak variation, no matter how well documented, is not proof of speciation, and variation within the species does not explain creation of finches in the first place, much less any other kind of "new being."

[32] National Academy of Sciences (NAS), *Teaching about Evolution* (Wash. D.C.: National Academy Press, 1998), p. 27. Note also in the same publication, the National Science Teachers Association (NSTA) in its "Position Statement on the Teaching of Evolution" proffers the following definition: "Science is a method of explaining the natural world" (Ibid., p. 124). The reader should take these definitions with a grain of salt, however, as they are offered up only when Darwinism is not under attack. Once challenged,

As an objective definition, Definition 1 is consistent with *the* scientific method, unencumbered by any philosophical bias as to whether an explanation might involve either or both unintelligent natural or intelligent causes.

By Definition 1 intelligent design is science, (and is, in fact, consistent with *the* scientific method). Tracking the language of the NAS, for example, intelligent design posits that explanations of true design in nature can be logically inferred from confirmable data (data of design in nature, which even Darwinists recognize, but deny is true design). And the results are obtained through observation (of design) and experiments (e.g., to test if design can be detected and/or duplicated by natural processes) that can be substantiated by other scientists.

Definition 2, on the contrary, imports an assumption as a limitation into what is "science," a limitation that is not required by any established definition of science or *the* scientific method. Definition 2 instead leads to *a* scientific method that is more limiting than *the* scientific method. Definition 2 inserts the limiting assumption of the philosophy of naturalism into an otherwise objective definition, and leads to *a* scientific method consistent with the assumptions of the philosophy of naturalism. This method is often referred to as *methodological naturalism*, and it is insisted upon as the only method of science by virtually all Darwinists.

Therefore, it is evolution-only Darwinists who attempt to re-define science to fit a philosophy of naturalism, the idea that natural causes and effects alone are sufficient to explain all natural phenomena. However, if one accepts the objective definition of science proposed by the NAS, state boards of education, and dictionaries,

---

Darwinism is (and can only be, it seems) defended with an insistence on the non-objective definition.

[33] *See*, Jonathan Wells, Ph.D, *Definitions of Science in State Standards*, Discovery Institute http://www.discovery.org/scripts/viewDB/filesDB-download.php?id=333. Generated in response to accusations that the standards adopted by Kansas in 2004 were "a radical re-definition of science," Wells' report shows that on the contrary, "Of the fifty states, nine include no definition of science or explicit description of scientific inquiry in standards accessible through the Internet. The standards of forty states include a definition of science or explicit description of scientific inquiry that is consistent with the one proposed in the Minority Report [i.e., a definition consistent with Definition 1 above]. **Only Kansas defines science as "seeking natural explanations."** Emphasis in original.

[34] *See*, Merriam-Webster online, www.m-w.com definitions for "science" as: [K]nowledge or a system of knowledge covering general truths or the operation of general laws especially as obtained and tested through scientific method. Definition of "scientific method" as: [P]rinciples and procedures for the systematic pursuit of knowledge involving the recognition and formulation of a problem, the collection of data through observation and experiment, and the formulation and testing of hypotheses.

intelligent design is, very definitely, science. Intelligent design is based upon empirical data, including observations of design in nature.

The question, then, with respect to how science is defined, becomes not what is knowable through observation, but, rather, what is permitted as an *inference* from such observations. That is, *what is the scope of possible alternatives with which to interpret our observations*? What *theories or hypotheses* is one allowed to entertain? For example, what if observations of living organisms show evidence of design?[35] What is one to do with such empirical evidence? Understanding how evolutionists answer this question leads to an understanding of why "science" as historically defined has been hijacked by science as practiced by importing and insisting on a *working* definition that includes a *non-refutable assumption* of only unintelligent causes for all effects with respect to the study of origins.

At least two options are available for honestly dealing with observable, empirical evidence of design: (1) openly consider the possibility that certain aspects of living systems are designed for a purpose; and (2) openly reject such a possibility regardless of the quantity and quality of the evidence of design.[36] Although Option 1 meets the *objective* definition of science (i.e., Definition 1 above), it is nevertheless rejected by Darwinists as not being "scientific" because it does not meet the unwritten altered working definition, which requires an assumption of only unintelligent causes for all effects. Option 2 meets the working definition (i.e., Definition 2 above), but eliminates intelligent design from "science," not by the evidence, but *by definition*. But when intelligent design is "defined out" of science the question of the cause of our origins is answered before being asked. To make matters worse, Option 2 is dogmatically insisted upon as if its acceptance were self-evidently required by "science."

Option 1 above is rejected because any hint at "intelligent design" sets off fears of a "divine foot" in the door and is taken as "creation science" which, regardless of the ultimate truth of the matter,

---

[35] As they inevitably do. Darwinists recognize one of their tasks as being to show how things that look designed are, in fact, not designed by an intelligent designer. As mentioned above, Darwinism can be considered the theory of "no design." We think we see design like we think we see the sun move around the earth; it is simply an illusion.

[36] Option 2 implies that scientists have sure knowledge of non-intelligent design, thereby deeming the question of intelligent design moot and, instead, referring to design as "apparent design" and all theories based on intelligent (non-human) design as "religious." For example, in the Introduction to the 1996 edition of *The Blind Watchmaker*, Richard Dawkins states: "Pretend as they will to scientific credentials, the anti-evolution propagandists are always religiously motivated, even if they try to buy credibility by concealing the fact." Attacking motives seems to be the only way evolutionists know to deal with opposition. Of course, who can blame them? Such a tack successfully plays to the public's reservations and the media's biases; and it conflates legal issues with emotional fears in a smokescreen of rhetoric designed (intelligently, for sure) to avoid actually having to refute scientific reasoning with scientific reasoning.

is simply anathema to modern science.[37] Supposedly, Option 1 is off limits to science because intelligent design requires an intelligent designer, which "person" or "persons" is unknown and unknowable to science—science simply can say nothing about such an entity. As stated in the "National Association of Biology Teachers Statement on Teaching Evolution":

> Explanations employing nonnaturalistic or supernatural events, whether or not explicit reference is made to a supernatural being, are outside the realm of science and not part of a valid science curriculum.[38]

Thus, the NABT contends that, *regardless of where the observations might lead,* explanations involving nonnaturalistic or supernatural events are outside the realm of science. Science, according to the NABT, is not to be objective, i.e., explanations based on empirical evidence. Rather, science *as practiced* is limited to only *naturalistic* explanations supported, if possible, by empirical evidence. One is certainly within reason to ask why. Why arbitrarily cut off one line of inquiry? The NABT continues:

> Evolutionary theory, indeed all of science, is necessarily silent on religion and neither refutes nor supports the existence of a deity or deities.[39]

Aside from the fact that an inference of design supports but does not entail a deity, three replies can be made to this position. First, although evolution's claim of no-design does not refute the existence of an impersonal deistic creator, it is not "silent" on whether life was designed and made for a purpose. This is because evolution is an

---

[37] For example, the National Science Teachers Association (NSTA) Position Statement on the Teaching of Evolution states: "'Creation science' is an effort to support special creationism through methods of science. Teachers are often pressured to include it or synonyms such as 'intelligent design theory,' 'abrupt appearance theory,' 'initial complexity theory,' or 'arguments against evolution' when they teach evolution." Similarly, the National Science Teachers Association Position Statement on the Teaching of Evolution states: "Policy-makers and administrators should not mandate policies requiring the teaching of creation science or related concepts such as so-called 'intelligent design,' 'abrupt appearance,' and 'arguments against evolution.'" NAS, *Teaching About Evolution*, p. 125.

[38] Statement on Teaching Evolution, National Association of Biology Teachers (NABT). Adopted by the NABT Board of Directors on March 15, 1995. Quoted in NAS, *Teaching About Evolution*, p. 127.

[39] Ibid.

undirected and purposeless process. For this reason, organizations such as the National Science Education Standards seek to teach children that natural objects and phenomena lack the attribute of design. This is an issue critical to both theistic and non-theistic religion. Notwithstanding claims to the contrary, Darwinian evolution does take a position on religion, particularly theistic religions positing a personal creator.

Second, it is clear, upon slight reflection, that scientific "silence" is not "necessary" any more than would be silence on a theory of gravitational attraction, electro-magnetic fields, or any other unseen but detectable phenomena. Presumably, because any supposed deity cannot be observed, he, she, it, or they cannot even be acknowledged as a possibility. But gravity is just as unobserved; we can detect only its effects.

Third, a fundamental problem with the NAS statement is that it implicitly uses a definition of religion that is too narrow. To be "necessarily silent on religion" is not possible. Being vocal about the scientific theories consistent with Secular Humanism (a recognized religion) is not being "silent" about religion. According to the U.S. Supreme Court, for example, religion includes both theistic and non-theistic belief systems. A position that excludes supernatural causes and allows for only natural causes is not silent or "neutral" as to competing religions.

In her book, *Total Truth: Liberating Christianity from Its Cultural Captivity*, Nancy R. Pearcey speaks of the role religion plays in the broadest sense. As she says, "The upshot is that no system of thought is a product purely of Reason—because Reason is not a repository of infallible, religiously autonomous truths, ..."[40] Pearcey continues:

> If you press any set of ideas back far enough, eventually you reach some starting point. Something has to be taken as self-existent—the ultimate reality and source of everything else. For the materialist, the ultimate reality is matter, and everything is reduced to material constituents ... For the doctrinaire Darwinist, biology is ultimate, and everything, even religion and morality, is reduced to the product of Darwinian processes....[41]

Pearcey concludes:

[40] Nancy R. Pearcey, *Total Truth: Liberating Christianity from Its Cultural Captivity* (Wheaton, IL: Crossway Books, 2004), p. 41.

[41] Ibid.

> And so on. Every system of thought begins with some ultimate principle. If it does not begin with God, it will begin with some dimension of creation—the material, the spiritual, the biological, the empirical, or whatever... In this sense, we could say that every alternative to Christianity is a religion.[42]

It becomes clear, therefore, that the effect of the undisclosed use of Option 2 blocks from consideration any non-naturalistic explanation of origins, whether or not based on empirically derived, observable, confirmable data.[43] Any theories that imply intelligent design of living organisms, are ridiculed, rejected, and reviled. The passion that accompanies such reactions suggests a non-theistic religious fervor.[44]

[6] *"... they merely have the appearance of paintings. They are, in fact, only* apparent *paintings."*

If the reader is to grasp only one point from this entire book, it should be an understanding of the concept of "design," its detection, and its implications. Interestingly, both Darwinists and intelligent design theorists see overwhelming evidence of design in nature. Only Darwinists, however, feel compelled to take the odd position of attempting to explain that what they identify as "complexity and beauty of design"[45] in nature is not true design, but is only *apparent* design. While somewhat lengthy, this note explores one Darwinist's attempt at convincing the reader of the Darwinist position that design is actually not design.

Richard Dawkins' best-selling books *The Blind Watchmaker* and *Climbing Mount Improbable* are instructive in understanding the

---

[42] Ibid.

[43] There are exceptions. For a good discussion on this topic, including a historical development of the philosophical basis for "science," *see* William Dembski, "The Demise of British Theology," *Intelligent Design, the Bridge Between Science & Theology* (Downers Grove, IL: InterVarsity Press, 1999). *See also*, Larry Witham, *By Design: Science and the Search for God* (San Francisco: Encounter Books, 2003).

[44] Not only the passion, but also the dogmatism of Darwinists rivals that of their creationist adversaries. It is amusing, then, to read such dogmatic statements as that in the "National Association of Biology Teachers Statement on Teaching Evolution": "The diversity of life on earth is the outcome of evolution..." followed by another statement: "The models and subsequent outcomes of a scientific theory are not decided in advance, but can be, and often are, modified and improved as new empirical evidence is uncovered." Another favorite: "Science is not teleological: the accepted processes do not start with a conclusion, then refuse to change it, or acknowledge as valid only those data that support an unyielding conclusion." And finally: "Science does not base theories on an untestable collection of dogmatic proposals." (NAS, *Teaching About Evolution*, pp. 127-28.)

[45] Dawkins, *Blind Watchmaker*, p. 21.

Darwinist viewpoint on design.[46] Both books, according to Dawkins, share an underlying theme: explaining "Darwinism and design."[47] And although the subtitle to *The Blind Watchmaker* is *Why the Evidence of Evolution Reveals a Universe Without Design*, it is clear that the only purpose for such a book in the first place is the fact that we all live in a universe *that appears* designed. All around us is unavoidably observable evidence of complex design in nature. Is it objectively real or is the design we see just an illusion?

Referring to design as "apparent" design is a problematic starting point. If language means anything, to even think of design at all in the absence of a corresponding thought of a designer is illogical, and in any other context would be absurd. Like speaking of a beautiful oil painting as "an apparent painting," speaking of "apparent design" is something of an oxymoron.[48] *A thing is either designed, or it is not designed.* Like knowing someone is "apparently pregnant," knowing something is "apparently designed" imparts no useful knowledge regarding the true state of affairs. If it is *apparent* design, then it is not design at all. Darwinists would agree with this, and Darwinists such as Richard Dawkins attempt to convince us that the beautiful oil painting of nature is, so to speak, not a painting at all, but an *apparent* painting.[49] But even attempting to refute design is itself an admission of the undeniable presence of design. Like atheism, "apparent design" is defined by what it denies. Even Dawkins can't help admitting design, like when he boasts of his ability to "easily fill another ten entirely new chapters [of *The Blind Watchmaker*] on the perennially fascinating topic of evolutionary design."[50]

Evolutionary design? In a universe without design?

Design, as opposed to "order" or "pattern," by definition requires an intelligent designer. "Order" does not necessarily require an intelligent "orderer;" natural laws can impose order on things.[51] But

---

[46] Dawkins, *Blind Watchmaker*; Richard Dawkins, *Climbing Mount Improbable* (New York: W.W. Norton & Company, Inc., 1996).

[47] Dawkins, *Blind Watchmaker*, p. ix.

[48] Likewise, "intelligent design" is redundant when design is used for such concepts as specified complexity, as discussed elsewhere herein. There is no such thing as "unintelligent" specified complexity. It is either design or not, and design requires intelligence in the form of a designer.

[49] As noted elsewhere herein, Richard Dawkins makes a distinction between "apparent design" and "true *human* design." As his argument goes, he bases his determination of "true" design on his sure knowledge of a human designer. Thus, for purposes of argument, the above illustration is on point if the painter is unknown and unknowable—not unlike the cave painting of this book. Dawkins would be forced by logical necessity to describe such a painting as an "apparent" painting.

[50] Dawkins, *Blind Watchmaker*, p. ix.

[51] Although there is a sense in which the natural laws were designed. Nevertheless, taking natural laws as purely natural for the sake of argument, such laws only "create"

it is illogical and somewhat ludicrous to even use the language of design while simultaneously denying the existence of a designer. Again, it is like explaining the origin of a mysterious painted masterpiece by first denying even the possibility of a painter. Imagine if homicide investigators operated under the same constraints as biologists, where, *regardless of the evidence,* only theories involving accidental death would be permitted. Murder (i.e., the hand of intelligent agency) could never be established—all deaths would be ruled accidental, even if perhaps some were *apparent* murders.

Understanding how Darwinists such as Richard Dawkins attempt to co-opt the language of design by denying design is instructive in understanding how Darwinists think about this critical point. Since the explanation that follows is an attempt to concisely capture the main points without overwhelming the reader, the reader is encouraged to purchase and read Dawkins' books. By reading them critically and for real understanding, the discerning reader will come to the conclusion that "apparent design" in living things is just wishful doublespeak for "design" and Darwinism can't explain either.

Dawkins begins his book *The Blind Watchmaker* with a chapter entitled "Explaining the Very Improbable."[52] Differentiating between "simple things" such as rocks, and "complicated things" such as people, Dawkins correctly notes that, "The difference is one of complexity of design."[53] Recognizing that "Words are our servants, not our masters,"[54] Dawkins states:

> [T]hese notions of complexity and design, are so pivotal to this book that I must try to capture a little more precisely, in words, our feeling that there is something special about complex, and apparently designed things.[55]

---

repetitive, relatively simple order, as opposed to specified complexity exhibited by living organisms.

[52] Ibid., p. 1

[53] Ibid., p. 1. Immediately after this statement Dawkins makes his famous observation that "Biology is the study of complicated things that give the appearance of having been designed for a purpose."

[54] Ibid., p. 1.

[55] Ibid., p. 6. Note how Dawkins starts with "notions of complexity and design" and then ends speaking of "*apparently* designed things." Without putting too heavy a gloss on this, such wordsmithing is significant and must be watched closely. As noted elsewhere herein with respect to Darwinists' use of logic, this is a case of "begging the question." Whether something is *apparently* designed or not is the *question*, i.e., what we are trying to ascertain. Dawkins starts with the right terminology, "complexity and design," and then suddenly assumes the truth of the premise he is seeking to prove by switching terms to "apparently designed things." Such looseness with terminology immediately calls into

Looking at how Dawkins explains his use of the "pivotal" terms "complexity" and "design" is an eye-opening introduction to Dawkins' sleight-of-ink approach to "using the tricks of the advocate's trade."[56] He starts his discussion by implying he would explain his use of the "notions of complexity and design." However, in the pages following this announced intention, Dawkins fails to distinguish between complexity and design. Instead, he ambiguously conflates the two notions, usually just referring to "complicated" things.[57] For example, after several pages of rambling about complexity requiring that something be "specified in advance," Dawkins states:

> This has been quite a long, drawn-out argument, and it is time to remind ourselves of how we got into it in the first place. We were looking for a precise way to express what we mean when we refer to something as *complicated*.[58]

No—we were looking for insight into how Dawkins uses the terms "complexity" (which may or may not be synonymous with "complicated") and "design," both terms that he introduced as being "pivotal" to his book.

The distinction between "complexity" and "design" is more than semantic, and it is more than academic; it is critical to understanding both the scientific claims of intelligent design theory and the shortcomings of Darwinism. Random chance can produce highly complex, and highly *improbable* things. For instance, the chance placement of all the rocks and stones that make up Mont Blanc is both complex and highly improbable. However, such a *complex* arrangement is not *designed* in any sense of the ordinary meaning of the word. Disappointingly, however, Dawkins misses the point. He comes close by recognizing that "complicated things have some quality, *specifiable in advance*, that is highly unlikely to have been acquired by random chance alone."[59] But it is not "complicated things" (nor even "complex" things) that have this quality—it is *designed* things that have this quality! Inherent in the very concept of design is "specified in advance" complexity. Without "specification in advance" a thing is, by definition, *not* designed.

---

question whether words are, in fact, Dawkins' slaves—or are they not actually his master?

[56] Dawkins, *Blind Watchmaker*, pp. 6 and xiv.

[57] Ibid., p. 9.

[58] Ibid., emphasis added.

[59] Ibid., p. 9, emphasis added. Dawkins here is perilously close to using the language of "specified complexity" which, of course is a key aspect of intelligent design theory.

Dawkins recognizes that something "specifiable in advance" is crucial, but crucial to *what*? The constraints of his theory (and language, with words Dawkins must find to be very unwilling servants), permit him to admit only that it is essential to "complicated things." But what is the difference between a "complicated thing specified in advance," and a "designed thing?" Nothing, of course; but Dawkins tries a particularly crafty sleight-of-ink maneuver to avoid becoming a proponent of intelligent design theory.

First, Dawkins correctly notes the problem. Directly following the quote immediately above ("This has been quite a long, drawn-out argument ...") he writes:

> We were trying to put our finger on what it is that humans and moles and earthworms and airliners and watches have in common with each other, but not with blancmange, or Mont Blanc, or the moon.[60]

Ambiguity on the relevance of "complicated" things aside, Dawkins states the exact right question, i.e., what is the difference between, say, humans and Mont Blanc? Surprisingly, Dawkins makes an argument for intelligent design by saying that there is a difference, and it is that some quality in living things is "specified in advance":

> In the case of living things, the quality that is specified in advance is, in some sense, 'proficiency'; either proficiency in a particular ability such as flying, as an aero-engineer might admire it; or proficiency in something more general ...[61]

Why is this not intelligent design theory? Logically it is. *If something is complex and highly improbable but for having a quality specified in advance, it is, by definition, specified complexity – and specified complexity lends itself to no explanation but design.*

Dawkins thinks he has avoided proving actual design only because he thinks an *afterward-identified* attribute like flying can, in hindsight, be attributed to "proficiency" which, in turn, is deemed "specified in advance." Is a thing good at flying? Then proficiency for

---

[60] Ibid. Note again, however, that we were *not* trying to express what we mean by something being "complicated." Dawkins said he was going to "capture a little more precisely, in words, our feeling that there is something special about *complex*, and apparently *designed* things" (Ibid., p. 6).
[61] Ibid.

flying must have been "specified in advance." If this is not an explicit argument for design, then logically it is no different than throwing a number of darts at a wall, drawing a bulls-eye around each one, and then marveling at the unlikely complexity of it all. But if the targets were first drawn on the wall (i.e., specified) in advance, and a dart was later found embedded into the middle of each one, such complexity would quickly be recognized as an extremely unlikely chance event, and the improbability of such organized complexity would argue against accidental happenstance.[62]

Therefore, agreeing with Dawkins that *specified complexity* requires explanation, we must ask, is specified complexity likely to be chanced upon, or does specified complexity require something to be specified in advance, an actual pre-determined target, goal, or purpose?[63] By any reasonable use of language and logic, specified

---

[62] There's no doubt that many earnest Darwinists object to terms such as "accidental happenstance," but that is exactly what Darwinism is *and must by definition be, unless it is teleological in nature*. Dawkins, in particular, seems to try his best to get people to believe that Darwinism is not a chance process (not *pure* chance, as he says (Dawkins, *Climbing Mount Improbable.*, p. 80)). He criticizes one of "Britain's most famous physical scientists" (Sir Fred Hoyle), for suggesting that Darwinism is a theory of chance by stating flatly that this scientist (and another) "miss the point that Darwinism is *not* a theory of random chance" (Ibid., p. 75). He attempts to focus the reader's attention on the non-random process of natural selection, as if a non-random step in a process renders the whole process non-random. For instance, Dawkins states that "[Evolution] is a theory of random mutation plus *non-random* cumulative natural selection" (Ibid., emphasis in original). Dawkins then asks: "Why, I wonder, is it so hard for even sophisticated scientists to grasp this simple point?" (Ibid.) To try to sell his "simple point," Dawkins refers only to the random, chance nature of mutations reluctantly and deceptively. For example, he states: "Mutation, the first stage in the process, is random in the sense of not pushing towards improvement. All improvement is therefore, in the first place, lucky, which is why people mistakenly think of Darwinism as a theory of chance. But mistaken they are" (Ibid., pp. 85-86). Mistaken he is, if he thinks that thinking people will buy this pseudo-scientific claptrap. Stressing the "non-randomness" of natural selection to cover for the random, chance events of mutation is like stressing the payout at a casino to say gambling does not involve chance. It is simply and inexcusably ludicrous. Natural selection can be a non-random process (it is non-random death of already-existing things). But make no mistake, variation, by mutations or otherwise, is quintessentially and purely a random, chance process. Variation is unguided, accidental, and random, and it alone, regardless of how many "non-random" events are pinned to it, renders Darwinism a theory of unguided happenstance. Therefore, Hoyle is right—Darwinism is a chance process. It is difficult to understand why Dawkins would take a contrary position. Why would Dawkins want so badly to distance Darwinism from being a chance process? After all, to deny random chance is to affirm intelligent design. If Darwinism is not accidental happenstance, the only other option is *guided* or *directed* creation of complexity. Dawkins can't have it both ways—at least not for sophisticated scientists and other serious thinkers. Either Darwinism is a theory of blind, random chance, or it is a theory of visionary, deliberate choice.

[63] The terminology of "specified complexity" is being used here because this is what Dawkins is describing, albeit in terms of "complicatedness," i.e., "complicated things specified in advance."

complexity requires a specified-in-advance target, and this is what separates design from ordinary complexity—design is characterized by specified (in advance) complexity. That is, the designed thing (*e.g.*, eye, wing, etc.) is specified in advance, not the *need* for proficiency with the thing, such as vision or sight. Even invoking a vague notion such as proficiency as being "specified in advance" leads to design as an explanation—*if something, anything, must be specified in advance to reasonably explain the creation of a highly improbable thing, the thing* is *designed, not "apparently" designed.*

Surprisingly, in a failed attempt to show otherwise, Dawkins proves that, in fact, a specified-in-advance target *is* necessary for any specified complex thing to be likely, i.e., probable in the time frames necessary for evolution. Specifically, Dawkins builds on the time-honored example of the proverbial monkeys typing at typewriters. Dawkins writes:

> I don't know who it was first pointed out that, given enough time, a monkey bashing away at random on a typewriter could produce all the works of Shakespeare. The operative phrase is, of course, given enough time.[64]

Even Dawkins is reasonable enough to understand that the odds of such an event render it effectively impossible in the time spans currently understood for our universe.[65] To simplify his supposed explanation for how unguided evolution can nevertheless do the job, Dawkins limits the task to just reproducing one line of Hamlet: "Methinks it is like a weasel."[66] Further specifying a restricted keyboard having only the 26 letters of the English alphabet, capital letters only, and a spacebar, Dawkins calculates the odds of our tireless monkey randomly typing out the 28 characters and spaces in the right order to be about 1 in 10,000 million million million million million million. Even Dawkins must admit that this simplified task "would be

---

[64] Ibid., p. 46. In fact, it may have been Thomas Henry Huxley, a contemporary of Darwin's and one of his strongest supporters. Huxley is reported to have said that "six monkeys, poking randomly at typewriters for millions of years, could write all the books in the British Museum." James Perloff, *Tornado in a Junkyard* (Burlington, MA: Refuge Books, 1999), p. 69. Such truisms, however carelessly bruited about, become a part of the general consciousness as an irrefutable truth.

[65] In general, most probability calculations are made by anti-Darwinists, since Darwinists consider the subject irrelevant. Darwinists know that, since we are here, and we are a product of evolution by any odds, probabilities are at best a curiosity and at worst a distraction.

[66] Ibid.

a long time coming, to say nothing of the complete works of Shakespeare."[67]

Dawkins considers a one-time random event of typing METHINKS IT IS LIKE A WEASEL to be prohibitively improbable given the limited time available in our universe's history.[68] However, what Dawkins believes, and, indeed, what his books *The Blind Watchmaker* and *Climbing Mount Improbable* attempt to show, is that such a highly improbably event is not so unlikely if it comes about in many, many smaller steps, cumulatively over time—but, according to Dawkins, a much shorter period of time.[69]

How does Dawkins figure that many, many random, *blind* steps are more likely than a single random, blind step to get to a highly improbable event? (Remember, the watchmaker is supposedly blind.) The answer is utterly amazing. He devised a way to successfully shorten the number of iterations his monkey must endure to type

---

[67] Ibid., p. 47.

[68] Ibid., p. 142. Commenting on the same probability, Dawkins stated, "The problem is, of course, that we couldn't assemble that many monkeys. If all the matter in the universe were turned into monkey flesh, we still couldn't get enough monkeys. The miracle of a monkey typing 'Methinks it is like a weasel' is quantitatively too great, measurably too great, for us to admit it to our theories about what actually happens."

[69] Indeed, the gist of Dawkins' argument is that the probability of a highly unlikely chance happening is increased if the event is achieved in innumerable smaller chance happenings, what he calls "cumulative selection." He says of his book *The Blind Watchmaker* "The whole book has been dominated by the idea of chance, by the astronomically long odds against the spontaneous arising of order, complexity and apparent design. We have sought a way of taming chance, of drawing its fangs. 'Untamed chance', pure, naked chance, means ordered design springing into existence from nothing, in a single leap. It would be untamed chance if once there was no eye, and then, suddenly, in the twinkling of a generation, an eye appeared, perfect and whole. This is possible, but the odds against it will keep us busy writing noughts till the end of time. ...To 'tame' chance means to break down the very improbable into less improbable small components arranged in a series. No matter how improbable it is that an X could have arisen from a Y in a single step, it is always possible to conceive of a series of infinitesimally graded intermediates between them" (Dawkins, *Blind Watchmaker*, p. 317). Of course, anything can be *conceived*. God can be conceived; space aliens can be conceived. But what we are looking for is a reason to believe that "pure, naked chance," *multiplied many times over in inconceivably small increments* can create specified complexity. There is no known natural law or process for doing so. Even Dawkins must admit that some mechanism for *guiding* the "series of infinitesimally graded intermediates" is necessary. He states that cumulative selection will work if there is "sufficient time" and "also only if there is a mechanism for guiding each step in some particular direction, otherwise the sequence of steps will career off in an endless random walk" (Ibid., p. 318). Guiding implies contingency—the ability to choose among alternatives—and this can be done only by intelligent beings. Dawkins believes that natural selection can "guide," but as discussed elsewhere herein, surreptitiously sneaking "guidance" in the guise of "natural" selection is the parlor trick of sophisticated Darwinists who pretend to scientific reasoning.

Hamlet's line—by starting with a "*target phrase*."[70] The monkey (actually a computer monkey in Dawkins' example, but it doesn't matter) types a phrase, *and then compares it with the desired outcome, i.e., the target phrase*! Letters in the right position are kept, and the process continues, each time comparing the random phrase to the target phrase and keeping any correctly placed letters, until the monkey randomly has typed the last letter in the right place.[71]

The significance of this "proof" cannot be overestimated. In an overt attempt to show that natural, unguided, purposeless evolution can lead to specified complexity in a reasonable amount of time, Dawkins describes an unnatural, guided, purposeful process that starts with a target, i.e., a *goal* or a *purpose*! A target is nothing less than purpose and his "selection" process is nothing more or less than "guiding" according to the purpose. There is nothing natural, unguided, or purposeless in Dawkins' supposed proof of the power of cumulative selection. Think about it—*the watchmaker is not blind after all*! How would blind nature know which letters to keep with each iteration? Dawkins might answer, "whichever letters are advantageous for proficiency." Proficiency for *what*? Does nature know? Does nature have a plan? Does nature have a *target phrase*? If so, by whatever name it is called, that alone is classic evidence of *design*, not chance.

Dawkins must have suspected that he had just explained how intelligent design works because several pages later he quickly backpedaled by admitting:

> Although the monkey/Shakespeare model is useful for explaining the distinction between single-step selection and cumulative selection, it is misleading in important ways.[72]

One "important way," Dawkins notes, is that nature has no "distant ideal" with which to compare each generation. That is, "[e]volution has no long term goal."[73] How, then, is the monkey/Shakespeare model useful? Why even waste the paper and ink, not to mention the reader's time? Dawkins continues that, "there is no long distance target," and the "criterion for selection is always short-term."[74] The short-term

---

[70] Ibid., p. 47. The target phrase is the desired sentence, i.e., METHINKS IT IS LIKE A WEASEL. That is, the process starts out with the end in sight.

[71] Ibid., pp. 47-8.

[72] Ibid., p. 50. Although Dawkins apparently knows his supposed example of the power of cumulative selection was flawed, he chose to leave it in his book. For what purpose, other than to awe the true believers, is difficult to understand.

[73] Ibid.

[74] Ibid.

criterion is "simple survival or, more generally reproductive success."[75] However, this does not explain "blind" evolution. An "ideal" is an ideal whether distant or not, and a "goal" is a goal whether long term or not, and a "criterion" is a criterion whether far off or close. An "ideal" or a "goal" or a "criterion" is nothing more than a "purpose" whether called so or not, and to the extent such "ideals" etc. are involved, living things are *designed*—and evolution is not blind. Dawkins' ill-conceived "proof" of the power of cumulative selection actually proves the need for specified-in-advance targets.

Calling the target "survival" or "reproductive success" neither removes the requirement for specificity-in-advance nor gives an example of sufficient specificity. Sufficient specificity must have the capability of giving cumulative *directional* complexity; the ability to make choices from a range of alternatives. Although this premise is what Darwinism is built upon, there is no reason to believe that such a cumulative process is more probable than one single event. Taking Dawkins' example of the typing monkeys (and unavoidably mixing metaphors), after the first random set of letters, *which letters should blind nature keep*? Dawkins might say, "Whichever one or two letters confer a reproductive advantage on the organism." That is the only answer Darwinists can make, but the one or two letters are *meaningful* only in the final sentence, or at least some other intermediate *meaningful* sentence. That is, "reproductive advantage" gained by an eye (or even 5% vision, which still requires 100% of a working eye), requires an entire eye.[76] To make the argument that some portion of a (non-working) "eye-to-be" provides a "reproductive advantage" is the same as saying that changing a letter or two in the random letters renders the sentence more meaningful (or at least as meaningful) in some way, in addition to being part of the final sentence for an eye. But it is highly unlikely that there is any meaningful "short-term" sentence among the innumerable possibilities of sentences. In fact, most certainly, repeated,

---

[75] Ibid.

[76] In a flawed line of reasoning passed off as scientific thinking, Dawkins and others like to push the argument that "half an eye is better than none." For example, Dawkins states, "A simple, rudimentary, half-cocked eye/ear/echolocation system/cuckoo parasitism system, etc., is better than none at all. Without an eye you are totally blind. With half an eye you may at least be able to detect the general direction of a predator's movement, even if you can't focus a clear image" (Ibid., p. 41). The flaw is easy to see—half an eye is in no way equal to half vision (and there's no reason to believe it might be otherwise). In other words, it takes 100% of an eye of some sort to get vision of any sort. Like saying half an automobile engine is better than none, the question is *better for what*? Having 50% of an engine will get you exactly 0% mobility. Half an engine is likely to be more of a liability than a benefit. There is simply no reason to believe that half of an automobile engine would create itself for no purpose, and there is no reason to believe that half an automobile engine would have any other purpose but to be united with the other half to be useful. Certainly 5 percent of an engine would in no way produce 5 percent power. Dawkins fails to see this simple truth. In his mind, "An ancient animal with 5 per cent of an eye might indeed have used it for something other than sight, but it seems to me at least as likely that it used it for 5 per cent vision" (Ibid., p. 81).

random changes to the letters of a meaningless "sentence" will ensure the sentence remains meaningless. Even if the sentence was to start out close to being meaningful, constant random changes certainly will render the entire sentence *less* meaningful, if meaningful at all. Furthermore, using "reproductive success" as the criteria not only assumes there is such reproductive success possible with a minor change but that the odds are cumulatively over time better that such a criteria is met instead of any of the limitless other phrases and sentences that could be formed, any or all of which may actually confer a *disadvantage* on the organism. Keep in mind that a simple 28-letter sentence is not even near the order of specified complexity necessary to make the simplest living organism.[77] Unless nature knows the direction, i.e., the target phrase, there is no basis for making the choice of letters to keep. Making a hindsight judgment reduces natural selection to the tautology it is, appearing to explain all, but actually having no explanatory power at all.[78]

So rather than learning "how much more effective" cumulative selection is,[79] we are led back to the well-worn path of ordinary Darwinism—that of enumerable *random*, purposeless changes, each as unlikely as the next, nevertheless cumulatively building *directionally* in a continuously more complex manner by natural, i.e., unintelligent, undirected selection. If Darwinists want to believe that one of their leading proponents has "proven" that innumerable small, random changes over time can "blindly" produce specified complexity, then there is little that logic or reason can do to help. However, the serious reader would be well served to consider carefully Dawkins' monkey business, as well as his own words on this issue:

> I am aware that my characterization of a complex object—statistically improbable in a direction that is specified not without hindsight—may seem idiosyncratic ... But ... whatever we choose to *call* the quality of being statistically-improbable-in-a-direction-specified-without-hindsight, it is an important quality that needs a special effort of explanation.[80]

---

[77] Indeed, Dawkins himself estimated the that there is 'enough information capacity in a single human cell to store the *Encyclopaedia Britannica*, all 30 volumes of it, three or four times over" (Ibid., pp. 115-16). We should not feel all that special, however. Dawkins is quick to assure us that the information capacity of amoebas has as much information in their DNA as "1,000 *Encyclopaedia Britannicas*" (Ibid).

[78] The tautology can be stated as: "The fittest are those who reproduce, because those who reproduce are the fittest." Such a tautological statement is right by definition, and, therefore, cannot be falsified.

[79] Ibid., p. 47. This was the purported purpose of the monkey/Shakespeare example.

[80] Ibid., p. 15, emphasis in original. Note also an apparent typographical mistake between the first instance of "specified *not* without hindsight" and the second instance:

Agreed. And Dawkins has provided a laudable attempt at giving a "special effort of explanation." However, his explanations sound plausible only when he is actually describing intelligent design! In fact, we need not wonder what to *call* Dawkins' mysterious "quality." After all, the quality of "being statistically-improbable-in-a-direction-specified-without-hindsight" already has a name—*specified complexity*, and it is the hallmark of design. To deny this is more than idiosyncratic—it is simply unreasonable. Like the Wizard of Oz asking earnest seekers to "ignore the man behind the curtain," the wizard of Oxford wants us to accept his "special effort of explanation" while ignoring the typewriting-monkey's crib sheets. Only the most gullible would be impressed with the smoke and noise of his evolutionary typing machines.

Richard Dawkins is not alone in his attempt to position Darwinism as an "undirected directed" process. The urge to ascribe some kind of direction-giving force to offset the otherwise random process of Darwinian variation can hardly be resisted. Another preeminent evolutionist, Stephen Jay Gould[81], while ostensibly denying a directed process, couldn't quite bring himself to declare Darwinian variation truly random. Distinguishing Darwinism from an earlier theory called Lamarckism,[82] Gould explained in an essay entitled "Shades of Lamarck":

> This is the essential difference between Lamarckism and Darwinism—for Lamarckism is, fundamentally, a theory of *directed* variation.[83]

"specified-without-hindsight." If the first meaning is intended, then Dawkins' characterization is equivalent to "specified with hindsight." If the second meaning is intended, the characterization is equivalent to "specified in advance." It seems that either of the two options is logically problematic for naturalistic Darwinists. "Specified with hindsight" reduces "evolution" to a tautology with no explanatory power (like predicting the previous day's stock market based on today's data; few people will be impressed with 100% accuracy). "Specified in advance" is, of course, a basic tenet of intelligent design theory.

[81] Stephen Jay Gould was one of the great Darwinists of our time. Not only was he a Harvard paleontologist and past president of the American Association for the Advancement of Science, he was a prolific writer in the popular media, writing for lay readers and scientists alike.

[82] Lamarckism, named after French biologist Jean-Baptiste de Monet de Lamarck (1744-1829), is essentially the idea that organisms adapt to changes in the environment, and transmit the changes to offspring. Thus, if the environment turns cold, an animal may grow fur in response and pass on furriness to its offspring.

[83] Stephen Jay Gould, "Shades of Lamarck," *The Panda's Thumb* (New York: W.W. Norton & Company, Inc., 1980), p. 79, emphasis in original. Lamarckism, "in its contemporary usage," according to Gould, is "the notion that organisms evolve by acquiring adaptive

But earlier in the essay, in an explanation of Darwinism's two-step process, Gould wrote:

> Darwinians speak of genetic variation, the first step, as "random." This is an unfortunate term because we do not mean random in the mathematical sense of equally likely in all directions. We simply mean that variation occurs with no preferred orientation in adaptive directions.[84]

Surely people as smart as most Darwinists can come up with a more "fortunate" term, if not an acceptable euphemism, to avoid using the so-called misleading term "random." But even the clever phraseology "no preferred orientation in adaptive directions" does nothing to change the fact that Darwinism *must* be based upon *random* variation—truly and mathematically random variation. The only other option available is *directed* variation (with or without directed selection), which for Darwinists, in any sense, is a *most* unfortunate term; one that, at best describes Lamarckism, and at worst implicates an intelligent director.

Dawkins' complicated discourse on *complexity* needs simplifying. Because living organisms are clearly complex, any attempt at explaining their origin must include an explanation of the origin of such complexity. Further, an adequate scientific explanation requires recognition of what kind of "complexity" we are dealing with. Without oversimplifying the matter, suffice it to say here that two categories of complexity can be distinguished: *specified* complexity and *unspecified* complexity. Patterns made by the bits of a computer program and the bits of a shattered pane of glass can be equally complex. But while shattered glass exhibits unspecified complexity, information-rich things,[85] such as computer programs and living organisms, can

---

characters and passing them on to offspring in the form of altered genetic information" (Ibid., p. 77).

[84] Ibid.

[85] Information theory is a discipline of its own. In general, information theory deals with identifying and measuring information content. Living organisms exhibit vast amounts of information content. In fact, information is defined as "the attribute inherent in and communicated by one of two or more alternative sequences or arrangements of something (as nucleotides in DNA or binary digits in a computer program) that produce specific effects" (*Merriam-Webster's Collegiate Dictionary, Tenth Edition: 1996* (Springfield, MA: Merriam-Webster. Inc., 1996). The idea that the information in a computer program or in DNA could possibly be formed in an *unspecified* manner is a matter for statisticians, all of whom report back numbers that, in the time available in our universe, make such an occurrence effectively impossible.

reasonably be described only as exhibiting specified complexity.[86] What Dawkins is vainly trying his best to *not* describe is specified complexity, because that is the hallmark of design. By definition, specified complexity requires a "*specifier,*" i.e., a designer, one who can choose among equally possible alternatives. To imply that natural selection can in any way "specify" among alternatives is deceptive; such implications turn "natural selection" into a code word for an "intelligent selection" without full disclosure of the logical contradiction.

What Dawkins wants to portray, and what he wants the reader to think he has described, is *unspecified*-yet-nevertheless-meaningful complexity. Unspecified complexity is, by definition, the only kind of complexity an unguided process can accomplish. Unspecified complexity is not a hallmark of design. Unspecified complexity can be better described as "coincidental" complexity. That is, like a pattern made by the pieces of shattered glass, such complexity can be thought of as being a result of coincidence, constrained only by nature and natural laws, such as the ground and gravity. Thus, it's something of a coincidence that all the stones in Mont Blanc ended up where they are. Rather than co-opting the language of design, therefore, Dawkins and his kind should speak of "coincidental complexity." Perhaps their theory of accidental specificity could go by the name Coincidence Repeated to Organize Complexity (CROC).

In any event, based on Dawkins' somewhat convoluted explanation of the difference between specified complexity and unspecified complexity, it may be that we now need another theory of evolution called *Dawkinism.* Dawkinism misses Darwinism only by a couple of letters, and, therefore, may be a little behind or a little ahead of Darwinism on an evolutionary path of blind understanding. (How can we know, unless we have a target phrase?) The theory of Dawkinism, as clearly espoused by its namesake, is a theory of chance, *but not "pure" chance,* wherein random mutations provide variation on which natural selection can work if it knows a specified-without-hindsight, *but not "specified in advance,"* criterion, *but not a "long term" criterion,* and the process can be repeated over and over innumerable times to produce complex organisms that exhibit design, *but not "true" design.* The first book to accurately develop the theory could be called *The Peeking Watchmaker: Why the Evidence for Dawkinism Confirms a Universe with Design.*

[7] *"Our explanations can be based only on evidence – evidence to prove our theories."*

Much is written about the "overwhelming evidence" for evolution. For example, we are told that the fossils are "evidence for"

---

[86] Thus Dawkins' insistence on the "specifiable in advance" quality of evolution.

evolution; the genetic information of organisms is "evidence for" evolution; the anatomical similarities across species, the distribution of plants and animals, et cetera, et cetera—everything is "evidence for" the "fact of evolution."

Is there overwhelming evidence for evolution? Yes and no. As always, the answer to such questions depends on what meaning is attributed to the terms. The same statement can connote entirely misleading information if terms are not used *and understood* precisely and consistently. Thus, when someone makes a propositional statement that fossils are "evidence evolution occurred," it is important that we understand exactly what such a statement is intended to mean. We already know that the term "evolution" is essentially meaningless, but even given its full force to mean Darwinian descent with modification, it still remains to understand the meaning and use of the term "evidence."

Here is the point to always keep in mind when considering evidence: *"Evidence for" something is not the same as "proof of" something.*[87] Evidence, such as the fossil record, is indeed evidence for evolution in the sense of evidence "in support of" evolution. With this understanding one can see that the fossil record does support the notion of evolution; that is, it is consistent with a theory that predicts historical extinct organisms. But to *support* a proposition is far from *proving* the same, and this is where the confusion occurs. For evidence to "prove" a theorized historical event "happened," the evidence must not only sufficiently account for the event with no (or, at least a minimum of) anomalies, but it must also effectively rule out alternatives.

To illustrate, consider the competing theories of the 16th century with respect to the motion of the sun and the planets. A geocentrist in the tradition of Ptolemy could stand outside, look up at the sun traversing across the sky under his motionless feet and proclaim his observation as "evidence for the sun circling about the motionless earth," and he would be absolutely correct. His observation is evidence *in support of* his theory. However, a heliocentrist, considering Copernicus's more accurate observations, can use such observations as "evidence for the earth orbiting the sun," and he would likewise be correct. Both observers could rightly state that their respective observations are "evidence for," in the sense of "in support of," their respective theories. The point, again, is that evidence "for" either position is not "proof of" the position—neither observer can say that based on his observation he has *proved* his theory. Not until sufficiently detailed observations were made as evidence accounting for planetary motion without anomalies to support heliocentrism (and

---

[87] We know this in everyday experience; evidence the suspect was at the scene of the crime is not proof he did the crime. The suspect may or may not in fact be the criminal, but simply being at the scene of the crime is not enough to convict him.

sufficient to rule out geocentrism) could it be said that the heliocentric theory had been "proven".

Therefore, when considering the "evidence for evolution," it is important to always consider what actually is the nature and significance of the evidence. That is, what does the evidence *prove*, if anything, and what does it merely support.[88] And, more importantly, one should consider what competing theories the evidence can rule out.

To help sort through evidentiary claims of Darwinists the following "5-point Evidence Filter" is proposed. Using the 5-point Evidence Filter below will bring clarity and precision to bear on any claims for "proof" of a theory or proposition—leading to an objective "proof claim criterion." As applied to the proof claims of Darwinists, the 5-point Evidence Filter can be used as follows. When Darwinists present particular "evidence for" evolution, ask the following five questions,

1. Does this evidence actually support the theory of evolution (i.e., is the evidence consistent with what one would expect to observe if evolution were true)?
2. What inferences does the evidence actually support (i.e., what can reasonably or actually be inferred from the observations put forth as evidence, as opposed to what is purported to be proven)?
3. Does the evidence also support alternative theories, such as intelligent design theory (i.e., is the evidence consistent with what one would expect to observe if intelligent design were true)?
4. Does the evidence, taken as a whole with other evidence, weigh more heavily in favor of evolution or an alternative theory?
5. Does the evidence as a whole, taken together with other evidence, *rule out* a particular alternative theory?

To illustrate the point being made, consider the book *What Evolution Is*, written by long-time evolutionist and Harvard Professor Emeritus Ernst Mayr, billed as "one of the great shining figures of evolutionary biology."[89] Certainly Ernst Mayr was the longest living of Darwinists, having "been interested in evolution since before the 1920s ..."[90] Mayr wrote his book in the 21st century to "stress principles" and

---

[88] This is not to imply that "mere" support is somehow insufficient; it is simply to indicate that "support" is not to be confused with "proof."

[89] Ernst Mayr, *What Evolution Is* (New York: Basic Books, 2001); back cover, quote attributed to John Maynard Smith.

[90] Ibid., Acknowledgments, p. xvii.

not get "lost in detail"[91] in explaining what evolution "is." Apparently, detailed proof of evolution is hardly worth repeating. As Mayr states:

> [T]here is no longer any need to present an exhaustive list of the proofs for evolution. That evolution has taken place is so well established that such a detailed presentation of the evidence is no longer needed."[92]

Nevertheless, it seems that relatively few are convinced by such well-established proofs. Therefore, Mayr wrote his book for "three different kinds of readers."[93]

> First and foremost, it is written for anyone, biologist or not, who simply wants to know more about evolution. Such a reader is quite aware how important this process is but does not understand exactly how it works and how one can answer some of the attacks against the Darwinian interpretation. The second group of readers consists of those who accept evolution, but are in doubt whether the Darwinian explanation is the correct one. I hope to answer all the questions this kind of reader is apt to ask. And finally, my account is directed to those creationists who want to know more about the current paradigm of evolutionary science, if for no other reason than to be able to better argue against it.[94]

How anyone could be left in either of the first two groups is something of a mystery if evolution "is so well established that a detailed presentation of the evidence is no longer needed." It seems like even the borderline true believers of group two might benefit from a sprinkling of detailed evidence. Moreover, Mayr doesn't even have a category for those who are not Genesis-account creationists and simply find Darwinism scientifically untenable. Therefore, while not fitting neatly into any of Mayr's positions, intelligent design theorists align most closely with the third, and the following discussion will oblige Mayr with respect to his desire for the third kind of reader—those wanting to understanding the current paradigm of evolutionary science to be able to argue against it. That said, we can safely know that

---

[91] Ibid., p. xv.

[92] Ibid.

[93] Ibid., Preface, p. xiii.

[94] Ibid.

Mayr has let fly the best barbs from evolution's quiver of "evidence" for our consideration.

To make refutation easier, Mayr starts right off with a chapter (actually Chapter 2) conveniently entitled "What is the Evidence for Evolution on Earth?"[95] He begins this chapter assuring the reader that, "[B]eginning with Darwin in 1859, more and more facts were discovered that were compatible only with the concept of evolution."[96] Compatible *only* with evolution? That is the position Mayr must support for those readers-of-the-third-kind who wish to argue against it. The reader should pay close attention as Mayr continues:

> [T]he occurrence of evolution was supported by such an overwhelming amount of evidence that it could no longer be called a theory. Indeed, since it was as well supported by facts as was heliocentricity, evolution also had to be considered a fact, like heliocentricity. This chapter will be devoted to a presentation of the evidence that led to the adoption of the "evolution is a fact" conviction among scientists.[97]

Carefully consider Mayr's words. In his opinion, evolution is no longer a theory! It is "supported by facts ... like heliocentricity," and, is itself a "fact." Such an uncompromising position does indeed afford an opportunity for refutation as the reader waits for evidence so well supported that evolution is a fact just like heliocentricity. However, illustrating the frustrating (although somewhat amusing) tendency of Darwinists to effortlessly speak out of both sides of their mouths, Mayr immediately provides a disclaimer, reminding his eager readers-of-the-third-kind not to expect too much, because,

> Evolution is a historical process that cannot be proven by the same arguments and methods by which purely physical or functional phenomena can be documented. Evolution as a whole, and the explanations of particular evolutionary event, must be inferred from observations.[98]

Translation: In spite of what I just said about evolution being "as well supported by facts as was heliocentricity," *evolution can't be known or*

---

[95] Ibid., p. 12.

[96] Ibid.

[97] Ibid., p. 12, last paragraph.

[98] Ibid., p. 13, first paragraph.

proven in the same way as we know and have proven that the earth orbits the sun. The relative motion of the sun and planets can be observed directly, permitting the heliocentric theory to be confirmed by the scientific method. Evolution (at least macro-evolution) cannot be observed directly and cannot be tested by the scientific method; it can be *inferred* only from observations.[99] Now, there is nothing inherently wrong with inferences—inferences are necessary for theories involving unobservable and unrepeatable historical events. But the chicanery that Darwinists display in trying to pass off such duplicity on the issue of "proof" for the "fact" of evolution is equaled only by the incredulity exhibited in their inability to understand why most people don't believe them.

Not to be deterred by technicalities such as distinguishing between historical methods and the scientific method, Mayr proceeds in his chapter entitled "What is the Evidence for Evolution on Earth?" with the first example of "quite overwhelming"[100] evidence which is, predictably, the fossil record. According to Mayr, the fossil record is "the most convincing evidence for the occurrence of evolution."[101] Of course, as has been well documented in many other books,[102] the fossil "evidence" may be proof of "change over time," but it is far from being "overwhelmingly" supportive of Darwinism. Even Mayr must admit:

> Given the fact of evolution, one would expect the fossils to document a gradual steady change from ancestral forms to the descendants. But this is not what the paleontologist finds. Instead, he or she finds gaps in just about every phyletic series. New types appear quite suddenly, and their immediate ancestors are absent in the earlier geological strata. The discovery of unbroken series of species changing gradually into descending species is very rare. Indeed the fossil record is one of discontinuities, seemingly

---

[99] Later in the same book, Mayr takes a more measured approach. In an appendix addressing "Frequently Asked Questions," Mayr answers the question, "Are not the "facts" of evolutionary biology something very different from the facts of astronomy, which show that the Earth circles the sun rather than the reverse?" (Ibid., p. 276). His answer: "Yes, up to a point. The movement of the planets can be observed directly. By contrast, evolution is a historical process …" (Ibid.).

[100] Ibid., p. 13.

[101] Ibid.

[102] *See, e.g.*, Phillip E. Johnson, *Darwin on Trial* (Downer's Grove, IL: InterVarsity Press, 1991), Chapter 4; Jonathan Wells, *Icons of Evolution: Science or Myth: Why much of what we teach about evolution is wrong* (Washington D.C.: Regnery, 2002), Chapters 6 and 10; and Joe White, ed. *Darwin's Demise* (Green Forest, AR: Master Books, 2001), Chapter 4.

> documenting jumps (*saltations*) from one type of organism to a different type.[103]

Thus, Mayr clearly recognizes what Darwin recognized, and what is, indeed, no longer the trade secret of paleontologists: *the fossil record does not unequivocally support Darwinian evolution*, much less prove any kind of evolution! In an amazing display of selective adaptation to an academically advantageous view, the only reason Mayr can even attempt to use the fossil record as evidence of evolution is because of his first five words of the above quote—that is, even though the fossil record does not actually support evolution, in the "up is down" world of Darwinism it nevertheless does, *because it must*. This leaves Mayr scratching his head, wondering:

> This raises a puzzling question: Why does the fossil record fail to reflect the gradual change one would expect from evolution?[104]

This question is no different than the following one:

> Why does the movement of the planets fail to reflect the motion of the sun one would expect from geocentricism?

It is "puzzling" only because observed facts are being forced to fit a theory that may be wrong. Of course, Mayr can spend another several pages trying to salvage some semblance of scientific reasoning. He even hauls out old faithful, the *Archaeopteryx*, as an example of a missing link. Discovered almost 150 years ago and subsequently discounted as a missing link, it is still presumably the best and only example of something that *looks* like what should be one of billions of missing links.

Mayr continues, but what he is oblivious to is that he's already lost the honest readers—at least the careful ones. Unfortunately, Mayr is not alone; force-fitting facts to a pre-conceived theory to the point of looking silly doing so is what one has learned to expect from scientists committed to the "current paradigm of evolutionary science." After all, the essence of a paradigm is its tendency to blind all but the most

---

103 Ibid., p. 14, italics in original.

104 Ibid.

perceptive to its errors.[105] However, such blind adherence to the current paradigm does not lend itself well to serious readers of any of Mayr's three categories. He may be absolutely right, but his discourse sounds like he can't even convince himself. Nevertheless, to continue in our effort to argue against Mayr's proofs, let's refute them in a more systematic manner.

Even considering Mayr's clever cogitations to be exactly, 100% accurate, the question remains—is the fossil record "proof" of "the fact of evolution?" Applying the 5-point Evidence Filter with respect to the "fossil record" will help clarify.

1. Does this evidence actually support the theory of evolution (i.e., is the evidence consistent with what one would expect to observe if evolution were true)?

   Answer: No, if the theory is that of gradual evolution proposed by Darwin. However, if punctuated equilibrium is true as promoted by Stephen J. Gould, then perhaps yes.

2. What inferences does the evidence actually support (i.e., what can reasonably or actually be inferred from the observations put forth as evidence, as opposed to what is purported to be proven)?

   Answer: The existence of fossils shows that there are many extinct organisms, all of which appear in the fossil record fully formed. (The supposed links in lineages are inferences required by the theory of descent with modification but not otherwise implied by the evidence.)

3. Does the evidence also support alternative theories, such as intelligent design theory (i.e., is the evidence consistent with what one would expect to observe if intelligent design were true)?

   Answer: Yes. The existence of fossils of extinct organisms, each one of which has all the hallmarks of being intelligently designed, is consistent with what one would expect to find if intelligent design theory is true.[106]

---

[105] *See, e.g.*, Thomas S. Kuhn, *The Structure of Scientific Revolutions* (Chicago: The University of Chicago Press, 1996).

[106] Note that this is not to imply that the fossil record is consistent with the Genesis account of creation. It may or may not be, but for the purposes of intelligent design theory the question is moot; the only relevant inquiry is whether the fossil record shows

4. Does the evidence, taken as a whole with other evidence, weigh more heavily in favor of evolution or an alternative theory?

   Answer: Taken as a whole, the evidence weighs more heavily in favor of intelligent design. Consider, as admitted by all knowledgeable Darwinists, including Mayr above, that the fossil record does not support the continuity of gradual development that Darwinism predicts. However, a fossil record showing extinct organisms exhibiting the hallmarks of intelligent design is perfectly consistent with what intelligent design theory would predict.

5. Does the evidence as a whole, taken together with other evidence, *rule out* a particular alternative theory?

   Answer: Possibly. Taken as a whole, the evidence could be construed to rule out Darwinism. But whatever the fossil record shows, *it cannot rule out intelligent design theory*. As long as design can be scientifically detected, intelligent design theory cannot be ruled out as an explanation of the fossil record. But, and equally importantly, if intelligent design can be scientifically detected in the fossil record, the fossil record *does rule out Darwinism* as a theory—Darwinism is the antithesis of intelligent design.

The above analysis is not an anti-evolutionist-inspired straw man. Darwin's contemporary supporters often challenged Darwin as being "not sufficiently 'Baconian,'" or that he had violated the true path of induction.[107] Darwin was seen as not adhering to the scientific method by rejecting empiricist analogy from direct experience.[108] Rather, Darwin argued that a "consilience" is the key to confirmation.[109] Consilience is the idea that a theory's truth value is determined not based upon any direct evidence, but upon its explanatory power.[110]

---

organisms in which design can be scientifically detected. The reply to this inquiry is an unqualified yes.

[107] Michael Ruse, *The Darwinian Revolution* (Chicago: The University of Chicago Press, 1979), p. 238.

[108] Ibid., p. 235. The criticism came from Darwin's legendary defender, English biologist Thomas Henry Huxley (1825-1895).

[109] Ibid., p. 235-36.

[110] Ibid., p. 235. At the time, Darwin and Huxley viewed their differences as being *philosophical* in nature. (Ibid., p. 236).

Specifically with regard to the point above was the criticism of the lack of scientific rigor with respect to the fossil record. One of Darwin's contemporaries, physicist, geologist and Cambridge mathematics coach William Hopkins criticized Darwin's theory for not adhering to the "right" scientific methodology. As evolutionist Michael Ruse relates:

> Now, claimed Hopkins, the greatness of Newton's theory lies in the exactness with which claims deduced from his premises about physical causes correspond with facts in the world. But when we come to Darwinism theory, the situation is quite opposite. The phenomenal facts, for example, the fossil record, cannot be deduced from Darwin's claims about natural selection.[111]

In line with the Evidentiary Filter proposed above, Hopkins insisted:

> At best Darwin shows that the phenomenal facts *could* have been caused by natural selection—that there is nothing inconsistent in this—rather than proving that they actually have been so caused.[112]

Importantly, Hopkins recognized that the underlying difference between him and Darwin was philosophical. As Ruse continues:

> But, Hopkins contended, this "may be" philosophy will not do. Because it failed to satisfy his Newtonian ideal, Hopkins thus felt justified in rejecting Darwin's theory.[113]

The "may be" philosophical basis for the "fact" of evolution is so uncritically examined by Darwinists today that even honest, level-headed Darwinists can slip into a "support equals proof" mindset. For example, in his book *The Fifth Miracle, The Search for the Origin of Life*, evolutionist and mathematical physicist Paul Davies examines the question of life's origins, commenting:

---

[111] Ibid., p. 238.
[112] Ibid.
[113] Ibid.

> Today, biologists insist that Darwin's guess was basically correct: life on Earth has descended from a single common ancestor.[114]

What is the basis for this insistence? Davies continues:

> What makes them so sure? There are several excellent reasons to believe in a universal ancestor. For a start, every known organism shares a common physical and chemical system ... Perhaps the most convincing evidence for a common origin is that the genetic instructions are implemented using a universal code ... It is too much to believe that all these complex and highly specific features arose independently many times.[115]

Using the 5-Point Evidence Filter, Davies' "excellent reasons" can be show to be equally excellent reasons for attributing the observed evidence to intelligent design. By way of example, taking the "most convincing evidence" of the "universal code" of genetic instructions, we can apply the Filter:

1. Does this evidence actually support the theory of evolution (i.e., is the evidence consistent with what one would expect to observe if evolution were true)?

   Answer: Yes. The universal code of genetic instructions is consistent with what one would expect to find if the theory of evolution (i.e., descent with modification from a common ancestor) was true.

2. What inferences does the evidence actually support (i.e., what can reasonably or actually be inferred from the observations put forth as evidence, as opposed to what is purported to be proven)?

   Answer: The universal code of genetic instructions exists in living organisms. (The supposed derivations of code from one organism to another are inferences

---

[114] Paul Davies, *The Fifth Miracle, The Search for the Origin of Life* (New York: Simon & Schuster, 1999), p. 71.

[115] Ibid.

required by the theory but not otherwise required by the evidence.)

3. Does the evidence also support alternative theories, such as intelligent design theory (i.e., is the evidence consistent with what one would expect to observe if intelligent design were true)?

   Answer: Yes. The existence of a universal code of genetic instructions is consistent with what one would expect to find if intelligent design theory is true. Common ancestry is also consistent with intelligent design if you view the designer as a software programmer that is directing "mutation" rather then having heritable variations produced by chance or some unknown law.

4. Does the evidence, taken as a whole with other evidence, weigh more heavily in favor of evolution or an alternative theory?

   Answer: Taken as a whole, the evidence of a universal code weighs in favor of intelligent design. The very *existence* of a "code" weighs strongly in favor of intelligent design. Indeed, a recent study concludes that the code is the most optimal of a million other randomly selected possibilities.[116]

5. Does the evidence as a whole, taken together with other evidence, *rule out* a particular alternative theory?

   Answer: Possibly. Taken as a whole, the evidence of a universal "code" that, by definition, likely requires a "coder" could be construed to rule out Darwinism. However, whatever the universal genetic code shows, *it cannot rule out intelligent design theory.*

The same scientific analysis could be performed on all putative evidence "for" evolution, including each of Mayr's "proofs" of evolution. In the end, it becomes clear that all the "evidence" amassed for evolution is, at best, evidence "in support of" evolution. But the issue is not whether the evidence merely supports evolution's historical

---

[116] S.J. Freeland and L.D. Hurst, *The Genetic Code is One in Million,* Journal of Molecular Evolution, vol. 47, pp 238-248 (1998).

hypothesis. The issue is whether it is sufficient to enable one to conclude that it is the *best current competing explanation*. Thus, Mayr can deceptively end his chapter entitled "What is the Evidence for Evolution on Earth?" by referring to "irrefutable evidence in support of" evolution:

> As we have seen, whatever aspect of biology is studied, it provides irrefutable evidence in support of evolution ... Indeed, there is no other **natural explanation** other than evolution for the facts presented in this chapter.[117]

Here the slight of hand (which is largely lost on the public) is that the phrase "natural explanation" is used to exclude design. If Mayr removed the word "natural" and revised the phrase to read: "no other explanation derived using the scientific method," his conclusion would not follow because his evidence is also consistent with design theory.[118] His claim is true only when its conclusion of no-design is assumed as a premise. This kind of "science" is wholly the product of purely circular reasoning: define science as not allowing design, then say that evolution's claim of no-design is a scientific fact. In any event, whether as a matter of rational logic or adherence to the *true* scientific method, it is clear that all the "proofs" of Darwinism are nothing of the sort.

[8] *"You see, science, natural science, can consider only natural explanations for physical features of the world around us."*

The approach to science being espoused here is what some have referred to as "methodological naturalism" or "applied naturalism." That is, what is being described as science is the logical application of the philosophy of *naturalism* to the methods of science. Naturalism is a philosophical view that acknowledges only physical or material causes or forces in nature, explicitly denying the existence of any non-physical entity such as an intelligent cause.[119] With respect to science, and specifically biological evolution, naturalism is "the *doctrine* that cause-and-effect laws (as of physics and chemistry) are adequate to account for all phenomena and that teleological [design] conceptions of nature are invalid."[120] Sometimes referred to as scientific materialism

---

[117] Mayr, *What Evolution Is*, p. 39, emphasis added.

[118] *See, e.g.*, William Dembski, *Intelligent Design, the Bridge Between Science & Theology* (Downers Grove, IL: InterVarsity Press 1999).

[119] David Noebel, *Understanding the Times* (Eugene, OR: Harvest House Publishers, 1991) p. 115.

[120] *See*, Webster's Third New International Dictionary of the English Language, Unabridged, (1993), emphasis added.

(matter is all that exists), naturalism is a fundamental tenet of Secular Humanism.[121] As stated by Carl Sagan, 1981 Humanist of the Year: "The Cosmos is all that is or ever was or ever will be.[122]

A naturalistic approach to much of science is appropriate where a naturalistic claim can be tested and there is no rational reason to question its claim. For example, there is no reason to argue for an intervening intelligent cause when one seeks to explain why an apple falls to the ground from its tree. But a naturalistic approach becomes a problem when institutions of science mandate its use in areas where its claims cannot be effectively tested, as in the area of biological origins and where the observed data contradicts its claim.

Naturalism is often justified on the ground that science cannot study the supernatural—the code word for God. The problem is that the exclusion is applied to evidence of design, which scientifically supports (i.e., follows as an inference from observed phenomena) but does not necessarily entail God. The evidence is excluded because of its implications. Darwinists that hold to the philosophy of naturalism often defend naturalism because the supernatural, much less a God, cannot be proven to exist.[123] This may be true, but the converse is equally true: the absence of the supernatural, including a god or gods, cannot be proven.

In his book, *River Out of Eden: A Darwinian View of Life*,[124] Richard Dawkins addresses the "God" issue and implicitly agrees that Darwinism is a matter of faith. Writing about "nod provokers" at his public lectures, i.e., people who point out various challenges to his viewpoint, thereby eliciting "a murmuration of approving nods," he states that he is often challenged with this "reliable nod provoker:"

> Fundamentally, your belief in evolution comes down to faith, and therefore it's no better than somebody else's belief in the Garden of Eden.[125]

Dawkins' reply is important for its frankness:

---

[121] *See, e.g.*, Humanist Manifestoes I and II (Buffalo: Prometheus, 1980).

[122] Ibid., p. 119, quoting Carl Sagan, *Cosmos* (New York: Random House, 1980), p. 4.

[123] Most evolutionists—including Darwinists—would object, saying that their philosophical viewpoint says nothing about the existence of God; the subject is irrelevant. From their perspective, a supernatural being is simply not necessary to postulate—or deny. According to Darwinists, naturalistic theories such as Darwinism can account for life forms after biogenesis, and a natural mechanism of biogenesis surely exists, even if never discovered. Therefore, a God is at best superfluous, and, at worst counterproductive to scientific reasoning.

[124] Richard Dawkins, *River Out of Eden: A Darwinian View of Life* (New York: Basic Books, 1995).

[125] Ibid., p. 32.

> Every tribe has had its origin myth—its story to account for the universe, life and humanity. There is a sense in which science does indeed provide the equivalent of this, at least for the educated section of our modern society. Science may even be described as a religion, and I have, not entirely facetiously, published a brief case for science as an appropriate subject for religious-education classes (In Britain, religious education is a compulsory part of the school curriculum, unlike in the United States, where it is banned for fear of offending any of the plethora of mutually incompatible faiths.[126]) Science shares with religion the claim that it answers deep questions about origins, the nature of life, and the cosmos. But there the resemblance ends. Scientific beliefs are supported by evidence, and they get results. Myths and faiths are not and do not.[127]

Dawkins is certain that his science-based religion is superior because scientific beliefs are supported by the evidence while myth-based religions are not. No doubt many find this position entirely reasonable. But the challenge for the serious thinker (if not the educated section of our society) is to demand to see the evidence science has to account for life in the first place and how such evidence fares against the competing evidence that is otherwise ruled out of order by the unwritten rule of non-objective science: only *natural* explanations allowed. It has none.[128] Serious thinkers should ask to see the evidence that Darwinism can account for the creation of new species. It has virtually none.[129] Furthermore, serious thinkers should demand a definition of science that is consistent, *e.g.*, by insisting that

---

[126] As an aside, note that if "offending any of the plethora of mutually incompatible faiths" were the criteria for choosing curriculum, then *evolution* would not be taught in the United States.

[127] Ibid., p. 33.

[128] Darwinism makes no claims with respect to the origin of life. Darwin's theory of "descent with modification" requires a living, replicating organism from which to evolve a descendent.

[129] Evolutionist literature is conspicuously silent on any actual examples of speciation, i.e., new species being formed. Darwin had to imagine an example of natural selection, much less speciation, which is theorized to be a result of natural selection. Contemporary research can do no better with respect to speciation. Except for some ambiguous examples based on a hyper-technical definition of "species" of fruit flies, no examples of speciation can be found in the popular literature. Scientific papers claiming to show speciation can be found, but such papers use hyper-technical definitions, are based on ambiguous plant species, and/or are simply speculative. Again, there is no simple example of speciation documented in the popular literature.

scientists take into account scientific evidence for intelligent design in origins research just as they do in other areas, such as in the Search for Extraterrestrial Intelligence, and archeology, for example. Finally, serious thinkers should expect a precise definition of the term "evolution" so that any propositional statements made concerning it can be judged for validity. Once definitions for science and evolution are consistently applied, even Dawkins may find his scientific "beliefs" must undergo a shift.

[9] *"To simply resort to a 'someone painted them' explanation is convenient, but it ignores the facts, and is, in fact, an unprovable theory. We can simply have no knowledge of any such imagined person."*

In an interesting essay entitled "Adam's Navel," evolutionist Stephen Jay Gould examined, in a sense, Adam's navel.[130] Specifically, Gould considered an essay by nineteenth century British naturalist Philip Henry Gosse dealing with the theological debate over whether Adam had a navel. As Gould stated:

> He [Adam] was, after all, not born of a woman and required no remnant of his nonexistent umbilical cord. Yet, in creating a prototype, would not God make his first man like all the rest to follow?[131]

Gosse's essay, entitled "Omphalos" (Greek for navel) set forth a comprehensive theory to explain how God might create species by sudden creation at any point in their life cycle with the illusory appearance of preexistence. Adam's navel is just one of many examples of anomalies that one would expect if an intelligent designer had created the world with the appearance of having a continuity with the past—which is one explanation for an old-looking young earth. Thus, newly created trees would have many rings, a recently made earth would have features that make it appear much older, and Adam would have a navel. In this way the world would have the appearance of having an ordered history. As Gould said, "Thus, the earth might be but a few thousand years old, as Genesis literally affirmed, and still record an apparent tale of untold eons."[132]

Of course, Gould had no real interest in contemplating Adam's navel; he was sure there was never a real Adam. Gould's essay

---

[130] Stephen Jay Gould, *The Flamingo's Smile: Reflections in Natural History* (New York: W. W. Norton and Co., 1985), p. 99.

[131] Ibid.

[132] Ibid., p. 100.

elegantly ridiculing Gosse's essay as "spectacular nonsense" was written so that "we will understand better how science and useful logic proceed."[133] "Omphalos," according to Gould, is "desperately wrong," because "we can devise no way to find out whether it is wrong—or, for that matter, right."[134] In other words, Gosse's theory was "unprovable," a fatal flaw for a scientific theory.

As Gould elaborated, "Science is a procedure for testing and rejecting hypotheses, not a compendium of *certain* knowledge."[135] Gould continued:

> Claims that can be proved incorrect lie within its domain (as false statements to be sure, but as proposals that meet the primary methodological criterion of testability). But theories that cannot be tested in principle are not part of science. Science is doing, not clever cogitation; we reject *Omphalos* as useless, not wrong.[136]

These are bold statements about the nature of scientific claims. How does Darwinism fare under these standards? Is Darwinism simply "clever cogitation?" Should we reject Darwinism as useless, instead of wrong?

Such questions are being asked by leading intelligent design theorists. For example, in his book *Darwin on Trial*, Phillip E. Johnson considers the rhetorical questions of Darwinists with respect to why "a proper Creator" should have designed organisms the way he is claimed to have done.[137] Johnson correctly observed that the questions are misplaced, stating, "The task of science is not to speculate about why God might have done things this way, but to see if a material cause can be established by empirical investigation."[138] According to Darwinists, such as Gould, a material cause established by empirical investigation must be, by the rules of science, falsifiable.[139]

---

[133] Ibid., p. 101.

[134] Ibid., p. 110.

[135] Ibid., p. 111, emphasis added. Surely Gould can't really mean this. If there is one thing scientists like Gould are absolutely, 100% certain about, it is that life is a product of natural, unguided evolution. Gould himself has contributed significantly to the compendium of certain knowledge on this topic.

[136] Ibid.

[137] Phillip E. Johnson, *Darwin on Trial* (Downer's Grove, IL: InterVarsity Press, 1991), p. 70-71.

[138] Ibid., p. 71.

[139] The notion of "falsifiability" is part of a model put forth by Karl Popper (1902-1994), Fellow of the Royal Society, Fellow of the British Academy, and Membre de l'Institute de France. According to "The Karl Popper Web,"

But it seems that Darwinists find all empirical investigations as proof of the "fact of evolution." The concept of natural selection itself has been characterized by Darwinists to be a tautology—that is, true by definition and unfalsifiable.[140] Other tenets of Darwinism are likewise unfalsifiable. Consider Darwin's famous challenge:

> If it could be demonstrated that any complex organism existed that could not possibly have been formed by successive slight modification, my theory would absolutely break down."[141]

The statement is written as to require "proving a negative." As reported by Larry Witham, the futility of proving a negative was discovered by Michael Behe while attempting to show irreducibly complex systems as achieving "breakdown." Behe stated, "One would have to show the system could not form by any of a potentially infinite number of unintelligent processes."[142] Witham continues:

> Because that is impossible, Darwin's challenge throws science out the window; if this is the criterion of its truth, Darwinian theory fails to pass the test of being testable science.[143]

That is, it is unfalsifiable.

According to Gould:

(http://www.eeng.dcu.ie/~tkpw/index.html), falsificationism is the idea that science advances by unjustified, exaggerated guesses followed by unstinting criticism. Only hypotheses capable of clashing with observation reports are allowed to count as scientific. "Gold is soluble in hydrochloric acid" is scientific (though false); "Some homeopathic medicine does work" is, taken on its own, unscientific (though possibly true). The first is scientific because we can eliminate it if it is false; the second is unscientific because even if it were false we could not get rid of it by confronting it with an observation report that contradicted it. According to Phillip E. Johnson, the problem with this criterion is that it is impossible to test every important scientific proposition in isolation. Background assumptions have to be made so that detailed statements can be tested. Ibid., p. 122.

[140] Natural selection can be stated, in effect, that the fittest survive to reproduce, and those who survive to reproduce are the fittest. This statement cannot be falsified. *See*, Chapter 2 of Johnson, *Darwin on Trial*.

[141] Charles Darwin, *On The Origin of Species, A Facsimile of the First Edition* (Cambridge: Harvard University Press, 1964), p. 189.

[142] Larry Witham, *By Design: Science and the Search for God* (San Francisco: Encounter Books, 2003) p. 134.

[143] Ibid.

> Gosse's deep error lay in his failure to appreciate this essential character of scientific reasoning. He hammered his own coffin nails by continually emphasizing that *Omphalos* made no practical difference—that the world would look exactly the same [regardless of occurring after creation, or only appearing to have been created in time past].[144]

An analogous question for Darwinists, then, is what would the world look like if Darwinism is *not* true? Can there be a practical, detectable difference? If not, then for all of Gould's rectitudinous rambling, he is simply like one accusing another of not closing his eyes during prayer—sure his own cause warrants an exception to the rules. As Johnson observed:

> If evolutionary biology is to be a science rather than a branch of philosophy, its theorists have to be willing to ask the scientific question: *How can Darwin's hypothesis of descent with modification be confirmed or falsified.*[145]

What is amazing is that a scientist with Gould's intellect and insight can fail to recognize that Darwinism fits his criteria of theories that cannot be tested in principle. In fact, by Gould's definition, Darwinism is not part of science. Intelligent design theorists agree wholeheartedly with Gould that science is doing, not clever cogitation; and on that basis they could very well reject Darwinism as useless, not wrong.

In a remarkable resemblance, the pronouncements of Darwinism are like those of the ancient Oracle of the Greek city of Delphi, its own *omphalos* or "navel stone" reputed to mark the exact center of the world.[146] According to Norman Davies, professor emeritus at the University of London, the problem for scholars "lies in distinguishing the Oracle's real achievements from its limitless reputation."[147] Davies elaborates:

---

[144] Gould, The Flamingo's Smile, p. 111.

[145] Johnson, *Darwin on Trial*, p. 71, emphasis in original.

[146] Norman Davies, *Europe: A History* (New York: HarperPerennial, 1996), p. 112. According to Davies, the navel stone marked the meeting-place of two of Zeus's eagles, one sent from the east and one from the west.

[147] Ibid., p. 113.

> Sceptics point out that none of the alleged predictions was ever recorded in advance of the events to which they referred. The amazing powers of the Oracle could never be tested. A powerful cult, an efficient publicity machine, and a gullible public were all essential elements of the operation.[148]

Incredibly, according to Davies, "belief in the omniscience of the Delphic Oracle is almost as great among enthusiastic moderns as it was among the superstitious Greeks of old."[149] Thus, it appears likely that Darwinism, with its tales of "amazing powers" that cannot be proved wrong, may linger among future enthusiasts long after the less gullible have left the cult and learned to ignore its efficient publicity machine.

[10] *"There are many examples of natural phenomena taking a recognizable form. Take, for example, mountain formations that look in the distance like a person's face. It's simply a coincidence that our brains see a face when what we are really seeing is just some craggy rocks formed by wind and rain."*

Similar reasoning was used by Richard Dawkins in his book *Climbing Mount Improbable*, [150] albeit in an attempt to show the difference between coincidental and true "design". Dawkins starts his book by declaring, "I need to begin by clarifying the problem of apparent design in nature, its relationship to true, human design and its relationship to chance."[151] He goes on to describe "coincidences" such as "when we think we see faces, or animal shapes, in clouds or cliff profiles."[152]

Dawkins' books can be enjoyable to read; he has a gift for illustrating the beauty of nature and nature's various forms. But when it comes to drawing conclusions from the beauty he so clearly apprehends, it seems he is an amazingly uninhibited purveyor of loose and inconsistent ideas. The fact that he gets away with such mischief and retains the admiration of his peers makes one wary of both. In his discussion of "apparent design" and its relationship to "true, *human* design" (emphasis added) he once again goes over the top with the

---

[148] Ibid.

[149] Ibid.

[150] Richard Dawkins, *Climbing Mount Improbable* (New York: W.W. Norton & Company, Inc., 1996).

[151] Ibid., p. 4. Again, this statement illustrates the real problem Darwinists face: explaining how observable, empirical evidence of design is actually just evidence of *apparent* design.

[152] Ibid.

easy hubris of one convinced that noble pedagogical ends need not be hampered by lowly petty logical means.

Using rock formations as an illustration, Dawkins starts with the common-sense notion that "ordinary stones weather into such a welter of shapes, it is not surprising if occasionally we find one that calls to mind a boot, or a duck."[153] "The resemblances," Dawkins rightly concludes, "are accidents."[154] Dawkins also refers to a "craggy hillside" in Hawaii that, if looked at in the right light and from the right direction, bears a striking resemblance to John F. Kennedy. Again, he rightly declares, this is "a pure accident."[155] He then compares the accidental resemblance to Kennedy to the four presidents on Mount Rushmore, in South Dakota, and states that you couldn't "persuade a reasonable person" that it just happened to weather into the features of the four presidents. Then Dawkins makes a most remarkable statement:

> We do not need to be told that these were deliberately carved (under the direction of Gutzon Borglum). They are obviously not accidental: they have design written all over them.[156]

How is design written all over Mount Rushmore? Can one really tell? Is it truly obvious? Remarkably, almost inexplicably, Dawkins tells us that design can be "obviously" detected. In this case, we know because "the *sheer number of details* in which the Mount Rushmore faces resemble the real things *is too great to have come about by chance.*"[157] Dawkins goes on to stress about Mount Rushmore that "Its four heads are clearly *designed.*"[158]

Dawkins continues with a clue as to his real basis for certainty in detecting design: "A sculptor conceived them, drew them on paper, made meticulous measurements ..."[159] Beyond such *a priori* knowledge of a sculptor, Dawkins finds a sheer number of details too great to have come about by chance, as convincing criteria for finding something "clearly designed." In fact, he says of Mount Rushmore:

---

[153] Ibid., referring to a collection of stones in The Natural History Museum in London.
[154] Ibid., p. 5.
[155] Ibid. Consider again the proposal in Note 6 to adopt the language of "coincidental complexity" which is what Dawkins is describing.
[156] Ibid.
[157] Ibid., p. 6, emphasis added.
[158] Ibid., emphasis in original.
[159] Ibid.

> [W]e'd estimate the odds against its four heads being carved by accidental weathering as astronomically high—like tossing a coin forty times and getting heads every time.[160]

Consider how incredible such admissions are in a book that has one purpose: to show how highly improbable things that look designed can come about by natural, unguided processes. Apparently Dawkins finds the odds of forty metal heads in a row sufficient to explain four stone heads in a row; but even those "astronomically high" odds are insufficient to account for one living head in a lively row with another disputing its origin.

If "estimating odds" of achieving a "sheer number of details" is Dawkins' method of detecting design, then he is not far from being an intelligent design theorist. But, of course, he's not and he really doesn't believe what he says. To avoid having his own logic force him into the intelligent design camp he is compelled to explain a contrived category of objects, what he calls *designoid* (pronounced 'design-oid').[161] What are designoids? They are objects that *look* designed, so much so that most people think they *are* designed.

The interested reader should study Dawkins' concept of designoids because it is instructive as an example of what is called in logic "begging the question." Begging the question happens when one comes to a conclusion which essentially restates (and assumes the truth of) the premise(s). Dawkins explains his idea of designoids by examining a series of various living things that look surprisingly like other living things—like "living sculptures."[162] With each description of various marvelously detailed, highly improbable living things such as beetles that mimic ants, fish that look like seaweed, and plants that function like pots, Dawkins simply informs us that these are what he considers to be designoids. That is:

> [T]hey are living things that look to us as though they have been designed but have actually been put together by a completely different process ...[163]

But this conclusion begs the question: How do we know these things were put together by a completely different process and were not designed? Dawkins' concept of designoids simply assumes this to be

---

[160] Ibid.

[161] Ibid.

[162] Ibid., p. 7.

[163] Ibid., p. 12.

the case in a straightforward tautology: designoids are things that resemble other things; and, things that resemble other things are designoids.

Dawkins states that if one detailed thing resembles another detailed thing, it is either a designoid or a design. He states that "one of the ways we recognize both true design and designoid pseudo-design is that we are impressed by resemblances between objects and other objects."[164] (To be clear, this is not the standard relied upon by intelligent design theory. Intelligent design theory is closer to Dawkins' earlier statements correlating number of details and their probability to evidence of design, regardless of resemblance to other things.) By this reasoning Mount Rushmore could be a designoid, and logically would be if Dawkins didn't have *a priori* knowledge of a sculptor.

An astute reader searching for an explanation of how to tell the difference between a design and a designoid will be disappointed as Dawkins delves into detailed descriptions of intricate, highly improbable living things without once actually explaining how one can tell the difference. After pages of words describing theoretical concepts such as *convergence* and *cumulative finding*, the reader is left with no basis for making the distinction. Dawkins simply ends his analysis by stating:

> [Designoid objects can be produced] ... which may make us gasp with admiration at the perfection of its apparent design. But it is not real design, because it has been arrived at by a completely different process.[165]

Again, this begs the question—the conclusion restates and assumes the truth of the premise. It assumes there is a "different process," but this is exactly the question we are seeking to answer. In the end Dawkins reveals no way to distinguish between his designoids and "true" designs. He seems to know based upon whether or not he has sure knowledge of an intelligent designer (*e.g.*, Gutzon Borglum). By Dawkins' reasoning, in the absence of knowledge of a designer, design is necessarily *apparent* design.

Dawkins does make a misguided attempt at demonstrating the capability of his "different process." According to Dawkins, he "pioneered" his approach of writing a computer program that "breeds, by artificial selection, creatures called computer biomorphs."[166] The computer program starts with an intelligently designed common ancestor—a two-dimensional figure designed to resemble something

---

164 Ibid., p. 19.

165 Ibid., p. 28.

166 Ibid., p. 30.

that looks a bit like a small tree. The computer then runs a program written by an intelligent person to produce a variety of "randomly mutated offspring," one of which is chosen by being selected by an intelligent human being for "breeding."[167] The chosen mutant then becomes the "parent" for the next round. In this manner, explains Dawkins,

> As generations go by, the selector [i.e., an intelligent human being] can guide evolution in very much the same way as humans guided the evolution of domestic dogs, but much faster.[168]

After "whistling" through many generations, Dawkins found he could "*home in on* an 'insect' or a 'flower', a 'bat', a spider' or a 'spitfire' [airplane]."[169]

Dawkins' supposed demonstration of his "different process" of obtaining *in silico* objects that look designed is laughable.[170] That someone of Dawkins' intellect and stature would truly believe that his proposition of "accidental design" *in vivo* is demonstrated by a programmed computer with a designed starting point further guided by an intelligent being to "home in on" a design boggles the mind. If this is the best reasoning Darwinists can produce, intelligent design theorists can be assured it is only a matter of time before more rational minds will prevail.

[11] *"What I can say for certain is that the figures appear to have been made slowly over long periods of time, and there were no people around at any time during their formation. This, at least, is scientific fact."*

Much of what we observe on earth appears to have been made over long periods of time, and the assumption of Darwinists is that there was no person, or entity, or god, or whatever, around at any time during its (or our) formation. This is presented to schoolchildren as if it were settled fact. Fanciful pictures of apes transforming into people from one page to the next fill textbooks, handouts, and worksheets.

Interestingly, it seems most people find this "fact" hard to believe. Polls (in the United States) consistently show that very large

---

[167] Ibid., p. 32.

[168] Ibid.

[169] Ibid., emphasis added.

[170] *In silico*, as contrasted with *in vivo* and *in vitro*. The term has emerged as a way of describing computer-based simulations relating to animal biochemistry. *See, e.g.*, www.worldwidewords.org/ weirdwords/ww-ins1.htm.

majorities of people simply find it too incredible to accept that we are an accident of nature. Darwinists dismiss this objection to Darwinism, somewhat properly, as being an "argument from personal incredulity."[171] When considering how "such a powerful idea [can] go still largely unabsorbed into popular consciousness," evolutionist Richard Dawkins states, "it is almost as if the human brain were specifically designed to misunderstand Darwinism, and to find it hard to believe."[172]

Dawkins is on the right track by considering what the human brain might be *designed* for. But maybe, rather than "misunderstanding" Darwinism, the popular consciousness understands it only too well. To Darwinists' continual chagrin, the public sees Darwinism for what it is—a theory that relies on chance. Surprisingly, Dawkins seems to do his best to make people believe that Darwinism is not a theory of chance. Without explicitly denying chance, he nevertheless tries at every turn to dispel "that old canard" that Darwinism is a theory of pure chance.[173] By carefully choosing his words (such as modifying chance with "pure," not unlike referring to sewer water as "sewage, but not *pure* sewage") Dawkins wants to deflect the obvious, that Darwinism is at bottom a theory that relies on chance—and lots of it. That Dawkins is so shy about admitting this, and even attempts to imply the opposite, is surprising because if Darwinism is not a theory that relies on pure chance, it must be a theory that relies on *direction* of some kind in some amount. If that is

---

[171] Dawkins, *Blind Watchmaker*, p. 39. Referring to arguments from personal incredulity, Dawkins states that they include those "versions which do not rest simply upon ignorance or lack of ingenuity." It is true that incredulity alone is an insufficient basis for rejecting a theory. Intelligent design can also be difficult to believe.

[172] Ibid, p. xv. Wouldn't it be ironic if the human brain *were* mischievously "designed" (note the inevitable use of language of design), to misunderstand Darwinism.

[173] Dawkins, *Climbing Mount Improbable*, p. 80. For example, in *The Blind Watchmaker*, Dawkins says, "This belief, that Darwinian evolution is 'random', is not merely false. It is the exact opposite of the truth" (Dawkins, *Blind Watchmaker*, p. 49). This is not an isolated thought; Dawkins continually obfuscates on the topic of "chance" and "randomness." For example, he makes statements such as, "Nowadays, and unfairly, this [i.e., the idea that the body can "choose" to mutate in the direction of getting better] seems all the more absurd to us because we are brought up to believe that mutations are 'random'" (Ibid., p. 306). Of course, we are only brought up to believe this because it is a fact, an unfortunate fact for Darwinists, but a fact nevertheless. But such a fact cannot stand in the way of Dawkins, who knows, as any thinking person does, that "If mutations are random, then, by definition, they cannot be biased towards improvement" (Ibid). How does Dawkins deal with the contradiction? He simply (and shamelessly) redefines words to mean what he wants them to mean. He confidently states, as if the problem is solved, that "There is randomness and randomness, and many people confuse different meanings of the word" (Ibid). Confusion is inevitable when language is abused, resulting in contrived meanings of terms. The only reason to contrive a new meaning for an existing word having an otherwise well-known meaning is to suppress, confuse and conceal the truth. Such tactics are not the natural forte of science, but necessity is the mother of invention, and necessity is relentless on the matter of "guidance" for evolution.

the case, then Darwinism is simply another version of intelligent design.[174]

With respect to the quote above about Darwinism being "hard to believe," for example, Dawkins attributes this to people's misunderstanding on "the issue of 'chance', often dramatized as *blind* chance."[175] Dawkins continues:

> The great majority of people that attack Darwinism leap with almost unseemly eagerness to the mistaken idea that *there is nothing other than random chance in it.*[176]

Again, Dawkins' lawyerly choice of words is intended to be self-serving, but actually works against him. First, whether there is "nothing other than" random chance is beside the point. The fact is that at least one of Darwin's steps in the evolutionary process *is* a random chance event. One random event in a multi-step process renders the entire process random.[177]

Other Darwinists don't seem to have a problem with admitting the fact that chance is a significant part of Darwinism. Thirty-eight

---

[174] In fact, Darwinists can hardly resist the urge to attribute a "direction-giving force" to natural selection. In the introduction to the Facsimile Edition of Darwin's *Origin of Species*, evolutionist Ernst Mayr stated: "That natural selection is a direction-giving force, within the limitations of the evolutionary potential set for a given species by its genotype, has now been substantiated abundantly" (Darwin, *Origin*, p. xvi-xvii). However, as this statement acknowledges, natural selection at best can only "direct" within the bounds of an organism's species. But even this is not true "direction," which, if language means anything, requires contingency, i.e., an ability to choose among possibilities with an end in mind. Some Darwinists go so far as to attribute true direction-giving powers to natural selection to "guide" evolution. For example, Richard Dawkins states, "*Core Darwinism*, I shall suggest, is the minimal theory that evolution is guided in adaptively nonrandom directions by the nonrandom survival of small random hereditary changes." Richard Dawkins, *a devil's chaplain* (New York: Houghton Mifflin Company, 2003), p. 81. To attribute a "direction-giving force" to "guide" something among possible alternatives, regardless what it is called, is to speak of an intelligent agent.

[175] Ibid. This assertion is particularly interesting. What other kind of chance is there but "blind" chance? "Five percent vision" chance, perhaps? And to infer that there might be some other kind of chance, or that there might be some deficiency in the ability of blind chance, is especially fascinating in a book entitled *The Blind Watchmaker*.

[176] Ibid., emphasis added.

[177] Consider a gambler at a craps table trying to convince someone that he is sure to get rich, since the result of a lucky roll is a non-random, non-chance, *automatic*, cash award. Suppose he strenuously argues that gambling is not "pure" chance, and criticizes you for even thinking that there is "nothing other than random chance in his gambling." This gambler's reasoning is logically equivalent to that of Darwinists who think that because one step in the evolutionary process is non-random, somehow chance is minimized (if not eliminated) as a factor.

Nobel laureates, for example, recently stressed in a joint statement that "evolution is understood to be the result of an unguided, unplanned process of random variation and natural selection."[178] Ernst Mayr, a leading Darwinist of the 20th century, agrees. In his book *What Evolution Is* he addressed this issue in an Appendix of Frequently Asked Questions.[179] Mayr not only attributes the production of genetic variation to chance, but also says chance plays a part in natural selection. Specifically, in answering the question, "Where does chance (stochastic processes) enter the process of selection?" he assumes without doubt that the first step of heritable variation is a chance event, and further states:

> Chance plays an important role even at the second step, the process of elimination of less fit individuals [i.e., natural selection]. Chance may be particularly important in the haphazard survival during periods of mass extinction.[180]

Mayr understands the implications of even hinting that evolution does not rely on chance, and only chance. In a related question in the same Appendix of *What Evolution Is*, Mayr answers the question "Does any process in evolution require a teleological explanation?"[181] Teleology is the Great Destroyer of Darwinian evolution, and Mayr's answer reflects the reaction of Darwinists to such a thought: "The answer is an emphatic 'No.'"[182] In light of Mayr's position, the fact that Darwinists such as Dawkins downplay the role of chance so strenuously is a mystery. As Mayr's answer clearly apprehends, the only option to a completely random, chance process is a teleological process—a directed (i.e., non-random), purposeful (i.e., non-chance) process. This, of course, is the basis of intelligent design theory.

Dawkins is welcome to join the more reasonable side of the issue, but that surely is not his intent. Evidently, he just wants to deal with a one-horned dilemma by declaring Darwinism to be something along the lines of "random non-randomness." Like proud Custer on

---

[178] Multiple authors, *Nobel Laureates Initiative* (The Elie Wiesel Foundation for Humanity, September 9, 2005).
[179] Mayr, *What Evolution Is*, p. 281.
[180] Ibid.
[181] Ibid., p. 275. The term "teleological" means "exhibiting or relating to design or purpose, especially in nature. *Merriam-Webster's Collegiate Dictionary, Tenth Edition: 1996* (Springfield, MA: Merriam-Webster. Inc., 1996). Teleology is the study of design in nature—a doctrine holding that ends are immanent in nature, that phenomena can be explained by final causes. Ibid. In other words, any theory of origins that relies on "specified in advance" concepts or features is consistent with teleological theories.
[182] Ibid.

the oxymoronic Little Big Horn, Dawkins thinks that being able to shoot in any direction means he must be in a good position. Wanting desperately to believe that Darwinism is simply misunderstood, he continues lamenting the fact that Darwinism is seen for the chance process it is, by stating:

> Since living complexity embodies the very antithesis of chance, if you think that Darwinism is tantamount to chance you'll obviously find it easy to refute Darwinism![183]

Intelligent design theorists *do* find it easy to refute Darwinism. What intelligent design theorists find difficult to understand is that, simply because natural selection can be considered to be non-random, Darwinists are blind to the overriding role chance plays in the process of providing natural selection something to select.[184] But let Darwinists continue denying that life can be attributed to a chance process (purely or impurely). This is exactly what intelligent design theorists maintain, and there is room for everyone on the side of truth.

Perhaps, however, it is more likely that our minds are designed (to use Dawkins' term) to follow Aristotle's advice that one should always prefer the probable impossible to the improbable possible. Sure, some suppose, evolution may be *possible*, but .... Darwin himself recognized the highly improbable nature of his theory of descent with modification. As he stated in *On the Origin of Species*:

> Nothing at first can appear more difficult to believe than that the more complex organs and instincts should have been perfected, not by means superior to, though analogous with, human reason, but by the accumulation of innumerable slight variations, each good for the individual possessor.[185]

Darwin worked through the difficulty by a series of propositions of which he stated, "The truth of these propositions cannot, I think, be disputed."[186] That his "truths" have been and

[183] Dawkins, *Blind Watchmaker*, p. xv.

[184] Note the true statement: "Survival of the fittest requires arrival of the fittest." That is, natural selection has no creative power. Natural selection can only "select" among already created beings. This leaves Darwinian "creation" wholly to the unintelligent processes of unguided physics and chemistry.

[185] Darwin, *Origin*, p. 459.

[186] Ibid.

continue to be disputed goes without saying. What is important to understand is that even Darwin recognized his ideas as "difficult to believe." We can either, like Darwin, choose to believe the highly improbable is possible, or, like Aristotle, choose to believe the seemingly impossible is probable.

[12] *"Let's face it; the facts will win the day. We were right about the sun, weren't we?"*

In 1543 Copernicus published his book, *On the Revolution of the Celestial Spheres,* setting forth his theory, unorthodox at the time, that the sun is actually the center of the solar system and the earth is one of many orbiting planets. His heliocentric theory, supported by detailed observations and mathematical precision, was in sharp contradiction to a theory that had held for over 1400 years prior—Ptolemy's geocentric theory. Moreover, in the intervening years Ptolemy's theory had been adopted by the Catholic Church in Europe, it supposedly being consistent with the idea of man being the center of creation and the earth being immoveable.

Galileo Galilei confirmed and promoted Copernicus's theory and was prosecuted and persecuted by the Catholic Church for his views in a highly publicized series of confrontations that remains an embarrassment to the Church, which only recently attempted to close the case with a frank concession that the whole episode had been a mistake. This period of confrontation between Galileo and the Church, sometimes referred to as the "Galileo Episode," continues to linger in the popular imagination as a reason to be forever skeptical of religiously motivated ideas. Whether or not it should be so used can be debated, but the historical account of the acceptance of the heliocentric theory of planetary motion is instructive in how "science" really works.

First, the Galileo Episode is instructive in observing that facts are rarely assessed in a vacuum, i.e., in an isolated, objective analysis. Facts are necessarily interpreted in view of a *theory* that seeks to explain what is observed in a larger context. Thus, if your theory is that the Earth is the center of the universe, you will observe facts that support this theory. This reality of science is illustrated beautifully by James Burke in his book *The Day the Universe Changed.*[187] Burke begins his book by relating the following:

> Somebody once observed to the eminent philosopher Wittgenstein how stupid medieval Europeans living before the time of Copernicus must have been that they could have looked at the sky and thought that the

[187] James Burke, *The Day the Universe Changed* (Boston: Little, Brown and Company, 1985).

> sun was circling the earth. Surely a modicum of astronomical good sense would have told them that the reverse was true. Wittgenstein is said to have replied: "I agree. But I wonder what it would have looked like if the sun *had* been circling the earth."[188]

Burke continues:

> The point is that it would look exactly the same. When we observe nature we see what we want to see, according to what we believe we know about it at the time.[189]

Burke's observation is correct and his insight notable. The fact is, as much as we like to think we are objective in our conclusions, in actuality we are typically led to conclusions by pre-conceived notions we call theories. That, in fact, is the whole point of proposing a theory—it provides a framework for sorting, analyzing and cataloging facts into a coherent system of knowledge. The explanatory power of a theory depends on the number of facts it can account for without anomalies and contradictions.

Lest one think that the idea that facts cannot be interpreted objectively is a ploy intelligent design theorists concocted to cry "bias," consider the words of Darwinists on this subject. In an essay entitled "An Early Start" appearing in his book *The Panda's Thumb*, evolutionist Stephen Jay Gould considers how a particular theory of gradual origin of life (life, not species), "ever gained such strong and general assent." He asked, "Why did it seem so reasonable? Certainly not because any direct evidence supported it."[190] This statement is amusing, but Gould's insight is otherwise noteworthy:

> I am, as several other essays emphasize, an advocate of the position that science is not an objective, truth-directed machine, but a quintessentially human activity, affected by passions, hopes, and cultural biases.[191]

---

[188] Ibid., p. 11.
[189] Ibid.
[190] Gould, *The Panda's Thumb*, p. 225.
[191] Ibid.

In addition to Gould, consider the words of an anti-intelligent-design skeptic. Michael Shermer is the editor-in-chief of *Skeptic* magazine and author of several books, including the book *The Borderlands of Science* which explores where "valid science leaves off and borderland science begins."[192] One of the most intriguing things about this admitted skeptic is the unbounded skepticism and ridicule of all theories of origins—except one. In his books and in regular columns published in *Scientific American* Shermer is not at all skeptical of natural, unguided evolutionary theories. He is absolutely certain that Darwinism is true and Darwin is among the great scientists of all time.[193] This skeptic of skeptics and gifted writer wrote in an April 2001 *Scientific American* column entitled *Colorful Pebbles and Darwin's Dictum* of "one of the deepest single statements ever made on the nature of science itself, particularly in the understated denouement."[194] What statement did he refer to? A statement from none other than Darwin himself, from a letter written to a friend in 1861:

> About thirty years ago there was much talk that geologists ought only to observe and not theorise; and I well remember some one saying that at this rate a man might as well go into a gravel-pit and count the pebbles and describe the colours. How odd it is that anyone should not see that all observations must be for or against some view if it is to be of any service![195]

To this, Shermer agrees, adding his own observation: "The facts never just speak for themselves. They must be interpreted through the colored lenses of ideas: percepts need concepts."[196] He continues:

> Science is an exquisite blend of data and theory, facts and hypotheses, observations and views. We can no more expunge ourselves of biases and preferences than we can find a truly objective, Archimedean

[192] Michael Shermer, *The Borderlands of Science* (Oxford: Oxford University Press, 2001), front flap.
[193] For example, the front flap of Shermer, *Borderlands*, states of Shermer: "He examines the work of Darwin and Freud, explaining why one is among the great scientists in history, while the other has become nothing more than a historical curiosity" (Ibid., front flap).
[194] Michael Shermer, "Colorful Pebbles and Darwin's Dictum," *Scientific American*, April 2001, p. 38.
[195] Ibid.
[196] Ibid.

> perspective—a god's-eye view—of the human condition. We are, after all, humans, not gods.[197]

If only such an honest acknowledgement of "biases and preferences" would be stated as a caveat before promoting theories as fact.

Second, the Galileo Episode illustrates the tendency of smart people to adopt and hold dubious theories to account for anomalies in observed phenomena. That is, philosophers, thinkers, and scientists have often formulated elaborate, if not incredible, explanations to "save the phenomena."[198] For the medieval philosophers (including non-religious astronomers and scientists), Ptolemy's "epicycles" were accepted as accounting for the irregular motion of the sun and planets as they orbited about the earth. Epicycles were periodic "mini-orbits" within a planet's larger orbit that were theorized to explain the occasional reversal of direction observed for the planets at various positions of their orbit (about the earth). What strikes us today as contrived and implausibly complex looping motions of the planets became a matter of firmly accepted belief because it fit the observed facts consistent with the accepted theory at the time. That is, it worked. And it worked for 1400 years.

How could secular and religious astronomers and scientists be fooled for so long? The answer is that in Ptolemy's attempt to go "beyond the information given," as Jeremy Bruner[199] put it, his explanation, even though improbable, fit the observed phenomena as far as it went. As Howard Margolis states in his book *It Started With Copernicus*, an accepted theory with supporting observations, anomalies and all, breeds a certain amount of economy and comfort.[200] This economy and comfort of a theory gives it a particular staying power, something of an intellectual inertia, so that over time even wrong, cumbersome theories can be difficult to unseat. That is, over time we become accustomed to the oddities of a cumbersome, yet comfortable system. We prefer what feels comfortable, including what

---

197 Ibid.

198 The task of science, in simple terms, is to explain reality. "Reality," is often disputed, but it is generally agreed to at least include what is manifest to our senses. In philosophical terms, "phenomena" refers to those things that are evident or manifest to our senses. Philosophers, such as Plato, Aristotle, and on up to more contemporary thinkers such as Jean-Paul Sartre, have attempted to formulate philosophies that "save the phenomena." To "save the phenomena" is to construct a theory that explains reality with a minimum of anomalies. An anomaly is a datum that does not fit the pattern or cannot be explained by a current model. *See*, R. C. Sproul, *The Consequences of Ideas,* and Kuhn, *The Structure of Scientific Revolutions,* for more on the philosophical and practical implications of "anomalies" in nature.

199 Howard Margolis, *It Started With Copernicus: How Turning the World Inside Out Led to the Scientific Revolution* (New York: McGraw-Hill, 2002), p. 25; quoting Jeremy Bruner, *Beyond the Information Given* (Norton, 1973)

200 Ibid., p. 25.

fits with our habitual patterns of thought. When choosing between two theories that seem equally economical, we will always find ourselves preferring the one that fits more comfortably with other cognitive patterns in our repertoire. Margolis explains that what seems complicated when we are new to it comes to seem simple as we become familiar with its arguments. Soon, a particular theory is no longer questioned; it is simply something that "everyone knows" is right as a matter of social knowledge. According to Margolis, when something takes hold as social knowledge, it begins to seem automatically both more economical and more comfortable. Sometimes logically, sometimes merely by trained inattention, we learn to ignore aspects of it that might otherwise seem clumsy.[201]

An example of how this scenario is currently being replayed with respect to biological evolution involves the idea of *entropy* as it relates to physical systems. Entropy is the name given to a measurable quantity of a system and is, very simply, a measure of the disorder of a system.[202] For example, we could measure the "orderliness" of a drop of red dye in a beaker of water. At first all the red dye molecules are grouped together in the drop—a relatively ordered arrangement. Over time, however, the red dye molecules do not stay grouped together in their relatively ordered state, but become spread out in gradually increasing random disorder. In more time, the red dye molecules become completely and evenly spread throughout the glass of water. Over even more time very little additional "spreading" occurs as the dye and water molecules reach a state of equilibrium—and its highest entropy.

What is described above with respect to the red dye in water is simply the most *probable* outcome for the position of the red dye molecules. The red dye molecules *could*, by chance, end up in only one half of the water, for example, only in the top half. Such an arrangement would represent a lower entropy than if evenly spread throughout the water. Or the red dye molecules *could* re-group into a drop again, an even lower entropy state. Over time, any distribution of red dye molecules is *theoretically possible*; the uniform distribution—that is, the highest level of disorder (and entropy) within the constraints of the system—is simply the most *probable*. Entropy is a measure that reflects the probabilistic notion of energy and matter that predicts, in physical terms, that any spontaneous changes in a given system (*e.g.*,

---

201 Ibid., p. 27.

202 *See also*, Note 63. Admittedly, entropy can be difficult to understand, if only because its applicability in different contexts causes the concept to take different forms of explanation. Moreover, explanations of entropy often become unavoidably complex, involving other concepts such as *free energy* and *equilibrium*. But in essence the idea of entropy is simply the idea that over time all natural processes tend to randomness, or disorder. For a more detailed yet very readable discussion written for the lay person, *see*, George Gamow, *One Two Three…Infinity* (New York: The Viking Press, 1968).

water and dye) occur in the direction of increasing disorder, i.e., entropy always increases over time.

Without going into great detail (and there are great details), the "law of increasing entropy" is reflected in what is called the Second Law of Thermodynamics—one of the most fundamental of all physical laws and one that has application in almost every scientific discipline. The universality of the Second Law of Thermodynamics is unquestioned; it is sometimes referred to as the basis for our perception of time itself.[203] It is the basis for the bleak concept of "the heat death of the universe." The "Second Law" says that in the absence of directed energy or physical constraints systems always "wind down;" they always go from a relatively ordered to a relatively more disordered state; they always go from high energy to low energy; they always go from organized to disorganized. Always. Except ... some would have us believe, in evolving life systems.

On its face, the Second Law stands diametrically opposed to any theory that requires matter to go spontaneously from a simple, random form to a more complex, ordered form, including biological evolution. The stock reply from Darwinists to such an objection is that the sun is part of our "system" and it constantly supplies massive amounts of energy input, thereby averting the Second Law from stopping the incredible ordering necessary for evolution.[204] (Evolutionist arguments with respect to the Second Law are dealt with more fully in Note 63 below.)

It is true that increasing entropy (increasing disorder) is inevitable only for systems as a whole (such as the system of our universe), and local parts of a system can experience entropy decreases (as happens in ordering events such as crystal growth). It is also true that living systems do use energy from the sun to go from simple to complex forms, but such systems are designed, like engines in cars, to convert energy to order. Simply invoking the sun into the earth's system as a "cure all" for evolutionary complexity ignores the fact that raw, undirected energy is not known to be capable in itself (i.e., in the absence of an existing physical law or process) of providing order out of disorder, much less *specified* order, and further much less *specified complexity*. Energy must be directed as *work* to be useful.[205] In fact, the undirected energy of the sun has the opposite effect—in the absence of an imposed ordering principle (like photosynthesis), the sun's

---

[203] *See* Harold F. Blum, *Time's Arrow and Evolution* (Princeton NJ: Princeton University Press, 1968). Blum reports that English astronomer Sir Arthur Eddington (1882-1944) expressed the second law as "time's arrow" (Ibid. p. 16.).

[204] *See, e.g.,* the National Association of Biology Teachers (NABT) Statement on Teaching Evolution in which the NABT states as one of their "tenets of science" that, "Evolution does not violate the second law of thermodynamics: producing order from disorder is possible with the addition of energy, such as from the sun." Quoted in NAS, *Teaching About Evolution*, p. 127.

[205] The term "work" has a specific meaning in physics and engineering.

radiation tends to break down matter into less ordered states. Awnings fade, clothing dry rots, and wood decomposes (at least partially due to the sun, and certainly helped by the sun).

A more clever approach to overcoming the problem of explaining evolution in light of the Second Law has been made by the introduction of a relatively new concept: *negentropy*. Physicist Erwin Schödinger coined the term "negative entropy" in his book *What is Life*, describing it as the force or energy that halts the decay of entropy.[206] While the term is subject to confusing usage among scientists, biologists use it in the sense intended by Schödinger, which is "to identify the remarkable ability of the living system, not only to avoid the effects of entropy production -- as dictated by the second law -- but to do just the opposite, to increase organization, which intuitively, seems like the converse of entropy."[207]

Negentropy, like Ptolemy's epicycles, is a contrived, scientifically suspect idea that has as its only purpose to explain how evolution could happen despite the universal applicability of the Second Law. That is, it "saves the phenomena" by postulating what otherwise, in any other context, would seem odd and unnecessary. In short, the idea of negentropy is that of the Second Law being short-circuited, even reversed in certain circumstances.[208] For someone committed to naturalistic evolution such an explanation must, like the epicycles of old, seem eminently believable—despite any factual basis or experimental support. Such a negative entropy force to power life, some Darwinists suppose, simply must, in some form, be present. This and other theories using terms such as "emergence" or "chaos" have proved popular as ways of attempting to explain how nature can move to higher, ordered levels of specificity without any intelligent direction. In any other context such mental gyrations would seem false, and the efforts farcical. However, like perpetual motion enthusiasts that design brakes on their machines to control the inevitable increase in energy output, such "negentropy" enthusiasts continue building on paper that which has no basis in science.

---

[206] *See*, E. Schrödinger *What is Life?* (Cambridge: Cambridge University Press, 1944).

[207] Ibid. *See also*, A. Szent-Györgi, A. *An Introduction to Submolecular Biology* (New York : Academic Press, 1960), "It is common knowledge that the ultimate source of all our energy and negative entropy is the radiation of the sun."

[208] This is not to say there are not "anti-entropic" processes. If, for example, instead of using red dye in our example above, we had added oil to water, we would have observed a distinct ordering between water in the bottom half of the beaker and oil in the top half. This more ordered state is anti-entropic, but it is not a violation of the Second Law and needs no special explanations involving "negentropy." Examples of spontaneous ordering include crystal formation, pebble size distribution on beaches, and the like. Likewise, life processes such as photosynthesis are not "violations" of the Second Law. A complete exposition of the subjects of the Second Law of Thermodynamics, entropy, and other energy-related concepts is far beyond the scope of this book. However, as noted, the topic is treated more completely in Note [63] below.

Third, the Galileo Episode shows what happens when scientific matters are controlled by a ruling elite that attempts to censor any and all dissent—regardless of the supporting facts of a matter. In this instance the Catholic Church was in control, stifling all dissent while enforcing a scientific view consistent with its own. To hold a strong opinion on a scientific matter is both right and necessary. However, to use political force to stop any challenge of, or opposition to, the scientific matter, is both wrong and ultimately harmful to science itself. Today in the United States any attempt to introduce challenges to Darwinism into public education (e.g., what is popularly termed "teaching the controversy") is met with blunt, raw political opposition. This opposition is aided by the media, complete with references harkening back to Galileo, to quickly suppress any attempts to even imply that Darwinism has any shortcomings.

It is ironic that the ruling "church" of the mainstream scientific establishment would today object to doing what Darwin did in his book *On the Origin of Species*: airing objections to his theory. Darwin, for all the bedevilment heaped upon him in the minds of those who have never read his book, was at least honest in his approach. He knew that a theory is either true or it is not, and truth is not troubled by adversaries. Would that this might be the lasting legacy of the Galileo Episode.

[13] *"And we're in charge now, don't forget," Dr. Hawkins added, looking around at the group. "At least we have that going for us."*

Whether it's the president's bully pulpit, the preacher's sacred pulpit, or the professor's secular pulpit, the undeniable fact is that the person or group "in charge" can dictate the rules of engagement for debating ideas. Whether it's by outright intimidation, or by simply "shouting down" opponents, those in charge have a built-in advantage over those in dissent. This advantage is not necessarily bad, but it becomes unproductive when the advantaged party becomes tyrannical in its suppression of dissent.

An interesting look at this principle can be found in Phillip E. Johnson's book, *Defeating Darwinism by Opening Minds.*[209] Johnson summarizes the story line from the movie *Inherit the Wind*, particularly recounting an event Johnson refers to as "owning the microphone."[210] In the final scene a character referred to as "Radio Man" appears in the courtroom, carrying a large microphone, and begins to report directly to the public on the proceedings as they happen. As Johnson relates, Brady (the volunteer prosecutor and anti-evolution crusader), famed for decades as an orator with a huge voice, attempts to speak into the

---

[209] Phillip E. Johnson, *Defeating Darwinism by Opening Minds* (Downer's Grove, IL: InterVarsity Press 1997), Chapter 2.
[210] Ibid., p. 32.

microphone but can't master the technique. During Brady's final tirade the radio program director decides that his speech has become boring, and Radio Man breaks in to announce that the station will return to the studio for some music.[211]

Johnson observes:

> The microphone (that is, the news media) can nullify Brady's power by (in effect) outshouting him. When the creation-evolution debate is replayed in our own media-dominated times, the microphone-owners of the media get to decide who plays the heroes and who plays the villains.[212]

When advantage turns to tyranny, those "in charge" attempt to thwart all dissent. It is telling when science organizations such as the National Academy of Sciences advocate an outright ban on teaching any criticism of evolution. Weakness fears a challenge and tyranny simply bars any attempt, often labeling it "religious." Tyranny always uses the sword for that which the pen cannot defend.

[14] *"But in this case to let you present your personal beliefs in class would be in complete violation of our guidelines for teaching science."*

James Perloff, in his book *Tornado in a Junkyard,*[213] records many incidents of teachers and students being punished for challenging evolution in public (and, in some cases, private) schools. Some examples:

> In 1980, a biology teacher for 17 years was dismissed for teaching both creation and evolution as possible explanations for the origin of life.[214]
>
> In 1996, a teacher in Washington State was suspended for two days without pay for showing his eighth-graders a video on creation and arranging a speaker to present scientific facts favoring creation.[215]

---

[211] Ibid., p. 33.

[212] Ibid.

[213] Perloff, *Tornado in a Junkyard*, pp. 237-38.

[214] Ibid., quoting Jane Ingraham, "Darwin is Dead," *The New American* 2 (24 November 1986): 34.

[215] Ibid., quoting "Teacher's Creationism Lesson Causes Stir," *Philadelphia Inquirer*, 22 August 1996, quoted in *Creation Ex Nihilo* 19 (June-August 1997): 9.

> In 1996, after being threatened by an ACLU lawsuit, a high school physics teacher was ordered to stop encouraging students to consider how their own worldviews affected their outlook on the creation-evolution debate.[216]

> Tenured professor Dean Kenyon taught a course on evolution at San Francisco State University. But when he coauthored *Of Pandas and People,*, a book advocating intelligent design theory, he was censured and his course taken away.[217]

These and other stories chronicle a sad state of affairs for a discipline that otherwise prides itself on promoting free inquiry and thrives on curiosity. Science, presumably the most objective of all studies concerned with knowledge, has cordoned off one sphere of inquiry to be off limits to any question, any challenge, and any opposing views. Evolution has become the grand guarded Citadel of Science, with walls protected by Knights of the Darwinian Creed, and true believers ready to pour the hot oil of political and legal power on any barbarian knaves at the gate. No exception is made, not even for the most innocent of those who approach the gate mistaking the stronghold as a bastion of academic freedom. Evolution is, quite simply, off limits to any challenge that hints at evidence supporting non-naturalistic causes.

Clearly not thinking about Darwinism, the National Academy of Sciences at one time issued an *Affirmation of Freedom of Inquiry and Expression* that read, in part:[218]

> I hereby affirm my dedication to the following principles:

> ... That the search for knowledge and understanding of the physical universe and of the living things that inhabit it should be conducted under conditions of intellectual freedom, without religious, political or ideological restriction.

> ... That all discoveries and ideas should be disseminated and may be challenged without such restriction.

---

[216] Ibid., quoting Phillip E. Johnson, *Defeating Darwinism by Opening Minds* (Downers Grove, IL: InterVarsity Press, 1997), pp. 49-50.

[217] Ibid., quoting Henry M. Morris and John D. Morris, *Society and Creation* (Green Forest, AR: Master Books, 1996), p. 178.

[218] *See*, Perloff, *Tornado in a Junkyard*, p. 238.

> … That freedom of inquiry and dissemination of ideas require that those so engaged be free to search where their inquiry leads, free to travel and free to publish their findings without political censorship and without fear of retribution in consequence of unpopularity of their conclusions. Those who challenge existing theory must be protected from retaliatory reactions.[219]

There is little doubt that the NAS meant (and possibly still means) these words. However, the discipline of Darwinism has evolved into a scientific mutant of sorts—a strange blend of scientific phenotype animated by philosophic genotype. The religious zeal that accompanies defenses of evolutionary theory goes beyond any explanation based solely upon scientific interest. Perhaps, as evolutionist Richard Dawkins has suggested, science should be treated as a religious topic, which would at least make dogmatic fervor expected, and would have the added benefit to Darwinists of making actual scientific support somewhat superfluous.[220]

Frustrated intelligent design theorists, however, can take heart at the pace of technological innovation and discovery. Just as modern technology rendered stone castles and shining armor obsolete, new technology is making quick work of Darwinism as well. Although bravely marshaled in defense of Darwinism, advances in genetics, molecular biology, cellular biology, and the like, continue to compound the problems for Darwinists. The evidence simply continues to build for specified complexity that can reasonably be explained only by intelligent causation. Though elite Darwinists remain hunkered comfortably behind their moats of arrogance and crenulated walls of indifference, each new advance in biology gives them more reason to question their sophisticated pretenses. While most will be blind to the obvious, and remain content to impress others within the walls of peer-reviewed articles in prestigious journals, there are some honest souls who will peer over the ramparts to see more than ignorant hordes encamped below. The bravest of hearts will begin venturing out to make peace before the walls of the great Fortress Darwin crumble as indefensible, its rubble left as a testimony for historians to sort out as another of the shifting paradigms of science.[221]

[15] *"I want you to think of a real explanation for the figures on the cave wall, not one that relies on myths."*

---

219 "An Affirmation of Freedom of Inquiry and Expression," National Academy of Sciences Resolution (27 April 1976), quoted in Robert V. Gentry, *Creation's Tiny Mystery* (Knoxville, TN: Earth Science Associates, 1992), p. 7.

220 Dawkins, *River Out of Eden*, p. 33.

221 *See*, e.g., Kuhn, *The Structure of Scientific Revolutions*.

Not surprisingly, Darwinists consider any non-naturalistic explanation of origins to be a "faith based" "myth." For example, in his book *River Out of Eden*, Richard Dawkins writes of Darwinian selection in his typical fluid style:

> Never were so many facts explained by so few assumptions. Not only does the Darwinian theory command superabundant power to explain. Its economy in doing so has a sinewy elegance, a poetic beauty that outclasses even the most haunting of the world's origin myths.[222]

In actuality, never before were so many assumptions explained by so few facts. Darwin himself had to assume a hypothetical example of natural selection,[223] and the best contemporary examples are tired, if not highly touted, stories of peppered moths and changing finch beaks.[224] Even these meager examples, however slight or cyclical, cover for a barrage of assumptions Darwinists use to fool themselves into thinking they've shown that with enough weeks and months you somehow get beaks and moths in the first place.

Natural selection can, by definition, only preserve what is already created. Indeed, the full title of Darwin's major work is *On the Origin of Species by means of Natural Selection, or the Preservation of Favoured Races in the Struggle for Life.*[225] Thus, light-colored moths on

---

[222] Ibid., p. xi.

[223] Darwin, *Origin*, p. 90. In a section entitled "Illustrations of the Action of Natural Selection," Darwin wrote, "In order to make it clear how, as I believe, natural selection acts, I must beg permission to give one or two imaginary illustrations." In fairness, it must be mentioned that Darwin considered natural selection too slow to be observed. Earlier in *Origin* (p. 84) Darwin said of natural selection that we "see nothing of these slow changes in progress, until the hand of time has marked the long lapse of ages, and then so imperfect is our view into long past geological ages, that we only see that the forms of life are not different from what they formerly were." Of course, appealing to "geological ages" to excuse lack of evidence does not substitute for evidence.

[224] Much has been written about two famous examples of natural selection—the peppered moths of Great Britain, and the beaks of finches on the Galapagos Islands. Both are presented as classical examples of "evolution in action." For example, in 1975, British geneticist P. M. Sheppard called the phenomenon "the most spectacular evolutionary change ever witnessed and recorded by man, with the possible exception of examples of pesticide resistance" (Quoted in Wells, *Icons of Evolution*,, p. 143). Two points are noted about the various examples of "evolution in action" (including pesticide resistance). First, as a minor point, after 150 years of Darwinism it seems that impressive examples of natural selection would be much better documented. Of the two standard textbook examples, only one is even from the twentieth century. Second, proof of natural selection is not proof of speciation, i.e., proof of new species being created.

[225] The original title is significant, and it is significant that the last part has been dropped from popular references to the work. The notion of "preserving" is by definition not one of "creating." In Darwin's theory, natural selection can only *preserve* that which is already in existence. While Darwin believed that heritable variations provide the "creation" that can be preserved by natural selection, such a proposition is far from being confirmed, or

darkened tree trunks were "naturally selected" by birds so that their death was not random with respect to dark-colored moths, which were preserved. But no new moths were created. Nothing evolved into a more complex organism, a morphologically changed organism, or even a different organism. The overall population of moths simply changed in terms of the ratio of dark moths to light. Likewise, with the beaks of "Darwin's finches," certain birds having certain beak sizes were "naturally selected" to die non-randomly with respect to other birds with different beak sizes, which were preserved. But, again, nothing new evolved. No new birds or beaks were created.[226] The same reasoning can be used with regard to every example of supposed "evolution." The great assumption upon which all evolutionary facts rest is that evidence of slight, even cyclical, changes in existing traits is evidence of simple-to-complex evolutionary creation of information-rich complex organisms in the first place.

But the point of Dawkins' almost rapturous obsession with the "sinewy elegance" of Darwinism is to distinguish it over any of the world's origin myths. His point is well taken if his starting assumption is correct. That is, if naturalism is true, and there is, in fact, no supernatural, then non-naturalistic explanations, however haunting, *are* myths.

The question, therefore, is how do we or can we, know?

[16] *"Jenny held back tears."*

This chapter further illustrates what regularly happens in public school classrooms across the United States. A student given an assignment, the purpose of which, at least in part, is to get the student to think, is censored by the government. Well-meaning teachers, already burdened with bureaucratic rules and regulations, think it is against the law to challenge Darwinism at all, much less by allowing talk of a creator. In the end, pressure from teachers and administrators, who are in turn coerced by the threat of lawsuits, force students to capitulate to a government-endorsed, dogmatic view of science.

Such a scenario may strike one as odd or rare, but the sad fact is that it is neither. In public education the subject of evolution is simply off limits to any challenge of any kind. As Phillip E. Johnson

---

even theoretically supported. Random, heritable variations in the form of mutations are unlikely to provide the requisite specified complexity for new organs, not to mention that such mutations rarely confer an advantage on an organism.

226 Nevertheless, Darwin's finches are continually touted as "evolution in our time". Even though no new species were created, the National Academy of Sciences touts Darwin's finches as "a particularly compelling example of speciation." NAS, *Science and Creationism: A View from the National Academy of Sciences* (Washington DC: National Academy Press, 1999), p. 10.

states in his book *Defeating Darwinism by Opening Minds*[227], "When evolution is the subject, questioning whether the official story is true is enough to make you an enemy of education."[228]

Johnson gives as an example the account of a Denver area fifteen-year-old high school junior named Danny Phillips. Danny effectively challenged the showing of a video program that presented the "evolution as fact" story by writing a paper critical of the video—characterizing it as propaganda. School administrators were at first convinced Danny had a point, and tentatively decided to withdraw the program from the curriculum. In the end, the resulting media firestorm forced the Denver school board to decide against Danny.

Other examples from publicized cases can be given, but most do not make the papers. Students simply respond to intimidation, subtle threats (e.g., from grades), and ignorance as to their rights as they silently kowtow to the party line. Johnson continues:

> Danny's case ended like so many others; he lost because the power was on the other side. Self-styled 'civil liberties lawyers' threatened to bring an expensive lawsuit, and the school board capitulated to them.[229]

In Danny's case, the attention of the scientific community reached all the way to the president of the National Academy of Sciences, who personally wrote to Danny to help silence discussion on the topic. The reader should read Johnson's book, but his conclusion on this issue is worth quoting here:

> It's an absurd situation, isn't it? Educators aren't allowed to address the issues about which their students, and the general public, are most concerned. When teachers challenge their students to think about how their worldviews affect their understanding of the creation-evolution controversy, so-called civil liberties lawyers censor the teaching by threatening to bring a lawsuit that the school district can't afford to defend. The president of the National Academy of Sciences writes an essay so simplistic that it insults the intelligence of a well-informed high-school student. He urges a bright high-school student not to think for

[227] Johnson, *Defeating Darwinism.*

[228] Ibid., p. 35.

[229] Ibid., p. 51.

> himself but to trust the findings of a research community that thinks it can settle the question of our origins by defining finch-beak variation as 'evolution.'[230]

[17] *"To believe that people visited the cavern, and that a person made these marks for some purpose, possibly even as art, could be, according to the book, "useful for some as a social convention," but otherwise "has no basis in science."*

The National Science Teachers Association (NSTA) Position Statement on the Teaching of Evolution quotes the National Science Education Standards:

> Explanations on how the natural world changed based on myths, personal beliefs, religious values, mystical inspiration, superstition, or authority may be personally useful and socially relevant, but they are not scientific.[231]

The question, of course, is not whether or not an explanation is based on a personal belief, but is it *true*? An explanation may be both based on a personal belief *and* true. Personal beliefs are rarely based on nothing. If a personal belief has a rational basis, even if it is so-called circumstantial evidence, it can still correspond to that which is true for scientific purposes. If a naturalist scientist were asked if gravity is real, much less scientific, he or she would not hesitate in answering in the affirmative. If asked *how* he or she knows, the answer would be, in so many words, "because we can observe and measure its effects." Gravity is simply a name we give to an unseen, unheard, unfelt, tasteless and odorless "thing" that we know exists only by its effect on our surroundings. The same can be said for electrical fields and energy—both are well documented and characterized physical phenomena that are undetectable except for their effects on their surrounding.

---

[230] Ibid., p. 52. The reference to "finch beak variation" is in response to the suggestion by the president of the NAS that Danny read *The Beak of the Finch* by Jonathan Weiner (Jonathan Weiner, *The Beak of the Finch* (New York: Viking Books, 1994). *The Beak of the Finch* purports to show documented evidence of "evolution," but in fact only reiterates the well-known phenomena of heritable beak-size changes in the subject finch populations. Not only do such changes fail to impress non-Darwinists because they simply show changes within a species, not unlike the difference between breeds of dogs or races of people, the greatly-touted changes have since been shown to have been cyclic—there was no directional, permanent change. *See, e.g.*, Wells, *Icons of Evolution*, Chapter 8 for discussion of data indicating cyclic changes.

[231] NAS, *Teaching About Evolution*, p. 125-26.

Such an approach to science, i.e., believing in unseen phenomena based on measurable and detectable effects, is accepted as scientific without question for applications involving gravitational fields, electrical fields, and energy. Nevertheless, when it comes to origins research we see mainstream scientific organizations attempting to stifle debate by defining out certain lines of inquiry, arbitrarily deeming certain beliefs "not scientific."[232]

[18] *"A naturalist named Seth Neihil first published the theory of natural selective inclusion, which relies on small incremental changes that are naturally selected for inclusion into various forms called specialties."*

Darwin's theory of evolution is, as evolutionist Richard Dawkins said, "a bewilderingly simple idea."[233] Darwinian evolutionary processes require only two things: (1) heritable variation in living organisms, and (2) a way to preserve advantageous variations, i.e., natural selection. Variation is generally attributed to genetic variation (mutations) in the modern theory commonly referred to as "neo-Darwinism." Thus, with each generation of any given organism, random copying errors in the genetic code of certain individuals can result in new gene sequences, which, in turn, can result in advantageous traits in an individual possessing the trait. Traits are considered advantageous if they result in higher survival rates, as natural selection selects those not-so-advantageously-endowed for death. Darwinian evolutionary theory can be summed up as "natural selection operating on random heritable variations to preserve advantageous traits in living organisms," with "advantageous" generally understood to be those traits contributing to reproductive success. That is, variations in organisms that confer an advantage, particularly in reproduction, are preserved (as individuals without the advantage are "naturally selected" to die without the requisite reproduction). Darwin thus referred to his theory as "descent with modification."

Darwin's book *On the Origin of Species*,[234] does not (and does not purport to) explain the origin of life. As its name suggests, it simply theorizes a possible explanation for the origin of *species* from one or a

---

[232] Not surprisingly, the very tack that such organizations take violates their own creed. The quote in this section reads (deleting some optional words): "Explanations on how the natural world changed based on authority are not scientific." But students and the public at large are asked (demanded, actually) to unquestionably accept the explanation of evolution based on the authority of these "science" organizations!

[233] Richard Dawkins, *Mechanisms of Evolution*, in Campbell N.A., Reece J.B. & Mitchell L.G., "Biology," [1987], (Menlo Park CA: Benjamin/Cummings, Fifth Edition, 1999), p.413). (Cite found at : http://members.iinet.net.au/~sejones/ceqcmech.html).

[234] Darwin, *Origin*.

few original life forms.[235] Noting the great variety of domesticated animals that human breeders could produce by *selecting* for certain inherited traits, Darwin theorized that nature could do just as well. While human breeders select for such traits as appearance or size, nature "selects" by "natural selection" for reproductive success.

That natural selection can "preserve" only by destroying what is not preserved, and, therefore has no power to create, is self-evident to anyone but wish-projecting Darwinists. Serious Darwinists know this as well, but try to convince the public otherwise. This deceit can be shown by an example that also illustrates what appears to be a common tactic of evolutionist authors—ending one chapter of a book with a bullishly optimistic statement that the writer claims will be supported in the next chapter—only to do no such thing.[236] In *The Blind Watchmaker,* Richard Dawkins ends chapter 6 with the following words:

> The earlier chapters have disposed of other alleged stumbling blocks, and the next chapter takes up yet another one, the idea that natural selection can only destroy, never construct.[237]

---

[235] Even as an explanation, *Origin* falls short of explaining the origin of any species. As evolutionist Jonathan Weiner states in his book *The Beak of the Finch,* "The *Origin of Species* says very little about the origin of species. Darwin's full title is *On the Origin of Species by Means of Natural Selection, or the Preservation of Favoured Races in the Struggle for Life*. Yet the book does not document the origin of a single species, or a single case of natural selection, or the preservation of one favored race in the struggle for life" (Weiner, *Beak of the Finch,* p. 6).

[236] Another example of this phenomenon can be found in Jonathan Weiner's book *The Beak of the Finch.* In the book, which purports to show "evolution in our time" (subtitle), Weiner ends Chapter 13 with the words: "The apparent fixity of species once seemed the greatest argument against evolution, just as the apparent fixity of the earth once seemed a commonsense argument against Copernicanism." The reader is led to believe that Weiner is going to dispel "the greatest argument against evolution," and show that (against common sense) new species can be explained, if not documented. In the next chapter, optimistically (and deceptively) entitled "New Beings" Weiner quotes Darwin with respect to "that mystery of mysteries—the first appearance of new beings on this earth" (Ibid., p. 203). However, the reader anticipating learning how "new beings" are created, thus illustrating that the species aren't in fact "fixed" will be disappointed. Weiner simply ends up restating the question: "Why are there so many kinds of animals?" Ibid., p. 207. The best Weiner can do is answer: "Adaptive radiations like Darwin's finches are the essence of the answer" (Ibid.). If "adaptive radiation" such as variation in beak sizes is "the essence" of the answer, then the answer must be that Darwinism *cannot* explain the appearance of "new beings" and the fixity of the species is more than merely "apparent."

[237] Dawkins, *Blind Watchmaker,* p. 166.

Thus, Dawkins implies that he will "dispose" of the stumbling block that natural selection can only destroy. Eagerly reading into chapter 7 (creatively entitled "Constructive Evolution"), however, only confirms the obvious—that natural selection actually does destroy, and it can never construct! Specifically, barely a dozen lines into chapter 7, after noting how a sculptor only creates by "subtracting," Dawkins states: "Natural selection may only subtract, but mutation can add."[238] The rest of the chapter discusses "ways in which mutation and natural selection together can lead ... to complexity."[239] Maybe, maybe not; but in fact Dawkins agrees and confirms that natural selection *can* only subtract, which, at best, is only a euphemism for "destroy" in nature.

[19] *"There seems to be no more design in the variability of the wall figures than in the course which the wind blows."*

This quote (with *organic beings* instead of *the wall figures*) is attributed to Charles Darwin.[240] The full quote is as follows:

> There seems to be no more design in the variability of organic beings, and in the action of natural selection, than in the course which the wind blows.[241]

Charles Darwin, like his modern successors, can see what he wants to see. Quotes such as the one above are not unlike a hypothetical statement uttered by one standing beside Isaac Newton, gazing at the sky, and commenting on what "seems" to be:

> There seems to be no more design in the path of the sun about the earth, and in the action of angels to push it around its path, than in the course which an apple falls from a tree.

---

[238] Ibid., p. 169. The analogy to a sculptor is revealing. A sculptor is clearly an intelligent agent, and the sculpture is clearly an intelligently designed creation (unless, by Dawkins' earlier reasoning one does not know, or know of, the sculptor, in which case his logic must compel him to deny true design; the sculpture can only be apparently designed; perhaps it is a designoid). If Dawkins wants to believe that nature "sculpted" current specified complexity by subtracting in an intelligent manner like a human sculptor, he is welcome. Just as there is no law against relying on bad logic, there is no law against believing wrong conclusions.

[239] Ibid.

[240] Wells, *Icons*, p. 202.

[241] Ibid.

Newton would look, perhaps kindly, on the fellow and think that it is not what "seems" to be that matters. What is important to scientific discovery is to propose testable explanations based on observable events. Although Darwinists think this is what they are doing, in fact most of contemporary evolutionary theory is based on what "seems" to be to those who cannot see any alternative.

To see anomalies within an existing paradigm is actually a rare event in science.[242] The history of science is full discoveries, which, according to Thomas S. Kuhn, come about only with the awareness (and acknowledgement, not the dismissal) of anomalies. Kuhn states:

> Discovery commences with the awareness of anomaly, i.e., with the recognition that nature has somehow violated the paradigm-induced expectations that govern normal science.[243]

Importantly, Kuhn notes that until the anomalous is assimilated into the paradigm such that it has become the expected, the relevant theory will remain in crisis.[244] As Kuhn observes:

> Assimilating a new sort of fact demands a more than additive adjustment of theory, and until that adjustment is completed—until the scientist has learned to see nature in a different way—the new fact is not quite a scientific fact all.[245]

Kuhn's point is insightful. If a new fact is "not quite a scientific fact," what kind of "fact" can it be? In the area of origins research anomalous facts, if they are acknowledged at all, are incorporated into the theory not as scientific facts, but as philosophically beholden "facts" that prevent the scientist from ever learning to "see nature in a different way." Darwinists simply cannot see the world in any way except through the eyes of a philosophical naturalist, so that the anomalous remains a mystery. A mystery for a philosophically based

---

[242] The term "paradigm" is defined as "a philosophical and theoretical framework of a scientific school or discipline within which theories, laws, and generalizations and the experiments performed in support of them are formulated" Merriam-Webster Online, http://www.m-w.com/. It is entirely appropriate to speak of "the current paradigm of evolutionary science" as did Ernst Mayr in his book *What Evolution Is*. *See*, Note 7 above.

[243] Thomas S. Kuhn, *The Structure of Scientific Revolutions* (Chicago: The University of Chicago Press, 1996), p. 52.

[244] Ibid. *See, e.g.*, Chapter VII: "Crisis and the Emergence of Scientific Theories."

[245] Ibid., p. 52.

theory is more tolerable than an explanation that implicates a rival theory. Philosophically based theories are not necessarily wrong if acknowledged as such, but the ultimate explanatory power of a theory is only as great as the number of explanations its starting assumptions will permit.

[20] *"... the truth is not changed by what anybody thinks about it."*

In today's post-modern world, it is often assumed that "truth," like beauty, is in the eye of the beholder. The beauty of truth, however, is that it is what it is, period. No amount of "clever cogitation," to use Gould's term, can change what *is*.

Consider a typical courtroom scene where a plaintiff and a defendant are disputing the events of an automobile accident. Both sides present their respective versions of an actual, objective, historical happening. As each side argues its case, one or the other may be more persuasive and win, but the outcome of the case does not alter the facts underlying the dispute. It may be that the jury wishes to have one side or the other win for reasons that have little to do with the underlying facts. But no amount of "believing" that one side or the other is being truthful changes the actual truth of the matter.

A real-life historical example illustrating this truth is the familiar dispute involving the Catholic Church and Galileo over the earth's motion in our solar system, as discussed in Note 12 above. As each side strove to make its case, one side using totalitarian tactics against the other's incorrigibleness, the earth continued its peaceful path about the sun, undisturbed. The truth of the earth's motion was unchanged by what either side thought about it.

The cause of our human origins is, of course, an actual, objective, historical happening. The happening was either a completely natural, chance process, or it was guided (at least in part) by intelligence. There is no other option, and the truth of our actual origin is unchanged by what anyone thinks about it.

[21] *"The truth," Jenny practically shouted the word truth, "is that a painting needs no more evidence of a painter than a watch needs evidence of a watchmaker!"*

This statement recalls William Paley's famous statement in his 1802 book *Natural Theology* that a "watch must have had a watchmaker" who "designed its use." The phrase was also the basis for Richard Dawkins' best-selling book, *The Blind Watchmaker*, in which

Dawkins attempted to show, according to the book's subtitle, why the evidence of evolution reveals a universe without design.[246]

In studying intelligent design theory, it is important to distinguish between various branches of arguments asserting design. In espousing one line of argument, Paley wrote that "a watch must have had a maker" who "designed its use," and so the purpose of things often reflected the wisdom and beneficence of God.[247] Thus, Paley was arguing for design as a proof for God's existence, God being understood as the deity of historical Christianity. As Larry Witham states in his book *By Design*, this position is obviously a premise of some design theorists, just as the Darwinians usually begin with the premise that God does not exist or work in nature.[248] However, it is not the premise of all intelligent design theorists.

Another line of intelligent design theory avoids any suggestion of who or what the designer might be, and stays away from any discussion of purpose in design. In an effort to propose a purely scientific approach, such intelligent design theorists simply posit that design can be detected, and design is evidence of a designer. Quoting Michael Behe, Witham continues:

> This is object-based rather than grand-purpose-based design. We sometimes think we see the purpose, but it's the purpose of the machine. The bacterial flagellum has a purpose: to propel things through the water. What greater purpose the designer intended for the machine is not something we look at.[249]

In any event, as Witham observes, in trying to frame a scientific concept of design, the intelligent design movement has repeated a few basic themes over and over. One, and likely the most important, is that in science the question is not between finding natural causes or supernatural causes, but between natural or intelligent ones.[250] Thus, whether the watchmaker is natural or supernatural can be debated; but regardless, what can be determined scientifically with certainty is that the watchmaker is intelligent.

[22] *"On the Origin of Specialties."*

---

[246] Dawkins, *Blind Watchmaker*, title.
[247] Witham, *By Design*, p. 141.
[248] Ibid.
[249] Ibid., p. 142.
[250] Ibid., p. 143.

Everyone interested in understanding Darwinian evolutionary theory should read Charles Darwin's *On the Origin of Species*. Easy to read, the book clearly lays out Darwin's theory and gives the reader understanding and insight from the book that brought biological evolutionary theory to the forefront and continues to be a key reference in the field. Although many books have come along since Darwin's book was published in 1859, the basics of his original theory remain intact, and his book remains authoritative. According to the National Academy of Sciences, "Darwin's original hypothesis has undergone extensive modification and expansion, but the central concepts stand firm." [251]

As mentioned earlier, a fair reading of *On the Origin of Species* leaves the reader with a sense that Darwin was an honest scientist. His careful, balanced approach to the topic he knew was controversial is refreshing to experience. In fact, much of his book is dedicated to dealing with "difficulties" with his theory. In one sense he was the first advocate of "teaching the controversy."[252] For example, the first "difficulty" he dealt with was the fact that the fossil record did not support his theory of gradual change over time.[253] If current teaching standards would simply follow Darwin's approach, contemporary science curriculum would be much more palatable to those who find Darwin's "difficulties" particularly difficult.

[23] *"... you have no right to force your will on us – you cannot legislate morality!"*

In a nation where the rule of law regulates activities, virtually *all* legislation is directed at moral conduct. In a democracy, laws dictate the will of the people as to what "ought" to be. Thus, speed limits legislate the "morality" of proper highway behavior. Criminal laws dictate what is clearly understood to be unacceptable moral behavior. Regulations on corporations restrict the behavior of corporate executives. All such laws are the imposition of the will of the majority onto the whole. Therefore, the idea that you cannot legislate morality is simply unfounded. People who say such things are usually those who would like to behave contrary to the imposed morality of others.

What is really at issue is the *legitimacy* of the imposed restriction on behavior, which, in turn involves the *basis* for

---

[251] NAS, *Science and Creationism*, p. 10.

[252] "Teaching the controversy" has come into use as a term to describe the idea of at least letting public school students learn the evidence both for and against Darwinism. That Darwinists would object to such an idea with forceful religious fervor should be proof enough of the need for such an approach.

[253] Darwin, *Origin*, p. 172. Darwin attributed the lack of transitional varieties to the "imperfection of the geological record." The interested reader should investigate whether the record is any more "perfect" after 150 years.

determining the "rightness" or "wrongness" associated with the behavior. In a materialistic world (i.e., matter is all that exists) there is no transcendent basis for any moral judgment, and, therefore, legitimacy can always be questioned.

Whether or not we live in a purely materialistic world is a subject for philosophy, but without a doubt Darwin's theory of evolution helped propel Western society into the current post-modern culture that is alternately described as "humanistic" or "secular." Richard Dawkins, for example, boasts that "Darwin made it possible to be an intellectually fulfilled atheist."[254] In the Introduction to the Facsimile of the First Edition of Darwin's *On the Origin of Species*, leading evolutionist Ernst Mayr states that, although Darwin was no philosopher:

> Yet few writers in the last 200 years have had so profound an impact on our thinking. This holds for logic, metaphysics, and *ethics*. It has taken 100 years to appreciate fully that Darwin's conceptual framework is, indeed, a new philosophical system.[255]

Of course we should enquire as to whether this "new philosophical system," that impacts "ethics" is the best system. If making atheists "intellectually fulfilled" is a noble objective, then Darwin's system may be the right means to that end. But what if atheists are wrong? It is doubtful that most atheists would feel very "fulfilled" if they knew that their basis for knowledge rested on false premises. To be sure, the truth (at least the validity) of a proposition is only as good as its underlying premises. Therefore, when seeking legitimacy of ethical systems, the underlying premises, or starting point, of the inquiry must be carefully examined.

Letting atheists set the agenda in ethics without examining their premises in the light of reason is risky business. No less dangerous is letting atheists set the agenda in science in areas that bear directly on ethics. One should be wary of an unexamined agenda, particularly from an intellectually fulfilled practitioner that resists and resents all such examination.

[24] *"Do you deny that there is another person whose voice, were it to be considered, might have a very different opinion in the matter?"*

A logical implication of evolutionary thought is that humans are merely one of all the various species on earth, no more and no less

---

[254] Dawkins, *Blind Watchmaker*, p. 6.

[255] Darwin, *Origin*, Introduction, emphasis added.

special than say, an aardvark. Those misguided souls who were under the impression that humans are somehow different, maybe even in some way superior, are either shallow humanists, reluctant atheists, or, worst of all, religious creationists. Darwinists have coined a term for those so unenlightened: *speciest.* As Richard Dawkins writes in his book *a devil's chaplain,*[256] the term "speciest" applies to those to whom "it is simply self-evident, *without any discussion,* that humans are entitled to special treatment."[257] The term is intended to reflect an analogy to racism.[258]

How might a speciest behave? To illustrate, Dawkins relates a hypothetical scenario in which one is appealing for money to "save the gorillas" in Africa, and another replies suggesting that the appeal is misplaced—the money could be better spent on the "*human* children suffering" on the same continent.[259] Dawkins does not deny that a good case could not be made for the latter cause, but he challenges the effort, "trying to point the finger at the *automatic*, unthinking nature of the speciest double standard."[260] Thus, the "speciest double standard," with which most of us are no doubt afflicted, is evidence of "the speciest assumption" that "lurks" within us, leading us to conclude that "[h]umans are humans and gorillas are animals." [261]

The ruthless consistency Dawkins exhibits when taking his atheistic, evolutionary philosophy to its logical conclusion is admirable. If his philosophy is right, then he is also right that humans are mere *animals*, whose worth is determined no differently than would be the worth of an aardvark, which worth, according to Dawkins "is just its replacement cost to its owner—or, in the case of a rare species, to humanity."[262]

It should not be surprising, then, that the issues in the abortion debate—another polarizing issue in the United States—are closely related to the issues raised in the evolution controversy. Specifically, if the evolutionary assumptions are correct, and purposeless evolution is the cause of our existence, then humans truly are an accident of nature

---

[256] Richard Dawkins, *a devil's chaplain* (New York: Houghton Mifflin Company, 2003), p. 21.

[257] Ibid., p. 20, emphasis in original.

[258] Ibid., p. 21.

[259] Ibid., p. 20, emphasis in original.

[260] Ibid., p. 20, emphasis in original.

[261] Ibid., pp. 20-21.

[262] Ibid., p. 21. Thus, if gorillas were a rare species, their worth could be far more than that of human children, since, unfortunately for the children, their species is not rare. It is somewhat amusing that Dawkins finds "replacement cost to humanity" to be a valid basis for valuation. Such a statement reflects pure speciesm at its worst—making human-based value judgments about other species. Indeed, Dawkins fails to give any reason why worth to "humanity" should be a consideration. What is humanity, but an extension of an animal line? Why should "humanity" be given any more weight than "gorilla-ity" or "aardvark-ity"?

to be accorded no more respect than, say, a worm, a tern, or a fern. In such a case there is no moral imperative of any kind with respect to human life, much less an obligation to treat human life as sacred (or even special). Again taking the example of humans and gorillas, Dawkins finds it odd that:

> There is an unquestioned yawning gulf between them such that the life of a single human child is worth more than the lives of all the gorillas in the world.[263]

Specifically targeting the abortion issue, he states further with respect to the worth of an animal's life:

> But tie the label *Homo sapiens* even to a tiny piece of insensible, embryonic tissue, and its life suddenly leaps to infinite, uncomputable value.[264]

Such speciest thinking is not only wrong, according to Dawkins, but it characterizes what he calls the "discontinuous mind" which "is especially influential when it afflicts lawyers and the religious."[265] The discontinuous mind chooses to divide animals up into discontinuous species. But according to Dawkins, "on the evolutionary view of life there must have been intermediates, even though, conveniently for our naming rituals, they are today usually extinct."[266]

What "must have been" is the question, not the answer, and Dawkins' conclusions as to the worth of human babies is inextricably bound to his assumption that we have a common ancestor with the gorilla (as well as the other African apes, such as the chimpanzee and the pigmy chimpanzee).[267] He finds the "discontinuous gap between

---

[263] Ibid., p. 21.

[264] Ibid.

[265] Ibid.

[266] Ibid. Note as a point of interest some logical discontinuities in Dawkins' thought. First, it is not anyone's mind, discontinuous or otherwise, that "divides animals up into discontinuous species." Species are, by accepted definition, "discontinuous" and distinct. That is the fact to be explained and it is the evolutionary mind that seeks to prove they are not truly distinct, which it has so far been unsuccessful in doing. Second, saying that intermediates "must have been," begs the question. There is no proof (as Dawkins admits) of intermediates – therefore to say they "must have been" is an unproved proposition that assumes the truth of the "evolutionary view of life" premise. Therefore, not surprisingly, it is Dawkins' "evolutionary view of life" that is a result of a discontinuous mind, not vice versa.

[267] Ibid., p. 23.

humans and 'apes' that we erect in our minds" to be "regrettable."[268] Such a gap is simply "the result of an evolutionary accident."[269]

Does such thinking really matter beyond being an academic question? Dawkins concludes his causerie aptly titled "Gaps in the Mind" with the following:

> Ethical principles that are based upon accidental caprice should not be respected as if cast in stone.[270]

Dawkins' book is dedicated to his daughter for "her Eighteenth Birthday."[271] Presumably, were his daughter found being unexpectedly attacked by African gorillas Dawkins would have pause for concern as he mentally considered his options to settle this family dispute. First, he would necessarily consider the worth of the respective animals. On the one hand, the gorillas may be members of a rare species—worth a great deal to humanity. On the other hand, as much as he would resist the speciest assumption, the life of that other animal that once was a tiny piece of insensible, embryonic tissue, would suddenly leap to infinite, uncomputable value—worth far more than the lives of all the gorillas in the world. Yielding to the speciest impulse, he may even call out for others to help him save the life of one who is now helpless in the hands of what he suddenly recognizes as brute beasts. Would anyone come to his aid? Maybe. At least if any speciests happened to be around. And, of course, if it's legal. And only if one's ethical principles permit such an intrusion into another's private affairs. Maybe if his health was also in danger the choice would be easier. Yes, maybe.

Or, maybe not.

[25] *"For her part, Alex was finding that the college experience was helping her to be more open-minded and tolerant of all ideas. Who was she, she thought, to say what is right or wrong, good or bad, truth or falsehood?"*

As stated by Paul Lawrence Farber in his book *The Temptations of Evolutionary Ethics*:

---

268 Ibid.

269 Ibid., p. 26.

270 Ibid.

271 Ibid., p. vii.

> The theory of evolution developed at a critical time in the history of Western ethics. In the decade following Charles Darwin's publication of the *Origin of Species* (1859) philosophers, clergymen, scientists, literary figures, and other intellectuals faced the problem of trying to establish a new foundation for ethics.[272]

Darwin himself attempted to explain the origins of morality in his book *Descent of Man* (1871). But, as Farber explains:

> [Darwin's] ... efforts [as well as those who attempted to extend his ideas] consisted of fashioning evolutionary justifications for accepted moral principles, but they did so in such a transparently contrived fashion that their failure to gain acceptance is understandable.[273]

It should not be surprising that any purely naturalistic approach to ethics should be "transparently contrived." What other option is there except to contrive an approach, other than possibly "opaquely" contrived? In general, Darwinists acknowledge the value of deeming some behaviors or ideas as "right" and others as "wrong," and they usually assume that such ideas are simply products of communities, not concepts derived from transcendental truths. But even this justification for accepting that notions such as cruelty and falsehood are "wrong," and mercy and truth are "right" fails to address the real issue, which is *why* society should even embrace such things as right or wrong? Why should there even be the concept of right and wrong? Isn't it almost always the case that one person's right is another person's wrong? Shouldn't the evolutionary forces of survival sort out the relative merits of each in fostering "better," "fitter" organisms?

One evolutionist that attempted to answer the question of how to scientifically determine right and wrong was the nineteenth century intellectual Leslie Stephens. Stephens embraced Darwin's new ideas and realized that central to this new worldview was a new system of ethics to replace the religious sanctions that for centuries had provided a guide for human conduct.[274] According to Farber:

---

[272] Paul Lawrence Farber, *The Temptations of Evolutionary Ethics* (Berkeley: University of California Press, 1994), p. 6.
[273] Ibid.
[274] Ibid., p. 26.

> [Stephens's work] ... represents the most developed elaboration in the nineteenth century of the Darwinian position on ethics, and it best reflects the internal problems its supporters faced.[275]

Stephens recognized that even if certain notions of right and wrong were not controversial, nevertheless, their *justification* remained at issue. To Stephens, it was not so much the matter of morality, but the form of morality that needed explaining. In the end, Stephens had to conclude:

> But if we turn from the matter to the form of morality; if, instead of asking what actions are right or wrong, we ask, what is the essence of right and wrong? how do we know right from wrong? why should we seek the right and eschew the wrong?—we are presented with the most contradictory answers; we find ourselves at once in that region of perpetual antinomies, where controversy is everlasting, and opposite theories seem to be equally self-evident to different minds.[276]

"Opposite theories" that "seem to be equally self-evident to different minds" is to be expected when the only source for such theories is "different minds" in the first place. Farber closes his book on the history of evolutionary ethics with the following conclusion:

> In coming decades philosophers may judge the construction of ethical systems to be impossible. The entire field of ethics itself may come to be regarded as a pseudo-subject like astrology. But perhaps a different approach in ethics will prove to be fruitful. Historians, however, should not try to predict the course of philosophy, or they cease to be historians. At best they can point out a century-long frustration over the state of ethics and in the case of evolutionary ethics, the significantly interesting but dismal track record of this approach as a method of resolution.[277]

---

[275] Ibid., p. 27.
[276] Ibid., pp. 28-29, quoting from Leslie Stephens, *The Science of Ethics* (London: Smith, Elder and Co., 1882).
[277] Ibid., p. 175.

Indeed, all attempts will be impossible. Trying to explain mercy and truthfulness as being a result of purely natural processes subject to evolutionary forces of survival is like trying to explain the Mona Lisa as being the product of wind, rain and erosion. Even though intelligent design theory does not posit a particular transcendent deity, such a deity is a reasonable explanation for seemingly transcendent ideas of right and wrong. What Darwinists must admit is that, for all their drawbacks, at least religious systems give some basis for objective morals. As Henry W. Longfellow stated: "Morality without religion is only a kind of dead reckoning,—an endeavor to find our place on a cloudy sea."[278] Intelligent design, while not religious or beholden to any religion, is at least amenable to religious thought.

[26] *"Why isn't a painting evidence itself of a painter?"*

As discussed above, evidence of design must be explained away by those seeking to convince the public that design is in reality nothing but mere "apparent" design. But what appears to be design *is* design, and it is evidence itself of a designer. Efforts to explain design as being illusory are in themselves unwitting admissions that design can be detected. The argument is essentially defined by what is self-evident, yet denied. Because Darwinists are fixated on what they deny, perhaps their position should be termed *adesignism*, in the spirit of *agnosticism* and *atheism*.[279]

Given that both Darwinists and intelligent design theorists acknowledge that design can be detected, the issue becomes which is more reasonable, to attribute design to a designer, or to deny design, call it *apparent* design, and attribute it to a long series of cumulative random mutations, some selected, some not? In any other context, the answer to that question would be self-evident. For example, as discussed above, an archeologist examining an intricate bone carving would never dream of attributing the design to an accumulation of random errors of nature, even if there was a mechanism for selection. Common sense would rightly reject the idea as being effectively

---

[278] Noebel, *Understanding the Times*, p. 222.

[279] *See*, Roddy M. Bullock, *Proposing a New Lexicon for Understanding in the Evolution/Intelligent Design Debate*, http://idnetohio.cinti.net/pages/adesignism.php, 2005. "Many things observed in nature appear to be designed. But because true design necessarily requires intelligence, the question of nature's appearances is simply whether or not such observed designs are true, intelligent *design*, or are apparently-designed-yet-actually-unintelligently-caused *occurrences*. Because there is no English word for "apparently-designed-yet-actually-unintelligently-caused" with respect to observed events and objects, I propose the term *adesign*. Adesign is the absence of design. Adesign leads to adesignism, which, like its linguistic counterparts atheism and agnosticism, is a belief system defined by what it denies. The new terminology creates a clear contrast between design and adesign and captures precisely the heart of the origins inquiry: nature exhibits either some design requiring intelligence, or all adesign requiring no intelligence. There is no other logical choice."

impossible. But when evolution is the topic, common sense is turned on its head; the self-evident becomes mysterious, design becomes apparent design, and the impossible becomes the inevitable.

[27] *"The question that concerns scientists is not* whether *the markings were made by natural processes, but* how."

Richard Dawkins starts Chapter 11 of his book *The Blind Watchmaker* by boldly proclaiming:

> No serious biologist doubts the fact that evolution has happened, nor that all living creatures are cousins of one another. Some biologists, however, have had doubts about Darwin's particular theory of *how* evolution happened.[280]

Thus, with characteristic snootiness, Dawkins relegates any doubters to the non-serious crowd. But he is not alone. Eminent neo-Darwinist Ernst Mayr states:

> [T]here is probably no biologist left today who would question that all organisms now found on the earth have descended from a single origin of life.[281]

At least Mayr moderated his fatuousness with the qualifier "probably." Others, however, dispense with such patronizing pleasantries, not even giving some biologists, serious or not, any benefit of the doubt. According to the National Academy of Sciences:

---

[280] Dawkins, *Blind Watchmaker*, p. 287, emphasis in original. The reader should carefully consider these words. What Dawkins is saying is exactly what Darwinism says: *all* living creatures are cousins of one another. But far from simply imagining that humans descended from a common ancestor of the ape, this means that humans are "cousins" with crickets, dandelions, worms, fungi, stink bugs, sponges and apple trees. Think about it this way: Darwinism states that if you trace back each of your generations, i.e., parents, grandparents, great grandparents, etc., etc., far enough, you will eventually get to a *common ancestor* with, for example, seaweed. You and seaweed share a common great, great ... great grandparent. To doubt such an outlandish proposition is not to make a so-called "argument from personal incredulity"—it is simply to reject absurdity.

[281] Ernst Mayr, *One Long Argument* (Cambridge, MA: Harvard University Press, 1991), p. 24, quoted in Wells, *Icons of Evolution*, p. 31.

> Scientists no longer question whether descent with modification occurred because the evidence supporting the idea is so strong."[282]

Therefore, according to the NAS, without question Darwinian evolution happened and is responsible for the variety of life on earth. When answering the related question "Don't many scientists reject evolution?" the NAS replied with a curt "No."[283] According to the NAS, the only thing left for scientists to argue about is *how* evolution takes place.[284]

The NAS's assertion that *no* scientists reject evolution is simply and inexcusably wrong. Such an offhand dismissal betrays the supercilious tunnel vision by which dogmatic Darwinists view the world. A casual review of the relevant literature reveals many, many scientists who question evolution in general and Darwinism in particular. For instance, the Seattle-based Discovery Institute has published a paper entitled "A Scientific Dissent From Darwinism" that lists over 500 scientists from universities and research institutes around the country that are skeptical of the claims of Darwinism.[285]

In fact, there are many scientists (and even more than a few serious biologists) concerned with the veracity of the theory of evolution. Many of these men and women probably harbor strong doubts but think they alone see the emperor with no clothes. But as more and more teachers, professors, and fellow scientists hear others boldly expose the ruler's nakedness, no doubt they, too, will raise their voices in asking *whether* the king has clothes, not *how* his clothes were made.

[28] *"Why? ... To authenticate their psychological wish projections? Or maybe to personalize the unknown forces of nature in a misguided attempt to give some certainty in an uncertain world?"*

Sigmund Freud, considered to be the founder of psychoanalysis, was born shortly before Darwin published his theory. Freud attempted to account for religion's origin, particularly since atheism's claim that there is no God seemed to be carrying the day. The

---

[282] NAS, *Teaching About Evolution*, p. 56. The NAS should be given credit for at least not using the meaningless term "evolution" in such a bold statement. "Descent with modification" is not generic "change over time" but is specifically Darwinism.
[283] Ibid. Interestingly, this question is one of about a dozen in a section entitled "Frequently Asked Questions About Evolution and the Nature of Science," and it is the only question answered forthrightly. The remaining "answers" are studies in carefully-worded non-answers intended to convey meaning without technically being inaccurate.
[284] Ibid.
[285] *See*, http://www.dissentfromdarwin.org.

most common explanation at the time for the widespread phenomenon of religion was that it is rooted in a deep psychological need or wish-projection.[286] Freud concluded that the chief task of civilization was to protect and defend us against nature. But nature seems to mock human control. Therefore, to create a defense against the forces of nature, said Freud, one must personalize them.[287] The personalized powers of nature then become objects of religious devotion.[288]

Freud's thoughts on the matter continue to influence atheists and humanists (usually the same thing) today. One prominent humanist insisted:

> A Humanist has cast off the ancient yoke of supernaturalism, with its burden of fear and servitude, and he moves on the earth a free man, a child of nature and not of any man-made gods.[289]

By denying God and the supernatural, humanism exalts man as the supreme authority. Even Darwinists, to whom man is simply a higher animal, recognize that man is at least unique in his level of consciousness.[290] This uniqueness, even if accidental and purposeless, places man in a preeminent position among all the conglomerations of atoms in the cosmos. The implications of this line of thought should be carefully studied by anyone leaning toward such a view. For one, Humanists themselves recognize that the logical endpoint of such a

---

286 R. C. Sproul, *The Consequences of Ideas* (Wheaton, IL: Crossway Books, 2000), p. 194.
287 Ibid.
288 Ibid., p. 195.
289 Blanche Sanders, *The Humanist*, no. 5, 1954, p. 226, quoted in David A. Noebel, *Understanding the Times* (Eugene, OR: Harvest House Publishers, 1991), p. 61.
290 For example, in March 2000 the National Academy of Sciences sponsored a "Colloquium on The Future of Evolution." Only humans were invited. The published Proceedings include topics related to the impact humans are having on evolution, and what they should do about it. (PNAS, May 8, 2001, Vol. 98, No. 10, pp. 5477-5480; available at www.pnas.org). One topic, presented by Paul R. Ehrlich, Department of Biological Sciences, Stanford University, was entitled, "Intervening in evolution: Ethics and actions." Ehrlich's contention is that "There is no question that *Homo sapiens* … is … altering the course of evolution for millions of years in the future" (Ibid., p. 5477). What should we then do? The paper starts with the following declaration: "Biologists should help to guide a process of cultural evolution in which society determines how much effort, if any, is ethically required to preserve options in biological evolution" (Ibid.). *Biologists* should guide *cultural* evolution? Why should we entrust "cultural evolution" to biologists? For that matter, what is "cultural" evolution, and how might one go about guiding it? Perhaps by limiting public discourse on a particular topic, such as intelligent design? Thankfully, we (in the United States) still live in a free country where such claptrap can be both delivered with airs of significance and then (or perhaps simultaneously) politely ignored.

view is a life without meaning or purpose. As one humanist theologian[291] declared:

> [A]theism can offer no hope of personal immortality, no threats of divine chastisement, no promise of eternal recompense for injustices suffered, no blueprints to sure salvation ... A tragic view of life is thus an uneliminable ingredient in atheistic thought.[292]

Secondly, and related to the first point, atheistic, humanistic philosophy is "monistic." That is, it denies that the mind, or consciousness, is anything more than a purely natural phenomenon—molecules arranged just so. There being no part of humans to survive the body's demise, such as a soul, the end is the end and there is no afterlife.

Humanists can believe whatever they want, but here is the point to consider: as always, the truth of the matter is not changed by what anyone thinks about it. *If* humanists are correct in their thinking, then any fuss about "religion" hardly matters. But no amount of believing on their part will change the truth if, in fact, they are wrong.

[29] *"To do so would be tantamount to belief in fairies."*

This statement is adapted from a quote attributed to 1979 Nobel laureate Steven Weinberg. At "The Nature of Nature" conference in 2000, where atheists and theists gathered for three days of debate about God and science, atheist Weinberg said that "whatever naturalism is ... it is better than religion, which is tantamount to belief in fairies."[293] Similar sentiments are echoed from the highest orders of Darwinians. Richard Dawkins, for example, always a source of honest atheistic spouting, has stated things such as the following:

> Faith is the great cop-out, the great excuse to evade the need to think and evaluate evidence. Faith is belief in

---

[291] Humanists do have a theology. For example, *Critiques of God*, published by Prometheus Books in Buffalo, New York, contains 371 pages supporting humanist theology's denial of the existence of God. *See*, Noebel, *Understanding the Times*, p. 57.

[292] Ernest Nagel, "Philosophical Concepts of Atheism," *Critiques of God* (Buffalo, NY: Prometheus Books, 1976), p. 17, quoted in Noebel, *Understanding the Times*, p. 63.

[293] *See*, Witham, *By Design*, p. 153. According to Witham, Weinberg had said a few years earlier, "I am all in favor of a dialogue between science and religion, but not a constructive dialog" (Ibid.).

> spite of, even perhaps because of, the lack of evidence.[294]

> If people think God is interesting, the onus is on them to show that there is anything there to talk about. Otherwise they should just shut up about it.[295]

Interestingly (although a true believer would no doubt disagree), these two statements could be made with equal force by substituting "Evolution" for "faith" and "God," respectively. Evolution is "believed" in spite of any evidence beyond microevolution, i.e., change below the species level. And Darwinists have been remarkably successful in convincing the public that things they can only imagine are worthy of talking about. (Of course, Darwinists are not as successful in convincing the public such clever cogitations are worth *believing*).

All in all, Dawkins considers it not "too melodramatic to say that civilization is at war. It is a war against religious bigotry."[296] That is, one supposes, a war between atheist bigots and religious bigots. But not every atheist is a bigot, and not everyone in the evolution-intelligent design debate is interested in being at war. Nevertheless, Dawkins, who claims to "care about what's true," because he's "in the business of science,"[297] is troubled because:

> [T]here's a lot of ignorance about, and that's OK, we have to live with that. I see religion, however, as a kind of organised misconception. It's millions of people being systematically educated in error, told falsehoods by people who command respect.[298]

Once again, consider the quote above, substituting "evolution" for "religion," and the quote more accurately represents the truth.

---

294 Attributed to Richard Dawkins, from http://www.positiveatheism.org/his/quotes/dawkins.htm.

295 Ibid. Admittedly, and unfortunately, the source of these last two quotations cannot be confirmed. However, they come from a website billed as "Positive Atheism's Big List of Richard Dawkins Quotations," which is a website sympathetic to, and approving of, Dawkins. Certainly, the tone and content of the quotes is consistent with Dawkins' expressed views on religion and God.

296 Richard Dawkins, review of *Blueprints* by Donald C. Johanson and Maitland A. Edey, *New York Times*, April 9, 1989, sec. 7, p. 34.

297 Simon Hattenstone, "Darwin's Child," *The Guardian*, February 10, 2003.

298 Ibid.

[30] *"He wondered what motivation she might have for not readily embracing the only theory accepted by scientists such as himself."*

Many Darwinists simply cannot comprehend that honest, scientifically minded people could doubt the "fact" of evolution. What is incredible, however, is the arrogant, haughty attitude that takes offense at any criticism of evolution as being from simple-minded "cretinists," to use a term of derision suggesting that creationists are cretins.[299] That such drivel would emanate from sources such as internet postings can be expected, but terms of derision are cast about by the leaders in the field, in surprisingly self-aggrandizing fashion. For instance, in *The Blind Watchmaker*, evolutionist Richard Dawkins refers to "those who seek to deny the truth of evolution" as "backwoodsmen,"[300] and "redneck" creationists.[301] According to Dawkins, such backwoodsmen only "pretend" to "scientific credentials."[302]

In an effort to downplay a significant challenge to Darwinian evolutionary processes from *within* the fold of naturalism, i.e., Gould and Eldridge's theory of "punctuated equilibrium,"[303] Richard Dawkins pontificates on the only reasons anyone could possibly *want not to* "believe in Darwinism." In *The Blind Watchmaker*, he states:

> There are people in the world who desperately want not to have to believe in Darwinism. They seem to fall into three main classes. First, there are those who, for religious reasons, want evolution itself to be untrue. Second, there are those who have no reason to deny that evolution has happened but who, often for political or ideological reasons, find Darwin's theory of its *mechanism* distasteful … Third, there are people, including many working in what they call (often as a singular noun) 'the media', who just like seeing applecarts upset, perhaps because it makes good journalistic copy; and Darwinism has become sufficiently established and respectable to be a tempting applecart.[304]

---

[299] The term can be found on internet postings on the topic.

[300] Dawkins, *Blind Watchmaker*, p. x.

[301] Ibid., p. 251. The exact quote refers to two prominent Darwinists who have been "doughty champions in the fight against redneck creationism."

[302] Ibid., p. xi.

[303] For further information on Gould and Eldridge's theory, *see*, Note [43].

[304] Ibid., pp. 250-1, emphasis in original.

To summarize, Dawkins can conceive of only three reasons why one might "want not to have to believe in Darwinism": (1) religious reasons; (2) political or ideological reasons; and, (3) commercial reasons (i.e., to upset applecarts to sell journalistic copy). Apparently, then, it never enters Dawkins' well-exercised imagination that there might be people who simply find the scientific justification for Darwinism lacking, or, similarly, that there are people who find little evidence for Darwinism's claims of "atoms-to-Adam" human development.[305]

To not even acknowledge that anyone, much less credentialed scientists, could have legitimate doubts to the point of "not believing" in Darwinism smacks of blind conceit. Such egotism is actually a good sign for intelligent design theorists. History shows that the emergence of new scientific theories is generally preceded by a period of pronounced professional insecurity.[306] One defense mechanism for professional insecurity is self-serving, egotistical vanity. But vanity can put on airs only for so long; and well-guarded apple carts become irrelevant when, by preventing examination of the fruit, they are relegated to being a curbside curiosity.

[31] *"Chance plays a part, Jenny," Dr. Seeting said, seemingly tireless in his refutation of what he referred to as "that old canard." "We must have some process for fresh variation of lines and spots of color. Marks must be put on the wall somewhere in an ever changing variety. But the randomness of mark variation must be contrasted with the non-randomness of selective inclusion, the other side of the process. It is not necessary that variation be random in order for natural selective inclusion to work. Selection can still do its work whether variation is directed or not. Emphasizing that variation can be random is our way of calling attention to the crucial fact that, by contrast, selective inclusion is sublimely and quintessentially non-random."*

This lengthy passage is adapted directly from Richard Dawkins' book *Climbing Mount Improbable*.[307] In this passage Dawkins once again seems to be attempting to convince the reader that Darwinism is not a chance process. By pressing reluctant words into unwilling sentences that are strung together into a strained paragraph, he tries to get the unwary reader past the point and onto other more interesting topics. Like a theologian navigating a tricky bit of theological whitewater, Dawkins takes liberties with language and ideas, no doubt hoping the first-timers will close their eyes and hang

---

305 The term "atoms-to-Adam" is set off in quotes for emphasis, not to attribute the quote to Richard Dawkins.

306 *See*, Kuhn, *The Structure of Scientific Revolutions*, p. 67-68.

307 Dawkins, *Climbing Mount Improbable*, p. 80.

on while the true believers simply ignore the foamy spray and ride out the rapids.

Dawkins starts by declaring that "Darwinism is widely misunderstood as a theory of pure chance."[308] In answering if Darwinism did anything to "provoke this canard," Dawkins even admits that, "Well, yes, there is something behind the misunderstood rumour, a feeble basis to the distortion."[309] Thus, Dawkins believes that the fault lies with a "rumor" that has been "misunderstood," which has led to a "distortion." Then, remarkably, Dawkins immediately writes: "One stage of the Darwinian process is indeed a chance process—mutation."[310] Dawkins then goes on with the rest of the discourse from which the above selection was adapted.

Remarkably, it is clear that Dawkins believes, or at least insists the reader believe, that because only one of the two stages of the Darwinian process (mutation and natural selection) is a random, chance event, the entire process is *not* a chance process (i.e., not *pure* chance), and he even leads the reader to believe that Darwinism is *non-random*. He says:

> But Darwinians make the fuss that they do about the 'randomness' of mutation only in order to *contrast* it to the non-randomness of selection, the other side of the process.[311]

Dawkins then, thinks that Darwinians are simply making "a fuss" over so-called "randomness" of mutation just to emphasize how selection is non-random—such a non-random step in the process presumably renders the entire process non-random.

The reader should carefully consider Dawkins' argument on this point. Dawkins is one of the world's leading Darwinists and writer of popular books touting Darwinism. His exposition on this topic is simply another example of the kind of logical sleight-of-ink that goes

---

[308] Ibid. Note the qualifier *pure* as Dawkins seemingly tries to minimize chance as a factor. As noted above, this is not unlike referring to sewer water as "sewage, but not pure sewage." Although few people are saying Darwinism is *pure* chance, it must be acknowledged that a process that has as a critical factor even one chance element, is in effect reduced to pure chance. As alluded to elsewhere, imagine combining one part pure water and one part sewage and claiming it's not "pure" sewage—it may depend on one's definition of sewage, but it is nonetheless hard to swallow. In the case of evolution, such a denial of pure chance is very significant. What else can it be? Directed chance?

[309] Ibid.

[310] Ibid. Dawkins is referring to what is commonly referred to as "neo-Darwinism," in which genetic mutation is the mechanism for heritable change in replicating organisms.

[311] Ibid.

unchallenged by "mainstream" scientists even as they go about enforcing their dogmatic views on the rest of us.

In fact, as noted above, intelligent design theorists should take heart in Dawkins' vigorous denials of Darwinism being a chance process. If Darwinism is not a chance process, even in part, then it must be a guided or directed process. In reality, Dawkins wants to emphasize the non-random character of natural selection to detract from the implications of chance random mutations. As noted above, he admits that if Darwinism is actually a chance process it's easy to refute:

> Since living complexity embodies the very antithesis of chance, if you think that Darwinism is tantamount to chance you'll obviously find it easy to refute Darwinism![312]

Dawkins agrees, therefore, that "living complexity embodies the very antithesis of chance," in a book written to convince its readers that evolution is "blind" and the universe is without design. Apparently Dawkins wants to have his cake and eat it too. He wants a blind watchmaker, but not one that stumbles upon various parts only by chance.

[32] *"The Stanley-Harold Experiment, as it was titled, purportedly demonstrated how the 'building blocks' of cave specialties could have originated naturally."*

In a lecture at the University of Chicago in 1951 professor and Nobel Prize winner Harold C. Urey challenged any student to try to turn chemistry into life by use of a spark of energy. One of his doctoral students named Stanley Miller took the challenge and was surprisingly successful in obtaining some preliminary results in a matter of weeks.[313]

Miller filled a sphere with methane, ammonia and hydrogen to simulate what was thought to be Earth's primitive atmosphere. Simulated lightening in the form of an electrical spark resulted in a reaction that produced an astounding product: amino acids. That such a simple experiment could produce the "building blocks of life" in such a short time gave impetus to a new field of biology called "abiogenesis," the chemical origin of life. References to the Miller-Urey experiment are ubiquitous in evolutionist literature, including high school textbooks, even though scientists recognized today that the

---

[312] Dawkins, *Blind Watchmaker*, p. xv.

[313] Witham, *By Design*, p. 97.

simulated environment used by Miller bears little resemblance to the conditions on the early Earth.

In his book *Icons of Evolution*, biologist Jonathan Wells explores the scientific controversy over early Earth conditions.[314] As noted by Wells, most scientists now think the Miller-Urey experiment failed to simulate actual conditions on early Earth. However, this does not stop the experiment from being prominently depicted in every leading biology textbook from 1998 and later.[315] Wells suggests that biology texts should be required to carry warning labels, such as:

> **WARNING**: The Miller-Urey experiment probably did not simulate the Earth's early atmosphere; it does not demonstrate how life's building-blocks originated.[316]

In any event, whether or not "life's building blocks" can be synthesized under any conditions, it is clear that between amino acids and life there lies an expanse that the "chance and necessity" bridge cannot traverse, no matter how well designed and controlled the experiment. Consider a person one moment after death. Not only are life's building blocks present, but the entire "building" is there. What is missing is "life," whatever that is. Just as there is no scientific way to put life back into a dead "building" of a person, there is no known scientific principle that would suggest one could add life to any supposed building blocks synthesized in a test tube. As evolutionist Paul Davies recognizes in his excellent book *The Fifth Miracle*:

> To be sure, molecular biology has scored some dazzling successes, but scientists still can't quite put their finger on *exactly* what it is that separates a living organism from other types of physical objects.[317]

Therefore, while the Miller-Urey is scientifically interesting, it is not even close to demonstrating what textbooks wish students to believe, i.e., that science can (if it has not already done so) synthesize *life*.

[33] *"What's with the extra, useless body parts?"*

---

[314] Wells, *Icons of Evolution*.
[315] Ibid., p. 249.
[316] Ibid., p. 259.
[317] Davies, *Fifth Miracle*, p. 32.

A frequent criticism of creationism, if not intelligent design, is that living organisms seem to have many "useless" body parts, including so-called "vestigial" organs. Examples include birds that have wings but do not fly, pigs that have two toes that do not touch the ground, and human males with nipples that serve no apparent purpose. The question is posed, why would an *intelligent* designer, much less a divine creator, make such useless parts?

Darwinists use evidence of "vestigial" organs to support evolution because such supposedly useless organs would be expected in an unguided, purposeless process of development. Vestigial organs could be organs that began evolving but never completely evolved, or body parts that formerly had a useful function but are no longer needed as the organism evolves to a different state.

The question of "why" with respect to so-called "useless" body parts is misplaced as far as intelligent design theory is concerned. First, if there was an intelligent designer, he (or she, it, or they), could have had any reason for designing organisms as he did, and he need not justify his actions to us. Second, to deem certain organs as "vestigial" and now useless is somewhat presumptive. There are many, many examples of essential organs previously, and wrongly, thought to be useless.[318]

The larger question for Darwinists, however, is not *why* there are so-called vestigial organs, but *how* they got there in the first place. That is, even if the organs are vestigial, noting this fact does nothing to explain their origin. Darwinists are left with the same question, whether organs are nascent or vestigial: How did they evolve? In fact, vestigial organs, if they are in fact vestigial, are evidence in support of *devolution* rather than evolution.

Therefore, the question of *why* certain questionable organs may have been created can be a red herring in the evolution/intelligent design debate. Although answering the "why" of the matter may help to explain the "how" in an evolutionary context, it is misplaced in an intelligent design context. For intelligent design theorists the "how" is important but the "why" is only ancillary.[319]

---

[318] *See, e.g.*, Henry M. Morris and Gary E. Parker, *What Is Creation Science* (Green Forest, AR: Master Books, 1982) pp. 62-63, stating that "essentially all 180 organs once listed as evolutionary vestiges have quite important functions in human beings."

[319] Ancillary with respect to detecting design, that is. Of course, the "why" of the presence and design of organs can lead to fruitful research, and the why of the creation persons is of immense importance relating to our purpose on earth. For an in depth look at the issue of vestigial organs, *see*, Jerry Bergman and George Howe, *Vestigial Organs are Fully Functional* (Creation Research Society Books, 1993).

[34] *"Is that the work of an artist, or a lowly greasepaint monkey? It appears to me, Jenny, that if these figures are the work of an artist, this 'artist' is a curiously inept cobbler of species."*

This remark is based partially upon comments attributed to biologist Kenneth Miller.[320] In response to arguments for design, scientists often resort to ridiculing the kind of God that design might implicate, using epithets like "lowly grease monkey." Evolutionist Frederick C. Crews, for example, has dubbed the God of design a "curiously inept cobbler of species."[321]

Such comments are revealing for several reasons. First, in the case of scientists such as Michael Behe to whom Miller apparently directed his remarks, evidence of design is not sought as evidence of a God. Michael Behe is a molecular biologist, Professor of Biochemistry at Lehigh University, and author of the book *Darwin's Black Box*.[322] He came to the scientific conclusion that certain molecular "machines" show signs of what he calls "irreducible complexity."[323] Irreducible complexity is one form of evidence for intelligent design. The concept of irreducible complexity is a result of the observation that some biological systems are made up of parts, all of which are necessary for the functioning of the system. Each of the parts has no function apart from the whole, and removing any one of the parts causes the system to fail to perform its function.

Irreducible complexity is one of the strongest scientific responses to Darwin's famous challenge and "safe harbor provision":

> If it could be demonstrated that any complex organ existed, which could not possibly have been formed by numerous, successive, slight modifications, my theory would absolutely break down."[324]

---

[320] Witham, *By Design*, p. 135.

[321] Frederick C. Crews, review of *The Wedge of Truth*, by Phillip E. Johnson; *Icons of Evolution: Science or Myth? Why Much of What We Teach About Evolution Is Wrong*, by Jonathan Wells; *Darwin's Black Box: The Biochemical Challenge to Evolution*, by Michael J. Behe; *Mere Creation: Science. Faith and design*, by William A. Dembski; *Intelligent Design: The Bridge Between Science and Theology*, by William A. Dembski; *Tower of Babel: The Evidence Against the New Creationism*, by Robert T. Pennock; and *Finding Darwin's God: A Scientist's Search for Common Ground Between God and Evolution*, by Kenneth R. Miller, *The New York Review of Books* Vol. 48, No. 15 (October 4, 2001).

[322] Behe, *Darwin's Black Box*.

[323] *See*, *An Introduction to Molecular Machines and Irreducible Complexity*, http://www.arn.org/mm/mm.htm. The link to "Molecular Machines Museum" shows by illustration and animation many examples of irreducible complexity.

[324] Darwin, *Origin*, p. 189. This statement is referred to as the "safe harbor provision" for Darwinism since it requires proving a negative, not unlike affirming that "If it could be demonstrated that God does not exist ..." Such a demonstration is impossible.

Darwin himself stated, "But I can find out no such case."[325] Richard Dawkins has stated, "I do not believe that such a case will ever be found."[326]

Although Darwin's challenge remains an evolutionist's safe harbor since it requires proving a negative, Behe believes he has discovered at least one such case that apparently cannot have been formed by successive slight modifications. The hair-like bacterial flagellum is a combination of precision parts that operates in the cell like an outboard motor on a boat. If only one of the many precision parts were missing, according to Behe, the flagellum would not work. This is an example, as Behe explains, of a system that is irreducibly complex.[327] The bacterial flagellum is composed of several well-matched, interacting parts that contribute to the basic function, wherein the removal of any one of the parts causes the system to effectively cease functioning.[328] If it cannot be shown that all the parts of the flagellum could have evolved independently and assembled accidentally, Darwin's theory absolutely breaks down. To reasonable minds, Behe's discovery is compelling evidence of the "break down" (if not the total dismantling) of Darwinism.

Second, such statements highlight the deliberate agenda of Darwinists to frame intelligent design theory as just another attempt at "scientific creationism."[329] In fact, intelligent design theory does not theorize a designer. Although intelligent design theory is embraced by theists as "proof" of their God, such use is ancillary to the theory itself. Rather than oppose intelligent design theory on a scientific basis, detractors of the theory loudly and lazily simply denounce the theory as another form of "creationism" to be avoided at all costs. Such theological potshots, usually delivered with religious zeal, are in themselves very telling. Usually failure by those holding one theory to attack the substance of an opposing theory, choosing instead to resort to fear-mongering and innuendo, is itself an accurate indicator of the relative merits of both theories. In this case, Darwinists have yet to seriously challenge the scientific basis of intelligent design theory. Until they do, and successfully, one can only assume there *is no* scientific opposition to intelligent design theory.

---

325 Ibid.

326 Dawkins, *Blind Watchmaker*, p. 91. Interestingly, Dawkins followed this quote with the line: "If it is—it'll have to be a *really* complex organ, and, as we'll see in later chapters, you have to be sophisticated about what you mean by 'slight'—I shall cease to believe in Darwinism" Ibid., emphasis in original. Presumably, an organ such as an eye, or a wing, is not "*really* complex" enough for Dawkins. Sophistication in this case is displaced by sophistry—and in truth Dawkins will never find a sufficiently-complex organ; he is, it appears, first a Darwinist and second a scientist.

327 Behe, *Darwin's Black Box*, p. 39.

328 Ibid.

329 By the same logic the charge could be leveled that Darwinism is merely an attempt at "unscientific atheism."

[35] *"So, to talk about the odds, or even to try to calculate the odds of a present reality occurring is somewhat irrelevant* – it happened, *regardless of the odds."*

Darwinists rarely seriously consider the odds of evolution actually happening as Darwinism requires. This is not necessarily because Darwinists don't consider evolution, including origin of life, improbable. They simply consider the subject somewhat *irrelevant*. The reason it is irrelevant is because of methodological naturalism—a construct that rules out design by definition. As pointed out by intelligent design theorists, there are only three explanatory causes: design, law (or necessity) and chance (and chance is not strictly speaking a cause—it is a descriptive explanatory device). If design is ruled out by philosophy and law cannot explain the evidence, such as the sequence of nucleotide bases that provide the messages in DNA, the only possibility that is left is chance. If chance is all that remains, then one is compelled to accept chance regardless of the statistical likelihood of the occurrence.

In fact, any serious contemplation of chance results in the realization that all phenomena, even that "explained" by chance is a result of physics (law), or intelligent manipulation (design). "Chance" explanations immediately confront the law that as the complexity of a system increases its probability of being explained by chance decreases exponentially. For example, assume that we wish to write the word "Design" using a random process consisting of the pulling of letters from a brown bag containing only the upper and lower case letters of the alphabet, a period, a comma and a space, with each symbol returned to the bag after each pull. There are 55 symbols in the bag so that the chance of pulling the capital "D" on the first try is 1/55. Pulling the "e" is also 1/55, but to calculate the odds of pulling "De" in that sequence is 1/55 x 1/55 or 1/3,025. The chance of pulling the letters to spell "Design" is 1/27 billion. Mathematician William Dembski argues that anything less probable than 1 over 1x10 to the 150th is statistically impossible given all the probabilistic resources in the entire universe from the time of the big bang, assuming it occurred a billion times 20 billion years ago. In calculating the probability of a sequence of letters, spaces and punctuation marks using our brown bag example, one reaches Dembski's probability limit at the 87th symbol. But living systems are defined by linear sequences containing billions of symbols precisely spaced and ordered. For simple life to evolve by chance is, by all honest accounts, either impossible or so highly unlikely by any known laws as to be practically impossible.

In general, most of the calculations made today to try to calculate the odds of evolution, particularly spontaneous generation of life, are made by anti-Darwinists. Biblical creationists are particularly vocal on this point. However, occasionally non-creationist Darwinists take a stab at some calculations. Unfortunately, the results of such calculations can lead to situations that discourage further inquiry. For example, as recorded by Phillip E. Johnson in *Darwin on Trial*, when mathematicians reported some calculations to leading Darwinists at the Wistar Institute in Philadelphia in 1967, the result was "a rather acrimonious confrontation."[330] In fact, one of the mathematicians concluded there is a "gap in the neo-Darwinian theory of evolution" that "cannot be bridged within the current conception of biology."[331] Treated as a peccant purveyor of heresy, this hapless conveyor of facts received the ultimate insult when it was suggested to him that "Your argument is simply that life must have come about by special creation."[332]

The unwavering consensus of Darwinists when confronted with numbers that suggest the only reasonable cause of life to be "special creation" is to retreat to an "it had to happen naturally somehow" mindset. Indeed, at the Wistar Institute event referred to above, leading evolutionist Ernst Mayr observed:

> Somehow or other by adjusting these figures we will come out all right. We are comforted by the fact that evolution has occurred.[333]

It's good to be comforted, if comfort at any cost is a goal. However, comfort—like its frequent companion bliss—can be just another manifestation of either ideology or ignorance (or both). Like the pampered guests on the Titanic, rather than "be comforted" by imagining immunity from the dictates of nature itself, Darwinists should be vigilant to facts that can suddenly alter their paradigm of invincibility. Darwinists and intelligent design theorists agree that the odds of a complex organism sufficient to start the Darwinian evolutionary process assembling itself by chance are off the chart, and that's just the tip of the iceberg confronting Darwinism. Mathematicians and statisticians have floated some numbers for the simplest of evolutionary tasks, and the numbers invariably reach magnitudes that can be ignored only by those intent on being distracted by the band in the dark of night.[334] When confronted with

---

330 Johnson, *Darwin on Trial*, p. 38.

331 Ibid., p. 39.

332 Ibid., p. 39.

333 Ibid., p. 39.

334 Regardless of how the odds are calculated, the results take on magnitudes that make naturalistic evolution impossible (since there is no accidental life form to start with). For example, British astronomer Sir Fred Hoyle is reported to have concluded that "The likelihood of the formation of life from inanimate matter is 1 to a number with 40,000 noughts after it ($10^{40,000}$) ... It is big enough to bury Darwin and the whole theory of

the numbers for chance assembly (knowing that the lifeboats of "somehow or other" are wholly inadequate), Darwinists get a sinking feeling because, as Phillip E. Johnson noted, "Chance assembly is just a naturalistic way of saying 'miracle.'"[335]

[36] *"Given so much time, the 'impossible' becomes possible, the possible probable, and the probable virtually certain. One has only to wait: time itself performs the miracles."*[336]

"Chance, luck, coincidence, miracle." With these words Richard Dawkins starts Chapter 6 of *The Blind Watchmaker* entitled, "Origins and Miracles."[337] Dawkins recognizes:

> [C]umulative selection cannot work unless there is some minimal machinery of replication and replicator power, and the only machinery of replication that we know seems too complicated to have come into existence by means of anything less than many generations of cumulative selection![338]

Therefore, even the all-powerful engine of "cumulative selection" cannot pull off something like producing the minimal machinery of a living organism—something on the order of a "miracle."[339] Even Dawkins has to succumb, reluctantly for sure, to the fact that cumulative selection had to first get started, "and we cannot escape the need to postulate a *single-step* chance event in the origin of cumulative selection itself."[340] What can be the explanation? Is this, as Dawkins asks rhetorically, a fundamental flaw in the whole theory of

---

evolution." *See*, "Hoyle on Evolution", *Nature*, vol. 294 (November 12, 1981): p. 148, cited by Joe White, ed. *Darwin's Demise* (Green Forest, AR: Master Books, 2001), p. 32.

335 Johnson, *Darwin on Trial*, p. 106.

336 Recounted by Stephen Jay Gould from a lecture by Harvard biochemist George Wald. *See*, Gould, *The Panda's Thumb*, p. 218-19. The "miracle" referred to is the origin of life. The phrase is also attributed to a 1954 *Scientific American* article in Phillip E. Johnson, *The Wedge of Truth* (Downers Grove, IL: InterVarsity Press, 2000), p. 50, quoting George Wald, "The Origin of Life," *Scientific American*, August 1954, pp. 44-53.

337 Dawkins, *Blind Watchmaker*, p. 139. Interestingly, Chapter 6 of Darwin's *On the Origin of Species* is entitled "Difficulties on Theory."

338 Ibid., p. 141.

339 Dawkins reiterated that "Cumulative selection is the key to all our modern explanations of life. It strings together a series of acceptably lucky events (random mutations) together in a nonrandom sequence so that, at the end of the sequence, the finished product carries the illusion of being very very lucky indeed, far to improbable to have come about by chance alone, even given a time span millions of times longer than the age of the universe so far" (Ibid., pp. 139-40).

340 Ibid., p. 140, emphasis in original.

the blind watchmaker? If life, by all reasonable accounts, requires a miracle to get started, isn't this proof of a "far-sighted supernatural watchmaker?"[341]

We shouldn't be so short-sighted. According to Dawkins, who never met an impasse from which he couldn't imagine an escape, it is simply a matter of reducing the magnitude of the miracle.[342] Dawkins' thesis, consistent with his naturalistic philosophy, is that "events we commonly call miracles are not supernatural but are part of a spectrum of more-or-less improbable natural events."[343] In fact, Dawkins explains, "A miracle, in other words, if it occurs at all, is a tremendous stroke of luck."[344] Further, and in harmony with his odd theory:

> [T]he more thoroughly we can break large chance events up into a cumulative series of small chance events, the more satisfying to rational minds our explanations will be.[345]

Dawkins explains that the key is to consider "*how* improbable, how *miraculous*, a single event we are allowed to postulate.[346] He elaborates:

> What is the largest single event of sheer naked coincidence, sheer unadulterated miraculous luck, that we are allowed to get away with in our theories, and still say that we have a satisfactory explanation for life?[347]

Thus, Dawkins admits that we need some amount of "sheer unadulterated miraculous luck." A miracle by any other name is just as improbable, but rather than postulate how such a miracle might be explained, Dawkins reformulates the question to be one of how much "luck" we can "get away with?" And pondering "how much" is quite an effortless task if one's imagination is not constrained by what one can know by scientific reasoning. To help the rest of us, Dawkins starts

---

341 Ibid.

342 Ibid., p. 139. Dawkins states, "We can accept a certain amount of luck in our explanations, but not too much."

343 Ibid.

344 Ibid. Surprisingly, Dawkins does not refer to biogenesis as an "apparent" miracle.

345 Ibid., p. 141. That "rational minds" would be juxtaposed in the same sentence with the idea that a highly improbable chance outcome is *more likely* to occur by "break[ing] large chance events up into a cumulative series of small chance events" is something of an unlikely occurrence in itself.

346 Ibid., emphasis in original.

347 Ibid.

his explanation with "Now here is a fascinating thought."[348] The remainder of the explanation involves fascinating thoughts about whether our own planet is the only one that has life, or whether life abounds all around the universe.[349] Imagine, Dawkins begins, that Earth is not the only planet in the universe on which life can abound. For starters, imagine there are about "100 billion billion"[350] such planets. And, imagine that based upon the number of eligible planets we were to "postulate a certain ration of luck"[351] available for our own "fantastically, lucky, miraculous event" of the origin of life.[352] Further, imagine that "given our ration of luck, we can then 'spend' it as a limited commodity over the course of our explanation of our own existence"[353] (i.e., for our necessary "miraculous event" and further Darwinian development). Continue imagining that, once the cumulative selection machine is revved up, "we need to postulate only a relatively small amount of luck in the subsequent evolution of life and intelligence."[354] Now, our ration of luck we can allocate to origin of life is certain to be quite high because:

> [I]f we assume, as we are perfectly entitled to do for the sake of argument, that life has originated only once in the universe, it follows that we are *allowed* to postulate a very large amount of luck in a theory, because there are so many planets in the universe where life *could* have originated.[355]

After all this imagining, we shall be pleased to find:

---

[348] Ibid., p. 142.
[349] Ibid.
[350] Ibid.
[351] Ibid., p. 145.
[352] Ibid., p. 144. The ration of luck is based on the number of planets by the following reasoning: Dawkins considers the probability that life will originate on any randomly designated planet of some particular type. Ibid., p. 144. Dawkins explains, "Suppose that our best guess [of this probability] is some very very small number, say one in a billion. This is obviously such a small probability that we haven't the faintest hope of duplicating such a fantastically, lucky, miraculous event as the origin of life in our laboratory experiments" (Ibid.). Of course, "one in a billion" is hardly close to the actual odds. But science has little to do with reality when science is based solely upon "supposing" one thing or another.
[353] Ibid.
[354] Ibid., p. 146.
[355] Ibid., p. 144, emphasis in original. Note also that we must indulge Dawkins' "entitlement," for if Darwinists from Darwin to Dawkins were not permitted to "assume for the sake of argument" they would have no theory.

> This means we can, if we want to, spend virtually our entire ration of postulatable luck in one big throw, in our theory of the origin of life on a planet.[356]

Consider carefully these words of one of the world's leading Darwinists. Admittedly, he is faced with a rather intractable problem, and his valiant effort probably succeeds in satisfying any true believers who actually care. But Dawkins' fanciful speculations fall short of convincing a scientific mind.[357]

First, it seems that Dawkins' "100 billion billion planet" theory rests on a shaky premise. The theory assumes that the actual probability we can reasonably entertain to explain life starting on earth—the probability of *our* miracle—is somehow permitted to be quite high because there were so many other planets on which life could have started. However, this clever cogitation does not in any way lessen the actual probability of life in fact originating on earth. It merely permits Darwinists to sit more comfortably with astronomical probabilities, imagining that we can simply allocate a "ration of luck" out of the great pool of luck available, since each of the other eligible planets had the same probability as earth of "abounding with life"—leaving a certain amount of "unused luck," so to speak.

Consider a lottery involving, say, correctly choosing an exact sequence of 10^{130} numbers (written as one followed by 130 zeros), which, according to theoretical physicist Paul Davies is the number of amino acids sequenced in a small protein.[358] If only one person plays and wins, we would say that person was very, very lucky—we would

---

356 Ibid.

357 Dawkins should at least be credited with attempting to deal with the problem of explaining how an event can happen against astronomical odds. Admittedly he is dealing to a certain extent in "out of the box" thinking, including "speculations about relatively probable ways in which cumulative selection might have got its start" (Ibid., p. 147). But such speculations, even when offered as "fantastic thoughts," have a way of becoming lodged in the popular imagination as having answered and settled the underlying problem.

358 *See*, Davies, *Fifth Miracle*, p. 91. Few of us would pay to play in a lottery requiring choosing a correct sequence of 10^{130} numbers—to even write out the numbers at two digits per second would take about 10^{48} billion years, many times more time than even the most generous Darwinists would grant the universe has existed. Note that life as we know it requires hundreds of thousands of specialist proteins, not to mention the nucleic acids. (Davies, *Fifth Miracle*, p. 95.) According to Davies, therefore, the odds against producing just the proteins by pure chance are "something like 10^{40000} to 1" (Ibid.). Not surprisingly, Davies considered random self-assembly of proteins to be "a nonstarter" (Ibid., p. 91). Likewise, Nobel Prize winner Francis Crick (who along with James Watson, determined DNA's molecular structure), considered the chance synthesis of even a small protein of 200 amino acids so improbable that he concluded that "the great majority of sequences can never have been synthesized at all, at any time." Perloff, *Tornado in a Junkyard*, p. 67.

even consider such a chance event *miraculous*.[359] By Dawkins' argument, such a seemingly miraculous event could be explained by considering the possibility that it wasn't just the lucky person playing the lottery, but that many, many (possibly 100 billion billion) other people were playing at the same time. Since each person needs the same amount of luck to win, each person puts more luck into the system, i.e., adds to a "certain ration of luck." Using Dawkins' reasoning, because only one person can win the lottery, it follows that we are *allowed* to postulate a very large amount of luck available to the lottery winner, because there are so many players in the lottery that *could* have won. Thus, the winning person wasn't so lucky after all, he or she simply appropriated the greatly increased *amount* of luck *postulated* to win.

By Dawkins argument, it seems that *how much* luck a person has available to win the lottery (i.e., his probability of winning, and, correspondingly, the attribution of winning to luck or a miracle) is dependent upon how many players are in the game—the more people playing, the more luck a person has available to appropriate to win. In fact, of course, *the probability of winning is not changed by the number of players*.[360] The probability can be calculated, and it is not reduced by any number of independent players, even if one or more players happens to also choose the exact same sequence. Likewise, the probability of life spontaneously generating on earth can be calculated, and the number is not reduced by any number of independent opportunities for life, whether or not they actually occurred.

Second, Dawkins equates the concept of increasing the probability of an event happening once *somewhere* in the universe with decreasing the time required for an event to actually occur *on earth*. As Dawkins concludes:

> [T]he point of our 'numbers of planets' argument is that, even if the chemist [who, according to Dawkins, might say you have "to wait a long time" for life to start] said that we'd have to wait for a 'miracle', have to wait a billion billion years—far longer than the

---

[359] Or *suspicious*. Most people would rightly suspect collusion, or some other type of intelligent intervention.

[360] It may be true that the probability of *at least one* person winning is increased with an increase in the number of players. But the probability of any one of the players winning is not changed; an effectively impossible event would nevertheless be considered a miracle. Further, we are not interested in the probability of *at least one* person winning out of a hypothetical number of possible players (i.e., origin of life on *at least one* of a number of possible planets); we are interested in the probability of *any given* person winning (i.e., origin of life on *a given* planet, such as Earth).

> universe has existed, we can still accept the verdict with equanimity.[361]

Why?

> There are probably more than a billion billion available planets in the universe. If each of them lasts as long as Earth, that gives us about a billion billion billion planet-years to play with. That will do nicely![362]

Dawkins does not deny that something on the order of a miracle is necessary to explain life on earth. He just says the "miracle" can *appear* more likely when considered in the context of a greater pool of universal probabilities, even if they have to be imagined. In essence, Dawkins is simply suggesting more monkeys at typewriters. Surely, it would seem, if you have enough monkeys and enough typewriters, you need less "luck" to get a line of Shakespeare. But this is not true—the chances of any one of the monkeys typing Shakespeare on any given try remains unchanged. Similarly, having 100 billion billion eligible planets does not change the probability of life starting on earth one bit—it simply might reduce the amount of time one might have to wait for life to happen *somewhere* on one of the eligible planets. Proposing a greater universe (using the word as a statistical term) of possible locations for an event does not change the probability of the event happening at any one location.[363]

Dawkins is led to faulty imaginations by asking the wrong question—actually a question that already has an answer. He starts out by considering "*how* improbable, how *miraculous*, a single event we are allowed to postulate." But *how miraculous* is not at issue—we know this by any number of calculations that invariably return astronomical—even impossible—odds. Our task is to consider what mechanisms might be available to deliver against such odds. And not only must we deal with the astronomical odds of life forming at all, anywhere, we must deal with the odds that such an astronomically unlikely event

---

[361] Ibid. p. 145.

[362] Ibid.

[363] For example, having 100 billion billion people flipping coins does not change the probability for any one of the people to successfully achieve a given sequence of flips. It may impact the time one might have to wait for any one of the 100 billion billion people to be successful.

happened on our planet, instead of one of the billions and billions of other hypothetical possibilities.[364]

Even if Dawkins was correct in his imaginations of 100 billion billion life-supportable planets, the cold facts of the matter make even that number insufficient to render biogenesis probable anywhere in the universe. Evolutionist Paul Davies considered biogenesis in a similar vein in his book *The Fifth Miracle*.[365] Davies calculated the odds of producing the required proteins for life (not including the nucleic acids) by pure chance at "something like 10^{40000} to 1.[366] This is one followed by forty thousand zeros. Such numbers are beyond any reasonable comprehension or correlation to real life events. Imagine dealing a perfect suit at cards, not once, not twice in a row, but *a thousand times in a row*. Such a feat, according to Davies, is easy by comparison.[367] Davies recounts the famous remark of British astronomer Fred Hoyle that likened the odds against the spontaneous assembly of life to those for a whirlwind sweeping through a junkyard and producing a fully functioning Boeing 747.[368]

Clearly understanding the phenomenally improbable chances of life as we know it self-assembling, Davies soberly considerers the options, including possible extraterrestrial origin of life. Even considering this possibility, however, hardly helps the situation. As Davies elaborates:

> There are indeed a lot of stars—at least 10 billion billion in the observable universe. But this number, gigantic as it may appear to us, is nevertheless *trivially* small compared with the gigantic odds against the random assembly of even a single protein molecule. Though the universe is big, if life formed solely by random agitation in a molecular junkyard, there is scant chance it has happened twice.[369]

Moreover, Davies rightly rejects arguments such as "life was inevitable, because we are here,"[370] or "there is nothing special or

---

[364] Keeping with the coin-flipping example, we are concerned with one specific coin flipper being successful. The odds of our chosen coin flipper being successful are not affected by the success or failure of the others.

[365] Davies, *Fifth Miracle*.

[366] Ibid., p. 95.

[367] Ibid.

[368] Ibid., quoting Fred Hoyle, *The Intelligent Universe* (London: Michael Joseph, 1983), p. 19.

[369] Ibid., p. 95.

[370] Ibid., p. 94.

exceptional about our place in the universe,"[371] or "life on Earth is a fluke."[372] At one point he asks the same rhetorical question Dawkins asked, "So what is the answer? Is life a miracle after all?"[373] Davies believes it is not, but is honest in his assessment of the chances of known physical processes doing the job. The job of the scientist, Davies says, is to at least make an attempt to explain biogenesis as a normal physical process.[374]

Dawkins' meanderings would be acceptable if he, too, would focus on finding natural explanations rather than on somewhat unnatural imaginations. However, as alluded to above, Dawkins is to be commended for at least trying to provide an explanation of some kind for what seems like an impossible event. Aside from pointing out faulty premises, however, a philosophically more relevant point is this: What is the difference between imagining 100 billion billion life-supportable planets, and simply imagining something like an intelligent designer? Why cavort through mental ether space to reach one imagination, while simultaneously rejecting another? The answer from Darwinists is certainly to be something along the lines of "With 100 billion billion planets there is no need to invoke the supernatural. Everything is all tidied up cleanly and naturally." Maybe, but the question is, as always, which is more *reasonable* to believe? Given, say, that both explanations require one to suspend any reliance on sure knowledge and venture into the realm of speculation—why not consider a designer, possibly even a God. God (or a god-like intelligent designer) is certainly more likely to bring about a miracle on the order of creation than is something like chance on one planet out of 100 billion billion planets.

When it comes to origin of life, the speculations of Darwinists are often substituted for explanations. In themselves, speculations are not wrong. "Out of the box" thinking and postulating theories is what science should be about. But why are Darwinists' speculations of unseen, undetectable, unproven entities taken seriously, while those of anti-Darwinists are brushed off without a thought? Why is Dawkins' speculation of billions and billions of life-supportable planets considered a rational attempt at removing the need for a miracle, and speculation on an intelligent designer to do the same is considered irrational at best, and "religious" at worst?

[37] *"Should we treat it as an acheropite image of sorts – and turn this research center into some kind of* sancta sanctorum *for deluded pilgrims?"*

---

[371] Ibid., p. 96.

[372] Ibid.

[373] Ibid., p. 93.

[374] Ibid.

A mediaeval tradition has it that the Roman Emperor Constantine's mother, Helena, had the staircase from the palace of Pontius Pilate sent from Jerusalem to Rome in about 334 AD. Still in Rome, and permanently housed in the Basilica of St. John in Lateran, the staircase is said to be the one that Jesus Christ ascended several times the day he was sentenced to death. At the top of the staircase and on an opposite wall is a window through which a visitor can look into the *Sancta Sanctorum*, a small chapel known as "Rome's most venerated sanctuary."[375] The most well-known thing kept in the *Sancta Sanctorum*, visible through the iron-grated window, is a painted image of Christ, which is called the *acheropite image*, "which is to say, an image not painted by human hand."[376] The current painting, which is said to have been touched up over the years, sits on an altar in the apse of the chapel. Below the altar is an ark with "relics" including, it is said, the heads of the apostles Peter and Paul.[377]

[38] *"What is truth? The* truth *is that I, and you for that matter," he leaned in with emphasis, practically touching Jenny with a pointed finger, "can know for sure only what we can see, hear, smell, taste or touch. No more, no less."*

Once again, we encounter a basic philosophical notion related to naturalism, specifically a position known as *positivism*. Logical positivism is a 20th a century philosophical movement that holds characteristically that all meaningful statements are either analytic or conclusively verifiable or at least confirmable by observation and experiment and that metaphysical theories are therefore strictly meaningless.[378] Although various philosophies would expand on the notion of our senses being the only reliable source of knowledge (such as knowledge by use of intellect, including reflection and imagination), in evolutionary theory sensory experience is still held out as the only basis for scientific knowledge. Limiting facts relied upon to those available to the sensory perceptions is rational. However, limiting *explanations* of observable phenomena to include only so-called "naturalistic" explanations is a limiting philosophical choice.

Aside from being limiting with respect to something like origin of life research, such a position is not consistent with other scientific disciplines—scientists are not generally prone to quibble over quiddity. For example, as odd as it may seem, scientists do not know what the essence of energy "is." "Energy," is not a "thing" that can be touched, seen, heard, or felt directly. It can, however, be observed in its effects, which can be measured and quantified. Equally enigmatic, perhaps,

---

[375] n.a., *The Holy Staircase: History and Devotion* (n.p., n.d.). Booklet purchased at the Basilica of St. John in Lateran, Piazza San Giovanni in Laterano, Rome.
[376] Ibid.
[377] Ibid.
[378] http://www.m-w.com/dictionary/logical+positivism.

gravity remains a mystery to physicists in the sense that they do not know what it "is."[379] The effects of gravity can be measured, predicted, and relied upon such that we speak of the "law" of gravity. But gravity is simply a name we give to an unseen, unfelt, tasteless, silent, odorless "thing" of which we can detect its presence only by its sensory effects we refer to as a "force."

Postulating unseen causes to explain gravity's effects has scientific precedent. For example, to make calculations agree with what is observed in the motion of deep space craft such as the *Pioneer 10*, "astronomers have been forced into an embarrassing situation."[380] It seems that to make calculations work, astronomers "have to assume that immense amounts of invisible matter surround all galaxies. Astronomers simply call this mystery stuff dark matter."[381] In other words, the empirically observed data forces astronomers to "assume" something so immense it surrounds all the galaxies—something they can not, and can never hope to, directly detect.[382]

"Invisible matter," "dark matter," these are simply names given by scientists to detectable but invisible phenomena to explain empirically observed data. Scientists don't give a second thought to naming such phenomena and insisting it exists, even though there is no evidence beyond the observable data. We could just as well call gravity "God" and it would not change our knowledge of it or its observable effects. Apples would fall just the same; airplanes would fly just the same; and, the universal constant in the formula for gravitational attraction just might be even more appropriate![383]

So, it is not strictly true that methodological naturalists, including scientists that hold to a position of positivism, cannot speak of and study an entity that eludes direct sensorial detection. With concepts such as electrical and gravitational "fields," for example, there

---

379 Tim Folger, "Nailing Down Gravity: New Ideas About the Most Mysterious Power in the Universe," *Discover*, Vol. 24, No. 10, October 2003, 34. Folger examines the strange behavior of deep-space craft that throw into question traditional notions of gravity. He quotes Michael Martin Nieto, a theoretical physicist at Los Alamos National laboratory on the subject of "our understanding of gravity, the most pervasive force in the universe:" "We don't know anything," [Nieto] says. "Everything about gravity is mysterious" (Ibid., p. 36). Folger admits that "Nieto's words are strong for a phenomenon that seems so familiar."

380 Ibid.

381 Ibid.

382 According to Folger, "a handful of heretics have begun to question the wisdom of believing in something no one has ever seen." Interestingly, therefore, in astronomy one who attributes observed data to "something no one has ever seen" is mainstream, while those who would insist otherwise are "heretics" (Ibid.).

383 The formula for gravitational attraction includes what is called a "gravitational constant" that is denoted by an upper case letter "G". The gravitational constant is an empirically-derived number that, when used in the gravitational formula, makes all gravitational calculations involving gravitational force come out right.

is ample acknowledgement of the presence of invisible forces. Simply observing their effect on things that are observable is evidence enough to speak of them as if they were real. No one would doubt the existence of gravity because the evidence for gravity, or something that we call gravity, is all around us. Likewise, if there is evidence of design all around us, it should not require leaps of faith to ascribe the design to an unseen entity, by whatever name it may be called.

[39] *"We must stress that myths can speak to purpose and values, both of which are very important, of course, and these are areas that are not for science to inform. Science simply provides the only reliable way of knowing about the real world."*

Ernst Mayr, "the Darwin of the 20th century,"[384] notes with regard to Biblical creationist positions:

> More or less similar creation stories are found in the folklore of peoples all over the world. They fill a gap in mankind's desire to answer the profound questions about this world ...[385]

To begin with, it is unlikely that "more or less similar creation stories" to the biblical Genesis account are found in folklore.[386] But that is a bone for creationists to pick. What is more amusing for an intelligent design theorist to note is Mayr's patronizing conclusion in his dismissal of such "stories":

> We still treasure these stories as part of our cultural heritage, but we turn to science when we want to learn the real truth about the history of the world.[387]

Aside from the doubtful possibility that Mayr includes himself in the "we" that "treasures" the Genesis account of creation, this statement is wholly consistent with intelligent design theory. Intelligent design theory is a scientifically sound approach to detecting design, even

---

[384] New York Times, back cover of Mayr, *What Evolution Is.*

[385] Mayr, *What Evolution Is*, pp. 4-5.

[386] Or, as Mayr informs his readers, one of the "two stories of creation in the Bible (Genesis)" (Ibid., p. 4). There is only one "story" of creation in Genesis. Referring to the two slightly different recitations of the days of creation in Genesis, Mayr seems to be attempting to subtly suggest there are two mutually inconsistent creation "stories" in Genesis.

[387] Ibid., p. 5.

evidence for an intelligent designer— in virtually the same way as the Search For Extraterrestrial Intelligence (SETI) project is doing.

Unfortunately, of course, Mayr's statement above, which is the party line for philosophical naturalists, is not entirely true. When the science of origins research hints that the "real truth" may implicate an unseen, intelligent being, it (unlike the SETI project) suddenly becomes "religious" and cannot possibly be "science." Thus, it becomes abundantly evident that "real truth" is not the goal; the goal for methodological naturalists like Mayr is to ensure that in origin-of-life research only philosophically consistent ideas can be "scientific."

[40] *"In fact, many prominent scientists believe myths. But, when it comes to explaining the origin of natural phenomena, these scientists rely on empirical data, and not on personal beliefs."*

The National Association of Biology Teachers affirmed in their Statement on Teaching Evolution, "Science teachers can, and often do, hold devout religious beliefs, accept evolution as a valid scientific theory, and teach the theory's mechanisms and principles."[388] This statement is, sadly, the truth, since the immediate question that comes to mind is: Do science teachers have a choice?

In what is a telling and unfortunate commentary, for example, author Jonathan Wells, wishing to give credit where credit is due for his book *Icons of Evolution*, states:

> In addition to the people named above, other scientists at universities in the United States, Canada, and the United Kingdom helped with various parts of the manuscript, but prefer to remain anonymous. In several cases, they chose anonymity because their careers might suffer at the hands of people who strongly disagree with the conclusions of this book. For those scientists, public acknowledgment will have to wait.[389]

These anonymous people, no doubt being professional men and women of science, find themselves in an awkward position of choosing between "science" and the pursuit of truth—beholden to the prejudices of the former in practice and the attraction of the latter in conscience. That such people are not even free to identify with any position contrary to Darwinism is an indictment of contemporary "science" that cannot be ignored. Such people are, for sure, not alone.

---

[388] NAS, *Teaching About Evolution*, p. 128.

[389] Wells, *Icons of Evolution*, p. xiv.

Like modern-day "Galileo's," there is a growing crowd of scientists increasingly aware of and uneasy with the mendacious nature of science in the area of origins—aware that science and truth have taken divergent paths, and uneasy in the knowledge that in any contest between the two, one remains unchangeable.[390] Many professors and academics embraced evolution because it was deemed "scientific," and now doubts are surfacing, like a bride's suspicions on the eve of her wedding day that the man she chose for lucre instead of love might be a penniless imposter.[391]

In a more knowledge-based paradigm, science teachers, professors and other academicians would be free to teach all the evidence, both for and against evolution just as Darwin did—regardless of whether or not the evidence correlates with a "religious" belief. Starting in Chapter 6 (out of 14) of his book *On the Origin of Species*, appropriately entitled "Difficulties on Theory," Darwin met head-on what he knew to be valid and strong arguments against his theory. What is noteworthy is that Darwin did not find it unusual that people would have difficulty with his theory, much less oppose it with compelling scientific arguments. He accepted it as part of the scientific process that his theory would necessarily have to withstand the scrutiny of scientific challenges. To do otherwise would not be science, but enforced dogmatism.

Enforced dogmatism would, for example, by peer ridicule, legal intimidation, or threat of career-ending retaliation, put obstacles in front of anyone who wished to teach evidence against Darwinism. The subject would simply be off limits to any contrary viewpoints. No choice. No tolerance. No diversity. Just state-imposed, peer-regulated, and legally enforced silence on the topic. That's what enforced dogmatism would look like.

---

[390] The reader is reminded that the geocentric theory of the solar system, while being stressed by the Catholic Church for religious reasons, was accepted as "truth" by virtually all the scientists of Galileo's day, including university professors in the leading universities of Europe who had become mouthpieces for the enforced dogma of the day. *See, e.g.*, Burke, *The Day the Universe Changed*, p. 128. ("There was little help to be gained from the once-powerful universities, for they were now buttresses of theology, their teachers defenders of the faith … New scientific and technological discoveries had no place in college lecture rooms … The colleges had become echoing, dusty halls of irrelevant logic-chopping.")

[391] This metaphor, being too good to not reapply, is adapted from Wes Pruden, "Pruden on Politics," *The Washington Times* (April 12, 2004), <http://www.washtimes.com/national/pruden.htm.> The full quote from Pruden, relating to the Democratic nomination of John F. Kerry in the 2004 US presidential election is, "The party embraced John Kerry because he was deemed "electable," and now doubts are surfacing, like a bride's suspicions on the eve of her wedding day that the man she chose for lucre instead of love might be a penniless imposter."

[41] *"Theory does not mean 'guess,' and progression has been observed so many times that there is no longer a compelling reason to question it. In this sense, it is a fact."*

In *Teaching About Evolution and the Nature of Science*, published by the National Academy of Sciences (NAS), teachers are provided with answers to "difficult questions about evolution ... from parents and others who object to evolution being taught."[392] In a chapter entitled "Frequently Asked Questions," for example, the answer to the question "Is evolution a fact or a theory?" is the basis for the above excerpt.[393]

The answer to this question, like most of the "answers" provided by the NAS, is a classic example of information without informing. First, the NAS starts with the non-informing tautological statement: "The theory of evolution explains how life on earth has changed."[394] (The NAS states that "evolution" means "cumulative change through time"[395] so the above statement reduces to "The theory of cumulative change through time explains how life on earth has changed [over time].") Whether or not this circular reasoning has any bearing on determining if "evolution" is a theory or fact depends on the meaning of the term "change." If "change" means generally "change over time" or "change within the genetic limits of a species," such as the cyclic change observed in finch beaks on the Galapagos Islands, then evolution is, very possibly, a fact. If, however, "change" means "atoms-to-Adam" human evolution from a single cell, then the statement is simply not true—evolution is only *an* explanation, but not *the* explanation, or even a very good explanation, of how human life originated.

The NAS's answer is a good example of the pervasive use among Darwinists of "begging the question" as a substitute for explanation.[396] That is, the statement "The theory of evolution explains how life on earth has changed" assumes the truth of the premise and states a tautology as an explanation. Evolution is, by the NAS definition, life on earth changing, and life on earth changing is, by definition, evolution. Unless and until Darwinists are willing to define "evolution" in a scientific manner, and use the term consistently, little knowledge can be imparted to those seeking to understand.[397]

---

392 NAS, *Teaching About Evolution*, p. 55.

393 Ibid., p.56

394 Ibid.

395 Ibid., p.11.

396 In logic, "begging the question" is a form of circular reasoning in which one comes to a conclusion which essentially restates (and assumes the truth of) the premise(s).

397 As noted in other places herein, the lack of precise terms in the "evolution debate" has hampered constructive dialog. As it is, "evolution" means anything and everything, and, therefore, nothing. Do finch populations exhibit larger beaks? That's "evolution in

Knowing this, however, the NAS hedged its answer with a further non-informing statement that is true by a contrived definition: "Biological evolution is the best *scientific* explanation we have for the enormous range of observations about the living world."[398] The NAS defines "scientific" as encompassing "anything that can be observed or measured," and "science" is limited to "explanations [that] are restricted to those that can be inferred from confirmable data."[399] Even though intelligent design theory meets the requirements of this definition, it is nevertheless arbitrarily excluded from any consideration by Darwinists, so that evolution probably is the "best" scientific explanation for how life on earth has changed. But that does not mean it is the best explanation, or a right explanation, or a factual explanation, or a true explanation. It is like presenting a painting by a blindfolded painter as the "best" that painter could produce. Without disclosing that the painter was blindfolded, the information is likely to yield errant conclusions as to the painter's ability and the painting's meaning.

Second, beyond the circular starting point, the NAS aids us in resolving the fact/theory question by assuring us:

> [S]omething that has been tested or observed so many times that there is no longer a compelling reason to keep testing or looking for examples"—is a "fact."[400]

The surprising conclusion based on this dubious proposition, we are told, is that the "occurrence of evolution in this sense is a fact"[401]—surprising because scientists no doubt dream of Nobel Prizes while "looking for examples" of observed change beyond mere variation below the species level. Darwinists are in dire need of an example of speciation beyond such things as the artificial "speciation" of the hordes of hapless fruit flies intelligently malformed and distorted in hopeful laboratories the world over.

Interestingly, the NAS tackles a related question with equal equanimity: Can evolution account for new species?[402] Without answering yes or no,[403] the NAS refers the reader to Chapter 2 of the

---

our time." Do finch populations go back to original beak sizes? That's "evolution in our time."

398 Ibid., p. 56, emphasis added.

399 Ibid., p. 27.

400 Ibid., p. 56

401 Ibid.

402 Ibid., p. 57.

403 This prevarication in itself, of course, speaks volumes. The question is a simple "yes or no" question. The answer given is a study in evasion, which means the answer is clearly no.

same booklet to see some cases in which "evolutionary biologists have documented many cases in which new species have appeared in recent years."[404] However, in a display of the type of brazen arrogance that gives good scientists a bad name, there is *no* example of documented speciation in Chapter 2. What the NAS describes in Chapter 2 is *speculation* on fossils and DNA evidence. One section is optimistically headed "Evolution as a Contemporary Process," in which the reader is promised to see "the occurrence of evolution in 'real time'—how changes come about and result in new kinds of species."[405] (One wonders what new "kinds" of species might be.) Thereafter the NAS marches out the usual suspects, confident that a hopeful reader might confuse them as actually being new species: antibiotic-resistant strains of bacteria; poison-resistant rats; toxic metal-tolerant plants that have reduced interbreeding (duly noted as "an initial step in the formation of new species"[406]); and, of course, "Darwin's" finches.[407] Example after example paraded out in a routine line-up, but not a new species in the group.

The venerable NAS is forced to admit (in Chapter 2):

> [C]reation of a new species generally requires thousands of years, so over a lifetime a single human usually can witness only a tiny part of the speciation process.[408]

Fair enough, but this type of statement should be made with an explicit denial of any documented cases of speciation, not as a timid surrender during a charade of proof for speciation.

[42] *"Basically, while it's no longer a trade secret that we don't see as many intermediate forms as we might expect, we do see some transitional marks, such as those demonstrated in Lavoux Cavern."*

Stephen Jay Gould, Harvard paleontologist and prolific writer in the popular media, was one of the great Darwinists of our time. As a scientist and president of the American Association for the Advancement of Science, Gould whole-heartedly advocated natural, undirected evolution, but he also recognized that many of the facts did not support the gradual change theorized by orthodox Darwinism.[409]

---

[404] Ibid.

[405] Ibid., p. 16.

[406] Ibid., p. 17.

[407] Ibid.

[408] Ibid., p. 17-18.

[409] Darwin's famous maxim, *Natura non facit saltum*, "nature does not make leaps" was challenged even by his staunchest allies of the time, including Thomas Henry Huxley.

For one, even though he never doubted the "fact of evolution" he recognized that the fossil record did not support Darwinian gradual change. In particular, he noted, as many had before and after Darwin, that the geological record was, at best, ambiguous with evidence for gradual change. As Gould relates in his book *The Panda's Thumb,* the "conflict between adherents of rapid and gradual change had been particularly intense in geological circles during the years of Darwin's apprenticeship in science."[410] The geological record at that time "seemed to provide as much evidence for cataclysmic as for gradual change."[411]

Realizing that the evidence from the fossil record is fundamentally unchanged since Darwin's time,[412] Gould expresses frustration that paleontologists "have paid an exorbitant price for Darwin's argument" because:

> [T]o preserve our favored account of evolution by natural selection we view our data as so bad that we almost never see the very process we profess to study."[413]

It is in this context that Gould states:

> The extreme rarity of transitional forms in the fossil record persists as the trade secret of paleontology. The evolutionary trees that adorn our textbooks have data only at the tips and nodes of their branches; the rest is inference, however reasonable, not the evidence of fossils. Yet Darwin was so wedded to gradualism that he wagered his entire theory on a denial of this literal

---

See, Gould, *The Panda's Thumb*. Gould relates a letter from Huxley to Darwin stating his religious zeal for Darwinism: "I am prepared to go to the stake, if requisite ... I am sharpening up my claws and beak in readiness" (Ibid., p, 179). Huxley, however, was also concerned that, "You [Darwin] have loaded yourself with an unnecessary difficulty in adopting *Natura non facit saltum* so unreservedly" (Ibid.). Huxley knew what Gould knew, based on his observation that, "The fossil record offered no support for gradual change: whole faunas had been wiped out during disarmingly short intervals" (Ibid., pp. 179-80).

[410] Gould, *The Panda's Thumb*, p. 180.

[411] Ibid.

[412] The quantity of fossils, of course, has increased dramatically, but the pattern exhibited by the fossils continues to support sudden appearance and/or extinction without intermediate forms.

[413] Ibid., p. 182. Note this is another (and admitted) example of the well-known and documented pattern of scientists seeing what they expect to see based upon predisposition to a theory.

> record: "… He who rejects these views on the nature of the geological record, will rightly reject my whole theory."[414]

Such a candid statement is refreshingly honest. Although Gould and those of his ilk will never bring themselves to reject Darwin's whole theory, similarly honest assessments of the facts could make the teaching of evolution much more tolerable among those who already rightly reject Darwin's theory.

[43] *"While Hawkins was a staunch proponent of strict Neihilian gradualism, Silva had published his own theory, more consistent with observed wall markings, proposing relatively long periods of stasis interrupted by relatively short periods of rapid change."*

Darwin's theory of gradual descent with modification has been often challenged, but rarely as drastically and publicly as in Stephen Jay Gould's theory of "punctuated equilibria." Richard Dawkins referred to this as "a minor, if much publicized disagreement over quite *how* gradual or 'jumpy' evolution is."[415] Recognizing that the fossil record does not support gradual Darwinian evolution, Gould, together with leading evolutionist and paleontologist Niles Eldredge of the American Museum of Natural History, proposed a theory that postulated two things: *stasis*—for long periods most species undergo little observable change; and *sudden appearance*—when change does occur, change is rapid and concentrated in small, isolated populations.[416] Thus, long periods of "equilibrium" are "punctuated" by relatively quick episodes of evolutionary change. Gould's recognition that the fossil record supports his theory of punctuated equilibrium was the basis for his famous quote that, "The extreme rarity of transitional forms in the fossil record persists as the trade secret of paleontology."[417]

According to the booklet, *Teaching About Evolution and the Nature of Science*, published by the National Academy of Sciences (NAS), Gould considered the "trade secret" quotation accurate, but partial and "dishonest in leaving out the following explanatory material showing my true purpose—to discuss rates of evolutionary change, not to deny the fact of evolution itself."[418] The NAS booklet quotes Gould as defining his theory, known as "punctuated equilibrium" as follows:

---

[414] Ibid., p. 181, quoting Darwin from *On the Origin of Species*.

[415] Dawkins, *River Out of Eden*, p. 11.

[416] Gould, *The Panda's Thumb*, p. 182.

[417] Ibid., p. 180.

[418] NAS, *Teaching About Evolution*, p. 56.

> Punctuated equilibrium is neither a creationist idea nor even a non-Darwinian evolutionary theory about sudden change that produces a new species all at once in a single generation. Punctuated equilibrium accepts the conventional idea that new species form over hundreds or thousands of generations and through an extensive series of intermediate stages. But geological time is so long that even a few thousand years may appear as a mere "moment" relative to the several million years of existence for most species. Thus, rates of evolution vary enormously and new species may appear to arise "suddenly" in geological time, even though the time involved would seem long, and the change very slow, when compared to human life.[419]

That punctuated equilibrium was ever mistaken as a creationist idea is doubtful, and it was very much considered by Darwinists to be non-Darwinian. But Gould's point is taken—his was a theory about rates, not about denying the "fact of evolution" itself—in context he was unmistakably clear that he was not challenging *Darwinism*. To be sure, Gould stated, "In fact, the operation of Darwinian processes should yield exactly what we see in the fossil record. It is gradualism that we must reject, not Darwinism."[420]

Most notably, Gould is remarkable in his honest assessment of his (and any) theory on this subject. He states:

> I emphatically do not assert the general "truth" of this philosophy of punctuational change. Any attempt to support the exclusive validity of such a grandiose notion would border on the nonsensical. ... I make a simple plea for pluralism in *guiding philosophies*, and for the recognition that such philosophies, however hidden and unarticulated, constrain all our thought.[421]

Gould, who was himself constrained in his thought by his guiding philosophy, was nevertheless honest enough on this narrow point to let the data lead to a theory, and not vice-versa.

---

[419] Ibid., p. 57.

[420] Ibid.

[421] Ibid., p. 185, emphasis added. Note the recognition of "guiding philosophies." Gould recognized that even in the world of naturalism, guiding philosophies "constrain all our thought."

[44] *"It just seems that college has opened my mind up to new ways of thinking – better ways. Even last night, what's so wrong with just having a little fun. I mean, we weren't hurting anyone." ... "I know that maybe, somehow, certain things are wrong. I mean, I wouldn't kill anybody. But who's to say what is right and wrong for me? There is no 'ultimate authority' that can be invoked."*

Again, the question of the basis for right and wrong is raised. Indeed, it will always be disputed in a relativistic world. Like seeking to agree on north and south in the absence of a magnetic pole, the inquiry is inherently flawed.

The excerpt quoted above reflects the values of secular humanism espoused (or at least tacitly approved of) on most college campuses in the United States. Paul Kurtz, author of *Humanist Manifesto II*, points out the following assumptions to be made about what man's eventual moral goals are:

> If man is a product of evolution, one species among others, in a universe without purpose, then man's option is to live for himself and to discover new areas of significance and achievement.[422]

Understanding the obvious implications of such a logical conclusion, Kurtz is quick to add, however, that humanists are not calling for the abandonment of all ethical principles. Without giving a logically consistent basis, Kurtz maintains that humanists "advocate basic moral virtues—fairness, kindness, beneficence, justice, and tolerance."[423] How advocating man "to live for himself" is consistent with advocating "moral virtues" such as beneficence or tolerance remains to be adequately explained. Although humanists disagree on an authoritative basis for morals, there is one area in which they are unequivocal in their unanimity. As explained by Kurtz in *Humanist Manifesto II*:

> Short of harming others or compelling them to do likewise, individuals should be permitted to express their sexual proclivities and pursue their life-styles as they desire.[424]

---

[422] Noebel, *Understanding the Times*, p. 195, citing Paul Kurtz, *The Humanist Alternative* (Buffalo: Prometheus, 1973), p. 179.
[423] Ibid., citing Paul Kurtz, "Humanism and Ethics," *Free Inquiry*, Spring 1981, p. 48.
[424] Ibid., p. 195, citing the Humanist Manifesto II, p. 18.

The great failing of humanism is pursuing an ethic of freedom without developing a basis for moral responsibility. While ethical relativism reduces the inquiry to, "Do your own thing (as long as it appears it's not harming anyone else)," and anything can be "good" or "bad," we would do well to recall the admonition of Aristotle:

> One may go wrong in many different ways, but right only in one, which is why it is easy to fail and difficult to succeed—easy to miss the target and difficult to hit it.[425]

Without a guiding principle or principles, there is no way to know what the target looks like, much less a way to distinguish between wrong ways and right ways of hitting the target. Until humanism can articulate a rational basis for right and wrong, we are justified in questioning its unquestioned authoritative position in current cultural thought and practice.

[45] *"Jenny, the beauty of college is learning to be free. Free from authority, free from the past. I feel like I'm finally liberated, and I like it!"*

Alex here is expressing what author David Wells refers to as the premises of the now dead "brazen promises of the Enlightenment." Writing from an unmistakably Christian perspective in his book *No Place For Truth, or, Whatever Happened to Evangelical Theology,*[426] Wells chronicles the "sorry parade of human foolishness," that has "enticed" the West.[427] Including "freedom from God and freedom from evil" in the above list of failed premises on which the Enlightenment was built, Wells identifies the ideas that were passed off under the illusion that life can be remade. According to Wells, it is this illusion that "both feeds and feeds off the idea of progress."[428]

Darwin is recognized as one of the "apostles of this doctrine of progress."[429] His ideas about evolutionary progress soon crossed over to other disciplines, including politics. Marxists, for example, applied Darwinian ideas in illusions that, according to Wells "have a way of

---

[425] Ibid., p. 203.

[426] David F. Wells, *No Place for Truth, or Whatever Happened to Evangelical Theology?* (Grand Rapids, MI: William B. Eerdmans Publishing Company, 1993), p. 58.

[427] Ibid.

[428] Ibid.

[429] Ibid.

making short work of reality." [430] Once the short work is done, the collapse of values into complete relativism affects every area of life and society.

[46] *"What is that, some sort of manifesto?"*

The humanist position on morals, ethics, religion and philosophy has been articulated in two manifestos: *Humanist Manifesto I* and *Humanist Manifesto II*.[431] A reading of these short documents is instructive in understanding the philosophical implications of Darwinism, which is held as an essential tenet of Religious Humanism as formulated in the Manifestoes.

[47] *"You can make fun, Jenny, and I know we disagree on this, but the world around us – call it matter, energy, substance, or what you will, exists independent of our minds, or any conceivable mind. We can make the world what we wish. And I wish to be free."*

The secular humanist Corliss Lamont held the philosophy of naturalism, and the closely related philosophy of materialism. In his book *The Philosophy of Humanism*, Lamont wrote:

> Like Naturalism, Materialism relies first and foremost on scientific method, believes in the ultimate atomic structure of things, and finds in Nature an order and a process that can be expressed in scientific laws of cause and effect.[432]

Thus, a key tenet of naturalism is its denial of the supernatural. The practical effect of this position is a denial of anything immaterial, such as the soul, personality, and, of course, God. This presents a problem for humanists—how to account for the human mind. For years the mind/body question has perplexed the minds of Darwinists and humanists (usually of the same mind). Lamont's logically consistent conclusion was:

> The central pillar in the Humanist metaphysics is that the underlying and continuing foundation of the universe is not mind or consciousness, but matter in its

---

430 Ibid., p. 59.

431 Ed., Humanist Manifesto I and Humanist Manifesto II (Buffalo: Prometheus, 1980).

432 Noebel, *Understanding the Times*, p. 11, quoting Corliss Lamont, *The Philosophy of Humanism*, (New York: Frederick Ungar, 1982), pp. 37-38.

> multiple and changing modes. The truth of this position is not dependent upon the definition of matter in terms of any particular stuff. Our position is based simply on the proposition that objective reality—an external world—call it matter, energy, substance, events, electricity or what you will, exists antecedent to and independent of the human mind, a Divine Mind, or any conceivable mind.[433]

In philosophical jargon, humanists are "monists." Monism is the philosophical doctrine that claims all reality is one substance, either natural or supernatural (as opposed to Dualism that affirms the existence of both). So for monistic humanists, the mind, including consciousness, is simply a manifestation of the brain—just an extension of the natural world, and explainable in purely physical terms.[434]

Think about that.

[48] *"I also know that there is no absolute moral code; we just have to adjust our ethical standards in each situation according to our judgment. As we learn to do this, emphasizing fairness, kindness, and tolerance, the world will be a better place."*

Ethical relativism—the belief that there is no absolute moral code, and therefore man must adjust his ethical standards in each situation—is a result of one of the great outstanding conundrums for Darwinists. If all is matter, simply molecules in motion, what is the basis for ethical standards? Why even discuss ethics, and how can there be "standards" of anything, much less behavior? Do other living things concern themselves with ethics? Do plants have "ethics"? Do roses have sympathy for those pricked while enjoying their fragrance? These and other thorny questions have occupied the minds of Darwinists from the dawn of Darwinism. In what might be the dusk of Darwinism, the subject remains unresolved, and probably hopelessly confounded by the seemingly intractable concepts of moral obligations and moral standards.

Darwin was chiefly concerned with a naturalistic explanation for the origin of living things, including the human body; but he realized that such an explanation must at least be consistent with explanations for the human *condition*. If natural selection explained the development of human individuals, could it also be a sufficient foundation for ethics? In his book *The Descent of Man*, Darwin explored the issue and sketched an evolutionary explanation for ethics based on

---

[433] Ibid., pp. 118-119, quoting Corliss Lamont, *The Philosophy of Humanism*, (New York: Frederick Ungar, 1982), pp. 145-6.

[434] Ibid., p. 123.

"the development of the intellectual and moral faculties."[435] Basically, realizing that ethical considerations, e.g., looking out for the good of another, is a very "un-evolutionary" concept for individual survival, Darwin came to the (inevitable) conclusion that a moral sense would likely be of little selective value to the individual; indeed, it might prove harmful.[436]

Despite Darwin's reservations, the history of naturalistic ethical considerations includes attempts by many to provide a purely natural basis for arriving at what seems to be a universal moral code. At least certain virtues such as mercy, truth, and temperance are universally considered virtuous. Most thinkers came to the conclusion, as did Darwin, that the foundation of moral sentiment is rooted in the behavior of *groups* rather than individuals. Thus a group possessing ethical individuals would have a significant adaptive advantage. Even with this clever take on the situation, the discussion is currently stalled not far from its nineteenth century beginnings. Consider the words of Leslie Stephens, a leading nineteenth century agnostic and thinker, whose writings represented the most developed elaboration of his time of the Darwinian position on ethics, and continues to best reflect the internal problems Darwinists face:

> No moralist denies that cruelty, falsehood, and intemperance are vicious, or that mercy, truth, and temperance are virtuous. Making every deduction from this apparent unanimity—allowing that similar names may be interpreted in very different senses; that the general outlines of a moral code may be the same whilst its spirit varies widely; and that the moral codes accepted at different times and places have been as different as has ever been seriously maintained—yet it remains true that there is an approximation to unity. The difference between different systems is chiefly in the details and special application of generally admitted principles. It is not such as we might anticipate from a radical opposition both of method and principle. But if we turn from the matter to the form of morality; if, instead of asking what actions are right or wrong, we ask, what is the essence of right and wrong? how do we know right from wrong? why should we seek the right and eschew the wrong?—we are presented with the most contradictory answers; we find ourselves at once in that region of perpetual antinomies, where controversy is everlasting, and

---

435 Farber, *Temptations of Evolutionary Ethics*, p. 18.

436 Ibid.

> opposite theories seem to be equally self-evident to different minds.[437]

Thus, Stephens did not see notions of right and wrong as controversial; it was their justification that was at issue. Attempts at justification for a moral code continue to occupy the leading figures of professional philosophy. The general consensus remains, that even if evolutionary ethics explained the origins of a sentiment in man that could truly be called a moral sentiment, it did not furnish a justification for any of the moral prompting that followed from it. It was a mere black box that cranked out opinions. The attempt to justify those opinions with evolutionary arguments was no more than thinly disguised attempts to carry over other ultimate criteria, such as utility.[438]

For an excellent summary of the development of ethics in the age of Darwinism the reader is referred to Paul Lawrence Farber's book *The Temptations of Evolutionary Ethics*, from which much of the above information is derived. In closing his book, Farber, reflecting on his role as a historian of evolutionary ethics development, and tempted to predict the course of philosophy, makes a general statement regarding the role of historians:

> At best they can point out a century-long frustration over the state of ethics and in the case of evolutionary ethics, the significantly interesting but dismal track record of this approach as a method of resolution.[439]

[49] *"I think that there are not only things we can know, there is also something that we can't not know. Regardless how enlightened we become, or how much we try to deny it, we can't not know it."*

Political philosopher J. Budziszewski finds God's general revelation to humanity built into creation—it is something *we can't not know*.[440] Unlike the nonsense of being agnostic about teapots orbiting the Sun, we can try to deny the knowledge of our own design, but the objective truth of the matter resists our denials and affirms the validity of the proposition.[441] The same is true for other notions, such as design.

---

[437] Ibid., pp. 28-9, quoting Leslie Stephens, *The Science of Ethics* (London: Smith, Elder and Co., 1882).

[438] Ibid., p. 95.

[439] Ibid., p. 175.

[440] Johnson, *The Wedge of Truth*, p. 152.

[441] *See*, Dawkins, *a devil's chaplain*, pp. 117 and 149. With respect to belief in a supreme being, and science's inability to disprove such existence, Dawkins concludes the only option is to be agnostic to the idea. He borrows a point from Bertrand Russell that "we

As noted elsewhere, Darwinist Richard Dawkins opens his famous book *The Blind Watchmaker* with the comment that "biology is the study of complicated things that give the appearance of having been designed for a purpose." This same sentiment is expressed by Sir Francis Crick in his autobiography where he states that "biologists must constantly keep in mind that what they see was not designed, but rather evolved."[442] Intelligent design theorist Phillip E. Johnson states in his book *The Wedge of Truth*:

> Darwinian biologists must keep repeating that reminder to themselves because otherwise they might become conscious of the reality that is staring them in the face and trying to get their attention.[443]

Johnson, proposing God as creator, notes the irony that "at one level [the idea God created us] is a truth we can't not *know*, and at another level it is a truth that we repeatedly *deny*."[444]

[50] *"When it comes to conflicting ideas like ours, we may each be wrong, but we can't both be right. Do you understand what I mean, Alex? Some ideas are exclusionary with respect to others, not all ideas can be right."*

Aristotle's Law of Noncontradiction is the foundation for all logical thinking. The chief principle of logic, the law states, "Something cannot *be* what it is and *not be* what it is at the same time and in the same sense or relationship." Applied to propositional statements, the law says "A statement cannot be true and not true at the same time and in the same respect." For example, the statements "the grass is green" and "the grass is red" cannot both be true at the same time and in the

---

must be equally agnostic about the theory that there is a china teapot in elliptical orbit around the Sun" (Ibid., p. 149). In an attempt at *reductio ad absurdum*, Dawkins equates the two disprovable ideas, concluding, "But that doesn't mean the theory that there is a teapot is on level terms with the theory that there isn't" (Ibid.). The fallacy in Dawkins thinking is revealed by considering his reasoning. He states, "As a lover of truth, I am suspicious of strongly held beliefs that are *unsupported by evidence*: fairies, unicorns, werewolves, any of the infinite set of conceivable and unfalsifiable beliefs epitomized by Bertrand Russell's hypothetical china teapot orbiting the Sun" (Ibid., p. 117, emphasis added). Even for Dawkins, therefore, the issue turns on the question of *evidence*. While there is no reason to believe a china teapot is orbiting the Sun, there is every reason to consider evidence of design as possibly pointing beyond nature, i.e., pointing to an intelligent designer.

[442] Johnson, *Wedge of Truth*, p. 153, quoting Francis Crick, *What Mad Pursuit* (New York: BasicBooks, 1988), p. 138.

[443] Ibid.

[444] Ibid.

same sense. Likewise, the statements "we are the product of intelligent design" and "we are not the product of intelligent design" cannot both be true at the same time and in the same relationship.

Understanding the implications of the Law of Noncontradiction is crucial to clear thinking on any issue. The idea that somehow certain contradictory conditions can co-exist ("what's true for you may not be true for me") is pervasive in modern culture. Reflecting on the idea that truths can be personal to individuals reveals the flaw in such a concept. For example, language is based on the Law of Noncontradiction: words have meanings; a word cannot mean one thing and not mean the same thing at the same time. If this were the case, language, communication, and knowledge itself would be impossible.

Because true knowledge is based on the Law of Noncontradiction through the use of language, it is important that words with precise meanings be used consistently. This is why Darwinists benefit by not being precise in their use of the term "evolution." By keeping the definition imprecise (*e.g.*, change over time), using it to describe very different things (*e.g.*, micro vs. macro evolution), and insisting that more precise definitions be avoided, Darwinists prevent an increase in knowledge and understanding of the subject. Saying "evolution happened" has no meaning outside of a precise, meaningful definition of "evolution," and debating the issue otherwise is futile.

As alluded to earlier, if "evolution" means "change over time," then we are all evolutionists. However, if "evolution" is defined as "a natural, unguided, information-producing process capable of creating specified complexity," then virtually no one would be an evolutionist. Further, whereas the former is trivial, and the latter is crucial, not one of the currently vogue natural explanations even attempts to explain information creation. The only theory that can do so is intelligent design theory, but like investors not willing to cut their losses because they are sure their stock will rise, most Darwinists will continue to believe in "change over time" believing that specified complexity will somehow be explained. This is an inevitable consequence of the nature of scientific paradigms. As Thomas Kuhn has well documented, paradigms change slowly, over generations. Gradually, however:

> [A]fter gaining a few supporters, men who will develop it to the point where hardheaded arguments can be produced and multiplied ... what occurs is an increasing shift in the distribution of professional allegiances.[445]

---

[445] Kuhn, *The Structure of Scientific Revolutions*, p. 158.

Kuhn notes that no single argument can be (or should be) expected to persuade all those of one particular view. In fact, Kuhn states, "At the start a new candidate for paradigm may have few supporters, and on occasions the supporter's motives may be suspect."[446] But because "scientists are reasonable men," Kuhn observes that over time,

> Still more men, convinced of the new view's fruitfulness, will adopt the new mode of practicing normal science, until at last only a few elderly hold-outs remain.[447]

Thus, the inevitable consequence of deep-rooted scientific paradigms is that truth takes time, and truth-seekers must sometimes wait out the "few elderly hold-outs," which themselves must ultimately yield to death. Those on the right side of truth, therefore, should take comfort in the long view, knowing that between truth and those who oppose it, only truth remains eternally unchanged.

[51] *"Why can't you just believe a person made the picture, and let others believe it was made naturally? There's room for all beliefs if we just show tolerance instead of dogmatism."*

"Room for all beliefs" is an irrational position if by this statement one means that all beliefs are valid and all or any can represent the truth about reality. As discussed above, the Law of Noncontradiction requires rigor with respect to according status to ideas. It is simply impossible (rationally) for each of two contradictory ideas to be right at the same time and in the same sense. Two contradictory ideas could both be wrong, but they cannot both be right. Tolerance with respect to hearing ideas and considering their implications is a good thing. That is, there should be "room for all beliefs" in the marketplace of ideas. But to uncritically accept ideas as true simply to be "tolerant" is not wise.

Ironically, while most Darwinists probably embrace "tolerance" and eschew "dogmatism" in theory, in practice they recognize the inviolability of the Law of Noncontradiction and choose to be dogmatic in their insistence that evolution, and only evolution, be taught in public schools. The Law of Noncontradiction never comes into play in the absence of competing ideas. While patronizing students and parents by assuring them it's "OK to believe in God," Darwinists

---

[446] Ibid.

[447] Ibid., p. 159.

demand that such belief be absolutely silent—dogmatically and intolerantly silent—with respect to origins of living things. If necessary, a sea of law and politics will be put into motion in a tidal wave of overwhelming stifling opposition to enforce the silence. Such politically enforced dogmatism is a good sign. Only false theories and insecure theorists cannot withstand the challenge of opposition. But truth, like water coursing downhill against an earthen blockage, will always find a way.

[52] *"Nothing in science makes sense except in the light of progression."*

This quote (using *evolution* in place of *progression,* of course) was made by evolutionist Theodosius Dobzhansky at a 1972 meeting of the National Association of Biology Teachers in San Francisco, California.[448] Dobzhansky was a prominent geneticist who has been called the "godfather of the modern synthesis."[449]

Quotes such as this are instructive for several reasons. First, as with all propositional statements about evolution, the truth of the statement depends on the meaning given to the term "evolution." If the term is used in its broadest sense of "change over time" then the statement could be true, but it is irrelevant. Whether or not "change over time" is necessary to understand science is not in dispute.

Second, taking the meaning of the term "evolution" to mean Darwinian "molecules-to-man" origin of species (as intended), the statement is manifestly untrue. Obviously there are many, many things in "science" that make sense without the least consideration of biological evolution. The study of planetary motion for example, makes sense now without a thought of evolution, and *made* sense long before any idea of natural, unguided evolution was theorized. Other examples can easily be imagined. Presumably it would strike Dobzhansky as surprising to learn that meteorologists, polymer chemists, aeronautical engineers, and scientists of almost all shapes and sizes are able to limp through their days without once considering natural selection.

Third, statements such as Dobzhansky's are often uncritically taken as truisms and used to form curriculum, policy and standards for science education.[450] Although the statement only assumes the truth of

---

[448] Witham, *By Design*, p. 27.

[449] Ibid., p. 19. The "modern synthesis" refers to the synthesis between Darwinian theory and genetic theory, producing what is termed as neo-Darwinism.

[450] For example, the National Association of Biology Teachers Statement on Evolution (adopted by the NABT Board of Directors on March 15, 1995) starts out with Dobzhansky's quote. As an aside, in an example of the subtle intimidation that accompanies official positions on evolution, the short NABT Statement ends with a review of court cases, and a reminder to teachers "to be mindful of these court cases, remembering that the law, science and NABT support them" (i.e., the teachers who teach evolution).

biological evolution, such presupposition is passed off as a reasoned position to be taken seriously. Once this kind of thinking is internalized as fact, evolution can be pressed into service at every turn, whether warranted or not.

[53] *"[T]hat many and grave objections may be advanced against the theory of events with modification through natural selective inclusion; he did not deny this."*

Darwin began the last chapter of his book *On the Origin of Species* with the following assessment:

> As this whole volume is one long argument, it may be convenient to the reader to have the leading facts and inferences briefly recapitulated. That many and grave objections may be advanced against the theory of descent with modification through natural selection, I do not deny. I have endeavoured to give to them their full force. [451]

With this frank account Darwin proceeded to recapitulate the difficulties on his theory of descent with modification through natural selection. Thus, not once, but twice in his monumental work he faced the difficulties of his theory with, in his estimation, "full force." Darwin identified certain "cases of special difficulty," such as "two or three defined castes of workers or sterile females in the same community of ants" and "geographical distribution [in which] the difficulties encountered on the theory of descent with modification are grave enough."[452]

In particular, Darwin was troubled by the lack of intermediate "linking forms."[453] Linking forms are animals (or plants) that are partly one species and partly another, in-between species by "gradations as fine as our present varieties." According to his theory, there should be "an interminable number of intermediate forms" such that "it may be asked, Why do we not see these linking forms all around us?"[454] Darwin asked the right question: "Why are not all organic beings blended together in an inextricable chaos?"[455] Darwin had an answer, and his reasoning and conclusion can be considered on its merits. However, that he anticipated the controversy and asked the question is

---

[451] Darwin, *Origin of Species*, p. 459.
[452] Ibid., p. 461.
[453] Ibid., p. 462.
[454] Ibid.
[455] Ibid., p. 462.

notable. Without a doubt, if Darwin were alive and teaching today he would be the first to "teach the controversy."[456]

One of the areas of controversy to which Darwin would give his full force is the lack of linking forms (including *living* linking forms). Since his theory requires innumerable linking forms, Darwin asked, "Why is not every geological formation charged with such links?"[457] That is, Darwin noted that the fossil record did not exhibit intermediate forms, as his theory would predict. "Why does not every collection of fossil remains afford plain evidence of the gradation and mutation of the forms of life?" he asked.[458] Further, he continued, "We meet with no such evidence, and this is the most obvious and forcible of the many objections which may be urged against my theory."[459]

As stated before, Darwin's honesty as a scientist is to be commended. His reply to the lack of fossil evidence at the time was, "I can answer these questions and grave objections only on the supposition that the geological record is far more imperfect than most geologists believe."[460] And, he explained, "Only a small portion of the world has been geologically explored."[461] Thus, Darwin contemplated whether "in future ages so many fossil links will be discovered" that naturalists will be able to decide whether or not these doubtful forms are varieties.[462]

Is the fossil record substantially different today? The answer, in short, is no. In Darwin's time the "lowest known fossiliferous strata" was the Silurian stratum that contained fully formed trilobites and other animals. (In a related "difficulty," Darwin recognized that some of these animals, such "as the Nautilus, Lingula, &c. ... do not differ much from living species."[463]) With respect to the sudden appearance of fully formed animals in the earliest fossils, Darwin stated:

> Consequently, if my theory be true, it is indisputable that before the lowest Silurian stratum was deposited, long periods elapsed ... and that during these vast, yet

---

[456] "Teaching the controversy" is the terminology that has evolved to describe the notion of teaching the criticisms of Darwinism along with the theory itself. Those who advocate teaching the controversy would, along with Darwin, approve of teaching not only the evidence for Darwinism, but the evidence against the theory. Of course, such a proposition is seen as a radical idea by present-day Darwinists, who believe it is simply impossible to marshal any evidence against Darwinism, except for mythical "religious" evidence.

[457] Ibid., p. 463.

[458] Ibid.

[459] Ibid.

[460] Ibid., p. 464.

[461] Ibid.

[462] Ibid.

[463] Ibid., p. 306.

> quite unknown, periods of time, the world swarmed with living things.[464]

Did the world ever "swarm with living things" prior to the Silurian period? Unfortunately for Darwinists, it appears not—at least not before the explosion of life that appeared during the Cambrian. The present-day state of the fossil record is virtually unchanged in terms of evidence for "swarms of living things." Rather than answer as Darwin did ("To the question why we do not find records of these vast primordial periods, I can give no satisfactory answer."[465]), however, today's Darwinists obscure the fact that the fossil record does not provide evidence of gradual descent with modification.

Vague statements like, "Fossils argue that evolution has happened"[466] are intended to imply that the fossil record is proof positive for the uninterrupted lineage of living organisms from simple to complex. Worse, misleading statements such as this one from the National Academy of Sciences are clearly intended to deliver partial truth:

> Fossils found in rocks of increasing age attest to the interrelated lineage of living things, from the single-celled organisms that lived billions of years ago to *Homo sapiens*. The most recent fossils closely resemble the organisms alive today, whereas increasingly older fossils are progressively different, providing compelling evidence of change through time.[467]

The first sentence of this statement is unquestionably, obviously, and inexcusably wrong, as it is unsupported by any scientific evidence. There simply is no demonstration of an "interrelated lineage" from single cells to Homo sapiens. The second sentence is undoubtedly correct, but it is irrelevant to the point of biological "single cell to Homo sapiens" evolution. Whether fossils are "compelling evidence" for "change through time" is not disputed by anyone. The scientific inquiry is to determine what kind of change has taken place.

In fact, many of today's oldest known fossils come from the geological period before the Silurian, known as the Cambrian, named

---

[464] Ibid., p. 307.

[465] Ibid.

[466] Weiner, *Beak of the Finch*, p. 7. Without intending to be annoyingly repetitive on this point, I note that this statement is meaningless without a precise definition of "evolution."

[467] NAS, *Teaching About Evolution*, p. 16.

after rocks in Cambria, Wales.[468] The fossils of the Cambrian rocks, however, like those of the Silurian, exhibit fully-formed phyla and classes of animals.[469] In fact, so sudden is the appearance of fully-formed phyla in the Cambrian that this period is referred to as the "Cambrian explosion" or "biology's big bang."[470] Much has been written on the Cambrian explosion and its inconsistency with Darwin's theory. As evolutionary theorist Jeffrey Schwartz puts it, the major animal groups "appear in the fossil record as Athena did from the head of Zeus—full blown and raring to go."[471]

The fossil record is at best ambiguous, and at worst contradictory, with respect to the "fact of evolution." Whatever it shows, however, is far from "plain evidence of the gradation and mutation of the forms of life" as Darwin predicted and illustrated as a "tree of life" that illustrated branched relationships from a common ancestor. The Cambrian explosion remains a paradox, totally inconsistent with Darwin's tree of life.[472] Alternative theories going by names such as "molecular phylogeny" are equally unhelpful in supporting biological evolution as being consistent with the fossil record. It seems that the missing links are non-existent links, and the "tree of life" is actually a "mis-tree of life"—a mystery that the fossil record cannot solve. For a more complete treatment of the problems with the fossil record, as well as Darwin's "tree of life" and the topic of "missing links" such as *Archaeopteryx,* the reader is referred to Jonathan Wells' book *Icons of Evolution,*[473] as such pervasive and persuasive icons are difficult to treat adequately in the present work.

[54] *"[H]e is arguing that a mechanism that can produce stalactites can also produce cave paintings, it's simply a matter of the degree of complexity, which is really just a question of time."*

Although some Darwinists deny the importance or relevance of the distinction, there are two generally recognized categories of "change" attributed to "evolution." The first, so-called *microevolution,* refers to observed change below the species level. The second, *macroevolution,* refers to speciation, the creation of new organs, new features, and new and more complex DNA, in short, new beings. Documented evidence of microevolution—changes in peppered moth populations, changes in finch beak sizes, and changes in bacterial

---

[468] Wells, *Icons of Evolution,* p. 35.

[469] Ibid.

[470] Ibid., p. 37.

[471] Jeffrey H. Schwartz, "Homeobox Genes, Fossils, and the Origin of Species," *Anatomical Record (New Anatomist)* 257 (1999), pp. 15-31, as quoted by Wells, *Icons of Evolution,* p. 41.

[472] Wells, *Icons of Evolution,* p. 48.

[473] Wells, *Icons of Evolution.*

resistance—are all taken as "proof" of "evolution," and, by extension, proof of speciation by macroevolution.[474]

Many Darwinists believe that the distinction between micro- and macroevolution is simply a difference in complexity, and that macroevolution is just microevolution continued and extended over long periods of time. Some, such as Darwinian evolutionist Frederick Crews, believe, "[by] conceding the reality of microevolution, the intelligent design people have already conceded all of evolution."[475] Others simply conflate the two notions as if there is no distinction to be accounted for. In his Pulitzer Prize-winning book *The Beak of the Finch*, Jonathan Weiner, speaking candidly of the heretofore lack of evidence for "evolution" states:

> Fossils argued that evolution has happened. Logic argued that natural selection can make it happen. But neither bones nor logic could demonstrate the one leading to the other, natural selection causing evolution.[476]

Clearly implying macroevolution, an avowed evolutionist made this surprising admission only because he believes that documentation of finch beak variation is the "missing link" of evolutionary theory, i.e., the demonstration of "the one leading to the other." In fact, Weiner is sure that the cyclic beak size variation of Darwin's finches "is the best and most detailed demonstration to date of the power of Darwin's process."[477] Indeed it is! And this is the point of making the distinction between micro- and macroevolution. The former can be scientifically demonstrated, the latter cannot. No one doubts that finch beaks may vary in size from one generation to the

---

474 As noted previously with respect to language and terminology, to have a fruitful discussion on the topic the term "species" must be defined. In the evolutionist literature the term species carries somewhat of a hazy definition. Darwin himself considered it a loose term and balked at defining it. *See*, Perloff, *Tornado in a Junkyard*, p. 46. Note, however, that intelligent breeding by humans can produce remarkable variation in pigeons, dogs, cows, etc., but it has never produced another species by any definition. Even the demarcation of the thirteen "species" of finches (now virtually deified as "Darwin's finches,") is not without controversy. Prominent Darwinists have differed over the delineation of "species" of Darwin's finches. Are there really thirteen species, or one species with variation not unlike what we see in a single species of pigeons, for example? Evolutionist ornithologist David Lack, for instance, stated in a monograph entitled *Darwin's Finches*, "In no other birds are the differences between species so ill-defined" (Weiner, *Beak of the Finch*, p. 43).

475 Witham, *By Design*, p. 164.

476 Weiner, *Beak of the Finch*, p. 7.

477 Ibid., p. 9.

next, but such variation within a species is not proof of creation of the species in the first place.

Like anti-biotic resistance or peppered moth color variation, the heritable change observed in the beak size of Darwin's finches is quite possibly convincing evidence of the power of natural selection. However, even granting the point that such observations are compelling proof of natural selection says nothing about creation of the organism, i.e., the origin of the species. Natural selection, by definition, can preserve only what is already created. In the example of Darwin's finches, as related by Weiner, no new beaks were created. No new organs came into existence. No new morphology was exhibited. In essence, at best the changing beak sizes demonstrated the non-controversial, well-documented, reasonable, and expected change capable within the limits of a species' genetic make-up, i.e., micro-evolution. The leap from micro-evolution to macro-evolution can be made, but it is a leap of faith, not a leap of science.

[55] *"We need to convince Dr. Hawkins on the basis of facts alone, regardless of the truth of the Legend."*

Truth is truth—particularly the truth of the historical account of our human development. If books that purport to have revelation on the historical account, such as the Bible, represent truth, such revelation should be borne out scientifically. While the details of creation and the creator can, and will, be debated for years to come, intelligent design theory provides a platform for scientific inquiry independent of any so-called revelation knowledge. Conversely, however, it should not be surprising that if the Bible does reveal truth, it is consistent with the scientific findings of design in nature.

The fact that intelligent design theorists are overwhelmingly Bible-believing creationists is not surprising. However, this fact alone is somewhat irrelevant to the scientific validity of intelligent design. If intelligent design theory represents the truth of the matter with respect to the origin of living organisms, such a position is not related to the validity or veracity of books such as the Bible. To simply link "intelligent design" immediately and necessarily to the Genesis account of creation is a lazy evolutionist's way of avoiding the necessity of refuting scientific arguments for design. But not all Darwinists are lazy, and not all intelligent design theorists are interested in defending the biblical account of creation. Together, it may be that "facts alone" will convince some Darwinists to objectively consider the evidence for design in nature. The separate inquiry of finding out who the (or a) designer *is*, can be set aside for another discipline to handle, informed by science if necessary.

[56] *"The stereotypes, the clichés. It seems just mentioning the Legend conjures up images of naïve simpletons trying to turn back the clock of science."*

In his book *The Wedge of Truth,* intelligent design theorist Phillip E. Johnson writes that the "creation-evolution controversy is one of those subjects that has become standardized in the press."[478] As Johnson says:

> I sometimes have the impression that journalists just click on a 'bash creationism' macro in their word processors and sit back while the printer pours out a string of clichés: the Catholic Church persecuted Galileo, the Scopes trial in 1925 should have settled the matter, the Bible is not a scientific textbook, scientists agree that 'evolution has occurred,' mainstream religious leaders say that God and evolution are compatible, and the country will fall into ruin if evolution is not emphasized in the schools.[479]

Like other topics of controversy in today's "culture wars," progress in understanding will happen only when both sides agree on precise definitions of terms, consistently insist on rigorous logical reasoning, and openly admit to assumptions. In particular, with respect to things that no doubt have a historical basis, and therefore have objective truth qualities, the debate should be able to go beyond clichés and stereotypes.

[57] *"But to produce a meaningful picture you have to have some way of getting* complex, specified information."

The fundamental claim of intelligent design theory is that intelligent causes are empirically (i.e., scientifically) detectable, and that such detectable intelligent causes are necessary to explain the complex, information-rich structures of biology.[480] One of the primary empirical methods involves the ability to detect complex, specified information. Complex, specified information is an important key to identifying design in anything, whether it be inanimate objects or animate living creatures. Together with other concepts, such as "irreducible complexity," complex specified information can be detected, tested, and falsified by the scientific method to determine with a high probability that something was designed.

---

[478] Johnson, *Wedge of Truth*, p. 73.

[479] Ibid., pp. 73-4.

[480] *See*, Dembski, *Intelligent Design*, p. 106.

Thus, the importance of the concept of complex, specified information lies in its scientific underpinnings. As intelligent design theorist William Dembski stated, "Science, we are told, studies natural causes, whereas to introduce God is to invoke supernatural causes."[481] However, as Dembski notes, a contrast that forces the debate in terms of "God or science," is the wrong contrast—one that focuses on the wrong distinctions. Dembski, a leading proponent of the scientific concept of complex, specified information, states, "The proper contrast is between *natural causes* on the one hand and *intelligent causes* on the other."[482] Dembski elaborates:

> Intelligent causes can do things that natural causes cannot. Natural causes can throw scrabble pieces on a board but cannot arrange the pieces to form meaningful words or sentences. To obtain a meaningful arrangement requires an intelligent cause. Whether an intelligent cause operates within or outside nature (i.e., is respectively natural or supernatural) is a separate question entirely from whether an intelligent cause has operated.[483]

Of course, the idea of "intelligent causes" is not new, and while intuitively appealing, such ideas in the area of biological evolution have for years been successfully stashed away on the shelf labeled "religion." However, the contemporary theory of intelligent design differs from the intuitive philosophical notions embodied in past design arguments in one very important way. Specifically, contemporary intelligent design theorists have established precise methods for *scientifically* distinguishing intelligently caused objects from unintelligently caused ones.[484] To do this, Dembski, for example, sets forth what he terms the "Complexity-Specification Criterion" for detecting design.[485] While the interested reader is referred to Dembski's book *Intelligent Design* for details, the following very brief summary is provided here:

> Whenever we infer design, we must establish three things: *contingency, complexity* and *specification.* Contingency ensures that the object in question is not the result of an automatic and therefore unintelligent process that had no choice in its production.

---

[481] Ibid., p. 105.

[482] Ibid.

[483] Ibid.

[484] Ibid.

[485] Ibid., p. 127.

> Complexity ensures that the object is not so simple that it can readily be explained by chance. Finally, specification ensures that the object exhibits the type of pattern characteristic of intelligence.[486]

Each of these three characteristics can be used to compare objects and things to determine evidence of intelligent causes. Thus, while each of these three ideas has its respective conceptual framework and difficulties, together they represent a powerful alternative to the chance/law limitations inherent in Darwinism. Dembski states:

> When called to explain an event, object or structure, we have a decision to make—are we going to attribute it to necessity, chance or design? According to the complexity-specification criterion, to answer this question is to answer three simpler questions: Is it contingent? Is it complex? Is it specified?[487]

The systematic answering of the three questions above constitutes what Dembski refers to as the "explanatory filter."[488] An affirmative answer to the three questions, for example, "filters" out necessity and chance as sufficient explanations for an event or structure, leaving design as the only rational, scientific alternative (or, at least a highly probable, reasonable alternative).

Dembski explains that what makes the complexity-specification criterion work to reliably detect design is recognizing that the principal characteristic of intelligent agency is choice.[489] For an intelligent being to act is therefore to choose between a range of competing possibilities.[490] Given this characterization of intelligent agency, Dembski maintains, the crucial question is how to recognize it, i.e., how do we recognize that an intelligent agent has made a choice?[491] While difficult to summarize Dembski's work in the format of this book, the gist of the answer to this question is:

---

[486] Ibid., p. 128. Interestingly, Darwinists such as Richard Dawkins also recognize specified-in-advance complexity in living things. As discussed above, rather than reach the logically reasonable position of inferring intelligent design, Dawkins proposes an illogical "specified-in-advance-with-hindsight-but-not-really-specified-in-advance" idea that is difficult to follow rationally, and difficult to accept reasonably.

[487] Ibid., p. 133.

[488] Ibid.

[489] Ibid., p. 144.

[490] Ibid.

[491] Ibid.

> [T]o recognize intelligent agency we must observe an actualization of one among several competing possibilities, note which possibilities were ruled out and then be able to specify the possibility that was actualized.[492]

Elaborating, Dembski adds:

> What's more, the competing possibilities that were ruled out must be live possibilities and sufficiently numerous so that specifying the possibility that was actualized cannot be attributed to chance.[493]

Dembski's complexity-specification criterion and the related explanatory filter are powerful tools to explore the presence of design in living organisms. Based solely upon empirically observable characteristics of living things, the tools meet all the requirements for scientific investigation and theory. Of course, such a concept is not likely to receive a warm welcome in a humanistic scientific community, and, predictably, the theory is almost universally cast in terms of "religion" and proof of the "direct hand of God." For example, in the National Academy of Sciences booklet entitled "Science and Creationism, A View from the National Academy of Sciences,"[494] the NAS states the following (after discussing "molecular evolutionary data"):

> Molecular evolutionary data counter a recent proposition called "intelligent design theory." Proponents of this idea argue that structural complexity is proof of the direct hand of God in specially creating organisms as they are today.[495]

Such a statement is hardly worth countering; but for completeness as well as exposure of the kind of unscientific reasoning employed by the NAS, the statement must be answered. First, the term "structural complexity" is not part of intelligent design theory—"specified

---

[492] Ibid., p. 146.
[493] Ibid.
[494] NAS, *Science and Creationism: A View from the National Academy of Sciences* (Washington DC: National Academy Press, 1999).
[495] Ibid., p. 21.

complexity" is the correct term, both by definition and by usage.[496] Specified complexity, as discussed above, is a precise term that is used consistently by intelligent design theorists.

Second, intelligent design theory does not propose a "hand of God" either directly or indirectly.[497] This rhetoric is simply and transparently part of the agenda of Darwinists to paint any theory that might remotely implicate the supernatural as "religious." The terms "specially created" add a particularly sinister twist to the "hand of God" slight. "Special creation" is a term applied to the act of creation as recorded in Genesis, the Bible. Special creationist scientists are a particularly hated breed among the propagandists of the NAS.

Third, intelligent design theory does not postulate that organisms were created "as they are today." In fact, not even the vile "special creationists" argue for this position.[498] Intelligent design theorists readily accept the fact of microevolution producing variants within species. There is no requirement in intelligent design theory that organisms were designed by an intelligent cause in the same form as they are today.

Dembski's complexity-specificity criterion offers a powerful alternative to the chance/law constraints of Darwinism. As with any scientific theory, intelligent design, including Dembski's complexity-specificity criterion, is not without its objections. Dembski deals with long-standing objections to design in a lengthy Appendix to his book. Again, the interested reader, including any brave true believers, are

---

[496] Or, perhaps, "irreducible complexity," discussed elsewhere herein. Both terms are precisely defined and consistently applied in the intelligent design literature.

[497] On the contrary, as clearly articulated by leading intelligent design theorists, the theory seeks to locate design, not the designer. For example, Dembski states unequivocally, "As a result [of intelligent design being the study of informational pathways induced by intelligent causes], intelligent design presupposes neither a creator nor miracles. Intelligent design is theologically minimalist. It detects intelligence without speculating about the nature of the intelligence" (Dembski, *Intelligent Design*, p. 107).

[498] In fact, such an argument is inconsistent with, for example, the Genesis account of creation. If all the variation of *Homo sapiens* is descended from Adam and Eve, then even special creationists must "believe in" microevolution—that is, change at the species level. This obvious fact is apparently ignored for rhetorical purposes with Darwinists that have no interest in understanding, but simply want to cast any creationist position in the worst possible light. For example, the same mistake is made by leading evolutionist Ernst Mayr writing in his book *What Evolution Is*, where he characterized "creationism" by stating as one of its beliefs: "Everything in the world today is still as it was when it was created" (Mayr, *What Evolution Is*, p. 4). Apparently, Mayr failed to do a modicum of research to understand the creationist position. If he had done so, he surely would understand his statement to be demonstrably false. Not only is his statement not true with respect to living organisms, including humans, but it cannot even be said to be true with respect to the geological record. Not intending to fight the creationists' battles here, my point is simply that it is unfortunate that such respected Darwinists as Mayr and his kind at the NAS cannot take the time to accurately represent a basic tenet of those they claim are so dreadfully wrong.

referred to Dembski's book for a more complete treatment of this important concept. What one will discover is that even if life could get started by a miracle of happenstance to achieve the first instance of specified complex information, it is highly unlikely that further, increasingly specified complex information could come about by happenstance. Such a happy event is like an elevator accidentally coming together—but only able to go up; it may work perfectly the first time, but then what? To continue going up it must have the cooperation of natural forces to build more floors, an elevator shaft, cables, and all the while without any direction-giving input and without falling back down to the bottom. Such a scenario seems highly unlikely, to put it mildly.

[58] *"Well, I don't know if you* can't *[get a good copying error]. I guess anything is possible, but such a thing is called a permutation, and it's almost always fatal to information ... and the results ... are called permutants. I've never seen a good permutant."*

Darwinism, and specifically neo-Darwinism, is built on the idea of heritable variation supplied by genetic mutation. Mutations are the changes in the sequence of nucleotide bases in DNA. Those changes supply the variation on which natural selection can work. A crucial observation is that neither the sequence nor the changes are dictated by any known chemical or physical law. The lack of any law that dictates the sequence was actually predicted by Watson and Crick and aided their discovery of the structure of DNA:

> So in building models we would postulate that the sugar-phosphate backbone was very regular, and the order of bases of necessity very irregular. If the base sequences were always the same, all DNA molecules would be identical and there would not exist the variability that must distinguish one gene from another.[499]

This lack of a law was also described much earlier by Jacques Monod in his famous essay *Chance and Necessity* as the principle mystery in biology:

> The ultimate ratio of all the teleonomic structures and performances of living beings is thus enclosed in the sequences of residues making up polypeptide fibers. ... In a sense, a very real sense, it is at this level of chemical organization that the secret of life lies, if indeed there is any one such secret. And if one were able not only to describe these sequences but to

[499] James Watson, *The Double Helix: A Personal Account of the Discovery of the Structure of DNA* (Touchstone 1968) pp. 52-4.

pronounce the law by which they assemble, one could declare the secret penetrated, the ultimate ratio discovered.[500]

Due to the lack of any law that dictates the sequence, and because he embraces methodological naturalism, Monod is compelled to attribute the order of the sequences to chance. Neither Monod nor other Darwinists test this conclusion by any statistical analysis because there is no other alternative if law does not play a role and design is ruled out by philosophy. Chance or random mutations are thought to be the product of a variety of influences, including radiation, foreign chemicals and copying errors. Darwinists claim that natural selection can preserve the variations that arise from such random changes to the software in germ cells to the extent that the changes manifest new function that confers a reproductive advantage on a mutated gene's "owner."

As a theory, the idea of genetic mutations providing a smorgasbord of choices for natural selection to browse among has a certain attraction. Patiently and coldly selecting between the weak and the strong, natural selection preserves organisms having "good" mutations while callously discarding the "bad." Over time, "selection pressure," like inanimate peer pressure, causes shifts in form and function.

In practice, however, the picture is quite different. Try as they might to make a case for mutation being a designer substitute, in real life mutations stubbornly resist Darwinian peer pressure, as they are almost always neutral, deleterious or fatal.[501] What seemed like a bonanza when first proposed as the main tenet of neo-Darwinism in the 1930's has turned out to be a bust for origin of species researchers. In 1937, Theodosius Dobzhansky wrote that "mutations and chromosomal changes ... constantly and unremittingly supply the raw materials for evolution."[502] Further research seemed to confirm this statement, including the discovery in the 1950's that DNA might transmit heritable traits. By the 1970's molecular biologist Jacques Monod announced that "the mechanism of Darwinism is at last

---

[500] Jacques Monod, *Chance and Necessity* (Vintage Books 1971), pp. 95-6, emphasis added.

[501] Depending on what one means by "good" and "bad" one can make the case that all mutations are bad. Examples such as the mutation for sickle cell anemia, for example, are touted as good mutations. However, even this mutation has its downsides.

[502] Theodosius Dobzhansky, *Genetics and the Origin of Species* (New York: Columbia University Press, 1937), p. 13, as quoted in Wells, *Icons of Evolution*, p. 181.

securely founded."[503] However, as molecular biologist Jonathan Wells[504] concludes:

> We now know that some DNA mutations are "neutral"—they have no effect at all. The vast majority of the rest are harmful. In the struggle for existence, natural selection would be expected to ignore the former and eliminate the latter. Only those rare mutations which benefit the organism could be favored by natural selection, and thus provide raw materials for evolution. Some mutations that affect biochemical pathways fit this description.[505]

Among the beneficial mutations Wells mentions are biochemical mutations—such as those leading to antibiotic resistance and sickle-cell anemia. However, as Wells notes, such mutations, considered beneficial in certain environments, nevertheless do not affect an organism's shape or structure. As Wells puts it, "evolution needs beneficial mutations that affect morphology."[506] Unless a mutation affects morphology—the shape of an organism—it cannot provide raw materials for morphological evolution.[507]

Darwinists do have one example of mutations leading to morphological changes. Although not naturally occurring, the intelligently and carefully designed four-winged fruit fly is becoming an increasingly popular icon of evolution.[508] Is this lab-created creature "proof" of evolution? The extra so-called "wings" lack flight muscles and are useless for flying so that they are of little benefit to the fruit fly.[509] Nevertheless, the useless wings do confer a benefit on Darwinists, and are, therefore, preserved by un-natural selection as a "proof" of the power of evolution. Such proof on the backs of captive,

---

[503] Horace Freeland Judson, *The Eighth Day of Creation: The Makers of the Revolution in Biology* (New York: Simon and Schuster, 1979), p. 217, as quoted in Wells, *Icons of Evolution*, p. 181.

[504] Jonathan Wells is a post-doctoral biologist and holds Ph.D.'s from both Yale University and the University of California at Berkeley (molecular and cell biology).

[505] Wells, *Icons of Evolution*, p. 181.

[506] Ibid., p. 182.

[507] Ibid., p. 178.

[508] The term "icon" is borrowed from, and used in the same sense as it is in Jonathan Wells' book *Icons of Evolution*. For a complete analysis of the development and importance of the four-winged fruit fly, the reader is referred to Wells' book.

[509] Imagine being bred with a pair of useless legs hanging about your waist. Other than being quite a curiosity for some and profit for others, you certainly would not consider yourself beneficially evolved.

malformed and handicapped fruit flies can earn Nobel Prizes for those doing pioneering work in this area.[510]

Even something like the extra wings on a fruit fly is not an example of evolution, however, if evolution means the natural process of morphological change. Like short ears on a dog, or long feathers on a bird, the extra "wings" on the four-winged fruit flies are simply a human-engineered modification of an existing body part. Such modifications are what Darwin referred to as artificial selection (albeit in this case by genetic modification), and no one denies the power of artificial selection.

Therefore, while Darwinists can glibly speak of mutations leading to new species, in fact the theory is very different from reality. For example, Richard Dawkins states:

> Evolution occurs because, in successive generations, there are slight differences in embryonic development. The differences come about because of changes (mutations—this is the small random element in the process that I spoke of) in the genes controlling development.[511]

In fact, Dawkins is exactly, 100% right. "Evolution" does "occur" because of mutations, but it is not the kind of evolution he has in mind. The random variations that mutations produce are almost without exception harmful to the organism possessing them. Paul Davies clearly recognizes the issue and comes closest to putting it in perspective:

> Mostly a mutation is damaging, just as a random word change in a Shakespeare sonnet would likely mar its beauty. But occasionally, quite by chance, an error might produce a positive benefit, conferring an advantage on the mutant.[512]

To stay with the metaphor, the idea that over time, millions of chance "word changes in a Shakespeare sonnet" would ever, by any means and in any length of time, do anything but "mar its beauty" is, again,

---

[510] Wells, *Icons of Evolution*, pp. 189-90. Without producing one morphological mutation that would benefit a fruit fly in the wild, German geneticists Christiane Nuesslein-Volhard and Eric Wieschaus shared the Nobel Prize for their work to find every possible mutation involved in fruit fly development.

[511] Dawkins, *Blind Watchmaker*, pp. 50-51.

[512] Davies, *Fifth Miracle*, p. 42.

not only "hard to swallow,"[513] it is beyond reason. It is easy for Darwinists to allege such a scenario, but alleging and supporting with evidence are two different things.

Reliance on the hypothetical "occasional beneficial mutant" is another example of an "up is down," "drawback-played-as-advantage" stance that Darwinists must insist upon by theoretical necessity, not because of any evidence for the position. But by and large people see the truth and reject the idea that random mutations can ever—much less consistently over time—help survival. As James Perloff writes:

> This explanation [that mutations produced all the changes that have brought the microbe up the ladder to a man] faces serious difficulties, however. Mutations are almost universally *harmful*. In human beings, they are classed as "birth defects." They often result in death or sterility. People today suffer from more than 4,000 disorders caused by gene mutations. Down's syndrome, cystic fibrosis, sickle cell anemia are familiar examples. But what benefits can we list from mutation? Who hopes his child will be born with one?[514]

To alleviate some of the obviously negative implications of gene mutation, Darwinists differentiate between so-called *macro*mutations and *micro*mutations, the former being the bad kind, and the latter being tiny and virtually undetectable, presumably the potentially good kind. Postulating micromutations, while perhaps avoiding the observable negative effects of macromutations, creates other problems for Darwinists. For one, a theory that relies on millions, probably billions of tiny, undetectable, undirected and random micromutations of the right type coming along when needed over

---

[513] Ibid. Davies acknowledges that the related idea of an instruction manual that writes itself simply by accumulating chance errors is "hard to swallow." But incredulity alone does not entirely account for why people don't "believe" in evolution. Many things are hard to swallow, but are nevertheless accepted as true. The constancy of the measured speed of light goes against all intuition and is hard to swallow, but it follows from observed and theoretical facts—i.e., it is *rational* and people "believe" it. What is at issue here is not what is "difficult to believe," but "what follows rationally from the facts." This is the gulf that Darwinists are apparently willing to trivialize.

[514] Perloff, *Tornado in a Junkyard*, p. 25, emphasis in original, citing Fritjof Capra, *The Web of Life* (New York: Anchor Books, 1996), p. 228 for the number of disorders caused by gene mutations.

millions of years, stretches one's capacity for reason.[515] As Phillip E. Johnson observes:

> The probability of Darwinist evolution depends upon the quantity of favorable micromutations required to create complex organs and organisms, the frequency with which such favorable micromutations occur just where and when they are needed, the efficacy of natural selection in preserving the slight improvements with sufficient consistency to permit the benefits to accumulate, and the time allowed by the fossil record for all this to have happened.[516]

It's one thing to believe such a scenario is probable. It's quite another to allege such a theory to be a "fact."

There is, however, a bigger problem associated with the idea of mutations, whether macro or micro. Regardless of the size of the mutation, and whatever the frequency, it is highly speculative and against reason to theorize that accidental, chance, random mutations can increase the *information content* of an organism's genome.[517] The reason birds have wings is because birds have *genes* that code for wing production with specific direction provided by the coded information. For a fish to develop wings, irrespective of how gradually, it first has to have genes for wings that provide the necessary information.[518]

Honestly conceding the existence and meaning of information-rich structures is likely the silver bullet for biological evolution, which is one reason why Darwinists support ideological gun control on the whole topic. Genetic information is virtually analogous to the software used to program and run computers. Both genetic information, such as in DNA, and a computer program have information content which permits the cell or the computer, respectively, to do something—to

---

[515] This is not an argument from incredulity. Rather than simply being unbelievable, the notion of many millions and billions of random, undirected mutations over billions of years accumulating to create the specified complexity of a human being, is beyond reason, i.e., it is irrational.

[516] Johnson, *Darwin on Trial*, p. 38.

[517] A *genome* is an organism's total genetic code, which, like a computer code, instructs the construction of the organism.

[518] *See*, Stephen C. Meyer, *The Origin of Biological Information and the Higher Taxonomic Categories*, Proceedings of the Biological Society of Washington, (117(2):213:239. 2004). Meyer argues, citing numerous studies by other scientists, that no current materialistic theory of evolution can account for the origin of the information necessary to build novel animal forms. Proposing intelligent design as an alternative explanation, arguing that new life forms such as those exhibited in the Cambrian explosion can only be attributed to a "remarkable jump" in complex specified information. No known material cause exists for such a jump.

purposefully carry out a task. To entertain the idea that random copying errors in a simple software program that, say, repeats the alphabet over and over, can be repeatedly replicated such that over time random copying errors produce a program such as the one that is now recording these words, is quite unreasonable, to say the least. Likewise, to think that some rudimentary genetic material gradually, over time, reproduced itself with innumerable, chance, random errors to produce the information-rich DNA of humans goes beyond scientific reasoning and reaches deep in to speculative wish projection.

The difference between the computer program analogy and Darwinism, Darwinists would say, is that the replicating computer program lacks a non-random selector, such as natural selection, to select out advantageous algorithms or subroutines. Natural selection is supposedly a "direction-giving force"[519] that acts as a great sieve to continually "sift" mutants, and acts like a "ratchet, locking in the advantageous errors and discarding the bad."[520] A self-replicating computer program does not have such a mechanism to "ratchet" it directionally toward more specified complexity. Therefore, say Darwinists, the analogy fails.

Does the analogy fail? On the contrary, it is exactly on the point of the "direction-giving" powers of natural selection to act as a one-way "ratchet" to higher specified complexity that *Darwinism* fails. Even Davies, who makes the most reasoned attempt at supporting the one-way sifting of natural selection, cannot in the end but restate the tautology of natural selection: "if it works, keep it; if it doesn't, kill it."[521] However, the following review of Davies' reasoning is instructive in understanding why natural selection, in fact, has no direction-giving capability, and to attribute any such power is simply to indulge in wish projection to avoid admitting evidence of design.

After relating the metaphor of randomly replicating Shakespeare and self-replicating instruction manuals, Davies writes:

> Some people find the idea of an instruction manual that writes itself simply by accumulating chance errors hard to swallow, so let me go over the argument once more, using a slightly different metaphor. Think of the information in human DNA as the score for a symphony. This is a grand symphony indeed ... By comparison, the DNA of the ancient ancestor microbe

---

[519] Darwin, *Origin of Species*, p. *xvi*; introduction by Ernst Mayr. Mayr's complete quote: "That natural selection is a direction-giving force, within the limitations of the evolutionary potential set for a given species by its genotype, has now been substantiated abundantly."

[520] Davies, *Fifth Miracle*, p. 42.

[521] Ibid., p. 43.

> is but a simple melody. How does a melody turn into a symphony?[522]

Thus, Davies clearly sets up the problem. And remember, this "slightly different metaphor" is supposed to shed light on what the previous metaphors lacked: how can one relatively simple, orderly, specified-complex thing turn into a more complex, highly ordered, specified-complex thing by simply accumulating chance errors? Davies continues:

> Suppose a scribe is asked to copy the original tune as a musical score. Normally the copying process is faithful, but once in a while a quaver becomes a crotchet, a C becomes a D ... Mostly these mistakes will spoil the balance or harmony, so that the score is of no further use; nobody would wish to listen to its musical rendition. But very occasionally the scribe's slip of the pen will add an imaginative new sound, a pleasing feature, a successful addition or alteration, quite by chance. The tune will actually improve, and be approved for the future. Now imagine this process continuing through trillions of copying procedures. Slowly but surely, the tune will acquire new features, develop a richer structure, evolve into a sonata, even a symphony.[523]

While not being brazen enough to explicitly suggest a "target phrase," (as does Dawkins; see Note 6 above) Davies' example suffers from the same fallacy as Dawkins' metaphor of typewriting monkeys. The metaphor suggests that there is some "target" or some "ideal" that is being "ratcheted" toward. Just what are the criteria for "a pleasing feature"? What might be a "successful addition"? And, most importantly, what could possibly be accidentally "approved for the future"? This imagery makes sense only if there is a criterion for comparison, with the ability to pick and choose among alternatives. But Davies assures us this is not the case:

> The crucial point about this metaphor, and it cannot be stressed too strongly, is that the symphony comes into being without the scribe's ever having the slightest knowledge of, or interest in, music. The scribe might

[522] Ibid., p. 42.

[523] Ibid., p. 43.

> have been deaf from birth and know nothing whatever about melodies.[524]

How, then, is this different from the randomly self-replicating instruction manual? Just like Dawkins' blind watchmaker, the deaf scribe can make only random changes with no thought of, nor ability to, choose among alternatives. Therefore, *by definition*, there are no criteria for "approving for the future" any sounds; there can be no basis for judging a "pleasing feature"; and there can be no improvement of the tune (what would "improvement" be, without an ideal?). This metaphor is no easier to "swallow" than that of instruction manuals randomly copying themselves.

Even Davies recognizes the flaw in the metaphor, and this is the key to understanding why Darwinian natural selection fails. Davies concludes:

> Where the metaphor fails is in the selection process. There is no cosmic musician scrutinizing the score of life and exercising quality control.[525]

Exactly! And for precisely the same reason the metaphor fails, natural selection itself fails as a direction-giving force! Unless there is some "cosmic musician scrutinizing the score of life" there is no *reason to believe* that random copying errors can produce a symphony.

It's not only that it is so improbable as to be impossible, the Darwinian mechanism, which operates just like a series of random environmental sieves, lacks any capacity to contemplate the future—to order events to attain a pre-determined goal or end. As the NABT explains, natural selection has "no discernable direction or goal." This is because it lacks a mind or some similar kind of information-processing systems that can make and implement choices. Furthermore, the laws of probability, the laws of thermodynamics, the laws of *nature*, all militate against such a happening. And yet it is exactly on this issue—the "selection process"—that Darwinism must be supported scientifically. But whether it is Dawkins' "target phrase"[526] or Davies' "approved for the future" musical score, any models, any metaphors, any examples—all rely on some kind of desirable end point to reasonably make the selection process viable. Otherwise, there is no reason to believe there is a "selection" process.

---

[524] Ibid.

[525] Ibid. The deaf scribe is a poor metaphor for another reason. A symphony is a polyphonic form; to write one would take multiple deaf scribe's errors, each building harmonizing instrumental themes.

[526] *See*, Note 6 above.

Like Dawkins' blind watchmaker and typewriting monkeys, Davies' deaf scribe metaphor provides a *reasonable* example only when "selection" is based on comparison with an ideal. Therefore, describing how a blind or deaf selection works based on a "target phrase" or a "pleasing feature" serves only to prove the opposite—that in fact, without an identifiable, specified in advance end point, it is *against reason* to believe that random copying errors *can*, much less *will*, produce increasingly precise specified complexity.

Davies and others insist that the selection criterion is "replication efficiency."[527] But replication efficiency does not explain how random errors can make Dawkins' "target phrase" or Davies' "symphony" without some way of choosing among alternatives. Copying nonsensical music faster and more prolifically has no bearing on the quality of the music. Typing random letters more furiously and easily guarantees more randomness, not Shakespeare. So "replication efficiency" in any metaphor makes scientific sense only with an implied intelligent agent to make selection among possible choices to achieve the requisite efficiency. There is no reason to believe that the metaphorical requirement is any less an actual requirement for living systems. Why, if the metaphors require a specified-in-advance criterion, does not the actual theory require the same? The fact that the metaphors break down is not an illustration of the power of natural selection, but rather a demonstration of the impotence of natural selection.

Again, with respect to the topic of speciation, for exactly the reason that all the metaphors fail, natural selection fails as a *scientific* explanation for speciation. Of course, whether speculation is required to explain speciation is beside the point for Darwinists. That "evolution occurred" is not in question, and the exact mechanism is a nicety to be debated among the professionals. One need not be a botanist to believe in flowers, and one need not understand all the intricacies of genetic mutations to believe in evolution. Certainly, suppose the true believers, any perceived difficulties will be worked out over time, in true evolutionary fashion.

Without question, a purely natural process should be amenable to metaphors; otherwise, what is being described is more likely amenable to metaphysics—a "magic step" process. Like creationism's much-maligned "God-of-the-gaps," Darwinism's unobservable "blind, deaf, mute-of-the-gaps" appears remarkably godlike in its ability to invisibly guide evolution along toward "more pleasing features." Deftly entailing the physical necessity for intelligent choice while denying the metaphysical need at the critical juncture of natural selection is the parlor trick sophisticated Darwinists perform in the company of the polite (if not fawning) ladies of science. But men not so

---

[527] Ibid. *See also*, Dawkins, *Blind Watchmaker*, p. 130. (What does "better" mean [in the context of mutational errors]? Ultimately it means more efficient at self-replication ...")

willingly fooled by such verbal legerdemain object when the trick leaves the parlor for the public as a *parlous* trick—passed off as fact before unbelieving eyes. And while successfully resisted for the moment, eventually these men will be successful in exposing natural selection for the unnatural speculation it is.

[59] *"Information. That's the fundamental difference ..."*

While the great Darwinists may be able to rationalize the odds of chance biogenesis, ignore the true implications of the Second Law of Thermodynamics, and dissimulate over true versus apparent design in morphology, *information* may be the linchpin that, once pulled, leaves such machinations smoldering on the scrap heap of scientific history. For if Darwinism is to legitimately maintain its status of being a scientific theory it must offer some natural (i.e., only unintelligent causes) explanation for the undeniable presence of *information* in biological structures. Of all the physical, observable characteristics exhibited by living organisms, information content (such as that in DNA) is the least amenable to rationalization or dissimulation (although it is often ignored).

"Information content" means the intangible property of an arrangement of elements that imparts a message or conveys meaning. This is known as the "semantic" character of information. The message formed by the specific, non-random combination of letters and spaces you are now reading reflects the semantic or meaningful character of this sentence.[528]

Meaningful information can be referred to as complex, specified information and is characterized as having non-repetitive, specified sequences of elements, or parts, such as letters in a sentence, colors in a painting, bits in a computer program, or chemicals in DNA. In each case the distinguishing factor over ordinary "complexity" is the contingent nature of the specific, non-random, meaningful organization of the elements.[529] As intelligent design theorist Phillip E. Johnson explains:

---

[528] Information theory is a complex subject, and a complete treatment is beyond the purview of this book. However, the definition provided for information is accurate and adequate, albeit simplified for the purposes of this discussion. For a more developed treatment, *see*, Dembski, "Intelligent Design as a Theory of Information," *Intelligent Design*, p. 153. *See also*, Meyer, *The Origin of Biological information and the Higher Taxonomic Categories*. *See also*, www.iscid.org, for indepth treatment from the International Society for Complexity, Information, and Design.

[529] "Contingency" is the idea that the source of information can admit multiple alternative possibilities, i.e., it could have been different. For example, a communication channel capable of transmitting only one signal cannot convey information. To convey information a communication channel must allow a multiplicity of distinct possible signals, any one of which might be sent. Ibid., p. 154. Again, information theory is a

> The important thing about DNA is not the chemicals but the information in the software, just as the important thing about a computer program or a book is the information content and not the physical medium in which that information is recorded.[530]

Very simply, there is no known unguided, purposeless process that can make non-repetitive, specified sequences of matter, particularly the long, detailed sequences of DNA, comparable in quantity and quality to an entire set of encyclopedias. On the Darwinian evolutionary machine the only two knobs that nature alone can turn are the knobs of *chance* and *law*.[531] Chance produces disordered, unspecified sequences that show no consistent patterns. Law produces the same simple pattern over and over again—highly ordered, repetitive sequences like crystals or snowflakes.[532] Law and chance, either alone or in combination, cannot produce the semantic character of information. Nevertheless, the sense that laws can sift chance and thereby generate complex, specified information is a deep-seated truism in the scientific community.[533]

In their own writings, Darwinists must tread carefully around the issue of information in living organisms. For example, in his book *What Evolution Is*, Ernst Mayr discusses the use of the German-coined term *Bauplan*, which is used by morphologists to refer to, in Mayr's words, the "basic designs" of all living animals.[534] According to Mayr:

> In German, the syllable *plan* in *Bauplan* means "map" or "blueprint"—not something that someone had planned. It is not a metaphysical concept.[535]

---

complex subject in its own right. For our purposes a general understanding of analogs such as written communications or programmed software is sufficient to grasp both the concept of information and Darwinism's shortcomings in explaining its creation.

530 Johnson, *Wedge of Truth*, p. 53.

531 Even "chance" is questionable as a "cause". Chance is not strictly a cause but a description of unguided physical results. As such, the causal "knobs" of Darwinism reduce to those of the laws of physics and chemistry.

532 Ibid.

533 Dembski, *Intelligent Design*, p. 168.

534 Mayr, *What Evolution Is*, p. 110.

535 Ibid. In fact, any number of German-English dictionaries confirms that *Bauplan* means "construction plan" or "floor plan" (*See*, www.dictionaries.travlang.com/GermanEnglish); "construction drawing" or "working drawing" (*See*, www.iee.et.tu-dresden.de); or, simply "ground plan" (*See*, John C. Traupman, *The Bantam New College German & English Dictionary* (New York: Bantam

Once again language becomes the enemy of Darwinists who cannot describe what they want to be true in terms different from what in reality is true. (How can something mean "blueprint" but not "something that someone planned"?) Such duplicity often is betrayed in their own writings, given that it's hard to keep all the stories straight when the unspoken objective seems to be to obfuscate. For example, in Appendix A of the same book, Mayr discusses "The Findings of Molecular Biology" where he states:

> The molecular discoveries of the greatest evolutionary importance are the following:
>
> 1. The genetic program (DNA) does not by itself provide the building material of a new organism, but it is only a *blueprint* (*information*) for making the proteins of the phenotype.[536]

"*Only* a blueprint?" Without the blueprint nothing meaningful would be built, regardless of how much building material was available. What cannot be concealed, even when trying, is that the "basic designs" of living things do, in fact, rely on a blueprint, and a blueprint is a carrier of *information*. Other Darwinists agree. Paul Davies states:

> DNA is nothing less than a blueprint—or, more accurately, an algorithm or instruction manual—for building a living, breathing, thinking human being.[537]

Thus, while Mayr may be concerned with the connotations of *Bauplan*, the term "blueprint" for DNA is quite accurate, and it has all the empirically observable signs of "something someone planned." Therefore, in the absence of a natural explanation it very well could be something that someone *did* plan; there is simply no known natural process that can create such information. As a rule, Darwinists seldom bring up the topic.

One Darwinist to whom the question of information creation was specifically proposed is Richard Dawkins. The incident provoked

---

Books 1981) p. 38.) In every case, and in every sense, the meaning is very definitely "something someone planned."

536 Ibid., p. 272, emphasis added. The acknowledgement of "information" here in the Appendix is, apparently, incidental. The term does not even merit an entry in Mayr's book's index.

537 Davies, *Fifth Miracle*, p. 41.

something of a minor international incident in the evolutionist/creationist community. According to Dawkins himself, as related in his book *a devil's chaplain,* in a chapter entitled "The 'Information Challenge,'"[538] he was "duped into granting an interview to creationists ..."[539] and:

> In the course of a suspiciously amateurish interview, they issued a truculent challenge to me to 'give an example of a genetic mutation or an evolutionary process which can be seen to increase the information in the genome.' *It is the kind of question that only a creationist would ask in that way ...*[540]

At this point in the interview Dawkins apparently paused for an extended period of time in which, according to Dawkins, he was "trying to decide whether to throw them out."[541] In any event, according to Dawkins, the edited version of the interview gave "the false impression that I was *incapable* of answering the question about information content."[542] Whether Dawkins was capable of answering the question at the time is somewhat irrelevant. What is relevant is: (1) why he would consider such a question as being only the kind of question creationists would ask,[543] and, (2) what his answer might be now, having had time not only to think about the question, but to include it in a book on the subject.

With respect to the "information challenge" being something only a creationist would ask, why is this? Either Darwinists have their collective heads in the sand on this issue, or they simply have no answer so they prefer the question not be asked—possibly they have a gentlemen's agreement to not broach the subject with one another. As we shall see, not all Darwinists have buried their heads, but, in fact, there are no good answers, so it may be that the unanswerable question is best not raised in polite company. Therefore, simply to ask the question to a Darwinist may make one truculent—if not virtually a creationist.

---

538 Dawkins, *a devil's chaplain*, p. 91.

539 Ibid.

540 Ibid., emphasis added.

541 Ibid.

542 Ibid., emphasis in original. Accounts vary on the actual sequence of events and the outcome. It seems that the pause and an eventual evasive answer to a completely different question led the creationists to claim Dawkins couldn't answer the question. The actual events and their meanings are not relevant to the present discussion, which gives Dawkins the benefit of the doubt.

543 Of course, Dawkins' objection may be with respect to *how* the question was asked (i.e., "in that way"). It seems more likely, however, that Dawkins, who is well accustomed to "truculent challenges," objected to the question itself.

With regard to Dawkins' thoughtful answer, it is less than helpful. He starts with a technical definition of "information" and a brief lesson on how information can be made more economical by deleting redundancy.[544] All this is very interesting, but irrelevant to the real issue. The question is not how to recognize information, redundant or otherwise; the question is whether there is any example of a genetic mutation or an evolutionary process which can be seen to increase the information in the genome—economically or not. As usual, part of the problem is that Dawkins wanders off into a discussion of a question that is not being asked. For example, after palavering on a few pages about some thought-experiment or another, he notes:

> But we can use the 'length of book' thought-experiment to agree upon what it would *mean* to ask the question whether information content increases over evolution, if only we had ancestral animals to look at.[545]

No, the question is not "whether information content increases over evolution." This is the question-begging fact in need of an explanation. What we want to know is *how* natural processes might accomplish such a task. In the end, Dawkins never gives us a clue. He finally surrenders to an unexplainably lame position of equating "information" with "complexity." He states:

> Perhaps the main lesson we should learn … is that the information content of a biological system is another name for its complexity. Therefore the creationist challenge with which we began is tantamount to the standard challenge to explain how biological complexity can evolve from simpler antecedents, one that I have devoted three books to answering, and I do not propose to repeat their contents here.[546]

Why then, did Dawkins treat us to the tutorial on information in the first place? His answer simply reflects his commitment to scientific materialism or methodological naturalism because he defines biological information as having only a syntactic content. This is also known as shannon information. His answer fails to respond to the challenge of Paul Davies mentioned above.

---

[544] Ibid., p. 92.
[545] Ibid., p. 101, emphasis in original.
[546] Ibid., p. 102.

First, to equate the information content of a biological system with only its "complexity" is to ignore its obvious semantic content. As we have seen, "complexity" can be either specified complexity or unspecified complexity, and the distinction is critical to the issue of information creation. Therefore, while information requires a form of complexity, complexity is not in information in its general sense. Presumably for Dawkins, where living systems are involved the entire discipline of "information theory" can be handled as "complexity theory" since information content is just "another name" for complexity. Slight reflection makes clear that such a retreat into this conflation of terms and concepts is simply a dodge—a poor substitute for simply saying "I don't have a clue."

Second, as we have seen, Dawkins has failed (in at least two books discussed above[547]) to explain how evolution can produce *specified* complexity. In earlier attempts to answer the question of biological complexity, Dawkins succeeds only in conflating "complexity" with "complicated things" and fails to address *specified* complexity at all, except to, in essence, describe intelligent design principles. No one is asking how natural process can produce generic "complexity." No one is asking how nature can produce order, patterns, or any other type of random complexity. What people are asking, and what Dawkins failed to answer in his book-length explanations, is *how can nature increase the information in the genome*?

In typically evasive fashion, Dawkins concludes:

> The 'information challenge' turns out to be none other than our old friend: 'How could something as complex as an eye evolve?' It is just dressed up in fancy mathematical language—perhaps in an attempt to bamboozle.[548]

Recall the original request: "to give an example of a genetic mutation or an evolutionary process which can be seen to increase the information in the genome." Evolutionary development and the question of the information content of the genome is related to, but is in no way the same as, the question of biological complexity. Perhaps Dawkins *was* bamboozled. But more likely he simply has no good answer—not even a wrong one.

One evolutionist who is not bamboozled, but nevertheless fails to provide a satisfactory explanation for information creation is Paul Davies. In his book *The Fifth Miracle*, Davies is not shy to acknowledge:

---

[547] Dawkins, *The Blind Watchmaker* and *Climbing Mount Improbable*.

[548] Ibid.

> There *is* a nonmaterial "something" inside living organisms, something unique and, literally, vital to their operation. It is not an essence or a force or an atom with a zing. That extra something is a certain type of information, or, to use the modern jargon, software.[549]

Further, unlike Dawkins, Davies recognizes the importance of using language with precision. Particularly with respect to the terms "information" and "complexity," he explains:

> I have explained how life, at rock bottom, has the same logical structure as a computer. This fact provides an opportunity to inject some precision *into the rather slippery notions of complexity and biological information* by appealing to the theory of computation ... Much of the bafflement about life is due to confusion in the meaning of terms like "order," "organization," "entropy," "chance," "randomness," "*information*," and "*complexity*." *These words are frequently employed in a slipshod or ambiguous way, without any proper definition.* In particular, "order" and "organization" are often conflated.[550]

Davies is simply underscoring what has been repeated throughout these notes: knowledge can be only as reliable as the language used to impart it, and language imparts useful knowledge only when used precisely and consistently. In Davies' discourse he correctly defines his terms, and then gets to the crux of the matter—the distinction between random complexity and information:

> Now, you might be thinking that, if biological organization is random, its genesis should be easy. It is, after all, a simple matter to create random patterns. Just take a jar of coffee beans and tip them on the floor. Surely nature is full of haphazard and chaotic

---

[549] Davies, *Fifth Miracle*, p. 40. To be clear, Davies is concerned with the creation of life in the first place, not Darwinian evolution. However, after the discourse that follows he acknowledges that the "ideas I have outlined above apply with equal force to the evolution of the genome over time" (Ibid., p. 120).
[550] Ibid., p. 115-16, all emphasis added.

> processes that might create a random macromolecule like a genome?[551]

Davies recognizes what anyone thinking seriously about this subject must admit: complexity is not "another name" for information. Random complexity of the type that nature can achieve (like spilled coffee beans) is not the same as specified complexity necessary for information-bearing things like a genome. The following quote is quite long, but the reader is encouraged to read it carefully, because Davies gets right to the heart of the matter:

> This [referring to the question ending the previous quote] is a good question, and it marks the point where we encounter the truly subtle and mysterious nature of life in the starkest manner. Fact one: the vast majority of possible sequences in a nucleic-acid molecule are random sequences. Fact two: not all random sequences are potential genomes. Far from it. In fact, only a tiny, tiny fraction of all possible random sequences would be even remotely biologically functional. A functioning genome *is* a random sequence, but it is not just *any* random sequence. It belongs to a very, very special subset of random sequences—namely, those that encode biologically relevant information. All random sequences of the same length encode about the same *amount* of information, but the quality of that information is crucial: in the vast majority of cases it would be, biologically speaking, complete gobbledygook.[552]

Again, Davies clearly comprehends the difference between "complexity" and "information," or better put, "complex" sequences and "information-bearing" sequences. Once more, an apology for the long quotations, but Davies' honest treatment deserves adequate consideration. He continues his logical thought process by expressing ideas usually uttered only by intelligent design theorists:

> The conclusion we have reached is clear and it is profound. A functional genome is *both* random *and* highly specific—properties that seem almost contradictory. It must be random to contain substantial

---

[551] Ibid., p. 119.

[552] Ibid., emphasis in original.

> amounts of information, and it must be specific for that information to be biologically relevant. The puzzle we are then faced with is how such a structure came into existence. We know that chance can produce randomness, and we know that law can produce a specific, predictable end-product. But how can both properties be combined into one process? How can a blend of chance and law cooperate to yield a *specific* random structure?
>
> To get some idea of what we are up against with this dilemma, it is rather like tipping out the coffee beans from a jar to make a *particular* random pattern. Not just any old random pattern, but a definite, narrowly specific, predetermined random pattern. The task seems formidable.[553]

Formidable indeed. An intelligent design theorist would say that not only do the properties of "random" and "highly specific" *seem* almost contradictory, they *are* in fact contradictory, at least for any known blind natural processes. An intelligent design theorist would provide Davies with the solution to his problem: the properties of biological information are not "random" and "highly specific," rather they are "aperiodic" and "highly specific." This is the property of the letters and words of this paragraph, for example. It is aperiodic and highly specific. The sequence of letters and spaces is not random. Randomness merely reflects the requirement of methodological naturalism.

But the formidability of the task of producing a definite, narrowly specific, predetermined random pattern is proportional to the means that one is allowed to postulate for its accomplishment. If intelligent design theory is permitted as an explanation, the formidability evaporates. But Davies is an evolutionist wedded to a philosophy-constrained theory, and he is not permitted the freedom to openly flirt with the possibility of intelligent causes for either genesis of life or origin of species. Nevertheless, he is apparently a sorely tempted evolutionist, as, remarkably, he concludes:

> Could a law on its own, without a huge element of luck (i.e., chance), do such a thing [i.e., biogenesis]? Can *specific randomness* be the guaranteed product of a deterministic, mechanical, lawlike process, like a primordial soup left to the mercy of familiar laws of

---

553 Ibid., p. 120, emphasis in original.

> physics and chemistry? No, it couldn't. No known law of nature could achieve this—a fact of the deepest significance, as we shall see in the final chapter.[554]

"No, it couldn't." These are the words of an intelligent design theorist. The fact that no known law of nature can produce the information necessary for biogenesis (Davies' immediate concern) is of equal importance to the production of information in general. However, although Davies agrees, he wants to believe there is an important difference between the two types of information generation. For example, he states:

> If you found the foregoing argument persuasive, you could be forgiven for concluding that a genome really is a miraculous object. However, most of the problems I have outlined above apply with equal force to the evolution of the genome over time. In this case we have a ready-made solution to the puzzle, called Darwinism. Random mutations plus natural selection are one surefire way to generate biological information, extending a short random genome over time into a long random genome. Chance in the guise of mutations and law in the guise of selection form just the right combination of randomness and order to create "the impossible object." The necessary information comes, as we have seen, from the environment.[555]

Here the linchpin is pulled, and the wheels fall off of Davies' otherwise flawless logic. Davies previously explained in concise, rational detail how and why law and chance alone *cannot* produce specific randomness, i.e., information. On what basis does he now rationalize that, given a miraculous starting genome, suddenly law and chance alone *are* sufficient to do just that, i.e., create more and more complex information? Is the chance and law of Darwinism somehow an exception to the general rule? If so, on what grounds? Chance and law are just that: chance and law, whether under the "guise" of something or not. Calling chance "mutations" and law "selection" renders neither suddenly able to do the miraculous. Even given a starting genome, there is no scientific basis for suddenly believing chance and law are transformed into a force sufficient to produce

---

554 Ibid., emphasis in original. In the "final chapter" Davies considers possibilities that he admits are "highly conjectural," such as the idea that life *always* existed, that is, it never came into existence in the first place. Ibid., p. 263.

555 Ibid.

additional information. This fact can be shown logically (as Davies did) and mathematically,[556] as well as empirically.

Even Davies, perhaps too honest for his chosen theory, can't help but believe that something more than law and chance are needed for Darwinian evolution. In the "final chapter" he referred to above, he considers:

> [E]mergent laws of complexity offer reasonable hope for a better understanding not only of biogenesis, *but of biological evolution too*. Such laws might differ from the familiar laws of physics in a fundamental and important respect. Whereas the laws of physics merely shuffle information around, a complexity law might actually *create* information, or at least wrest it from the environment and etch it onto a material structure.[557]

Why should Davies concern himself with conjecture about laws that "differ from the familiar laws of physics" to power the Darwinian machine? Davies informed us earlier that these "familiar laws" are the "one surefire way to generate biological information." Now we're told that they "merely shuffle" information around. Perhaps Davies' bold display of confidence earlier was a survival adaptation to his environment, surrounded (as he must be) by uncompromisingly fit Darwinists. But truth will find a way in honest minds, and Davies is close to the truth. He quotes a fellow evolutionist who staggered into the same truth on the subject of information creation:

> For example, Manfred Eigen has written, "Our task is to find an algorithm, a natural law that leads to the origin of information." Though acknowledging the crucial role played by molecular Darwinism, Eigen and his colleagues nevertheless see the need for it to be augmented by other physical processes which can be an additional source of biological information.[558]

It's hard to fault Davies, except that he is ultimately trying to fit a square peg into a round hole. "Augmenting" the square peg will only make matters worse. And searching for a "complexity law" or

---

[556] *See, e.g.*, Dembski, "Intelligent Design as a Theory of Information," *Intelligent Design*.
[557] Davies, *Fifth Miracle*, p. 259, emphasis added.
[558] Ibid., p. 260, quoting Manfred Eigen, *Steps Toward Life*, trans. P. Wooley (Oxford: Oxford University Press, 1992), p. 12.

"other physical processes" which can be "an additional source of biological information" is likely to be a fool's errand. Like searching for a "planetary law" or "other orbital process" to explain how the sun really, really does orbit the earth, the search will only prove fruitless. What is necessary is to put away the square peg, and find a round peg that matches the obviously round hole. The round hole is the origin of specified complex living organisms, and a round peg is a mechanism for creating complex specified information. As intelligent design theorist William Dembski states:

> The task of evolutionary biology is to explain the origin and development of life. The key feature of life is the presence of complex specified information—CSI. Caught up in the Darwinian mechanism of selection and inheritance with modification, evolutionary biology has failed to appreciate the information hurdles organisms need to jump in the course of natural history.[559]

Until Darwinists can secure the linchpin of information-creation, the lumbering Darwinian machine cannot hope to survive with scientific credentials. William Dembski considers what evolutionary biology would look like if information, rather than Darwinism, were taken as its central and unifying concept:

> First off, let us be clear that the Darwinian mechanism of selection and inheritance with modification will continue to occupy a significant place in evolutionary theory. Nevertheless its complete and utter dominance in evolutionary theory—that selection and inheritance with modification together account for the full diversity of life—will have to be relinquished. As a mechanism for conserving, adapting and honing already existing biological structures, the Darwinian mechanism is ideally suited. But as a mechanism for innovating irreducibly complex biological structures, it utterly lacks the informational resources.[560]

"Complete and utter dominance" has rarely, if ever, been relinquished without a struggle. So the struggle must go on, and such is the nature

559 Dembski, *Intelligent Design*, p. 180.

560 Ibid.

of scientific revolutions. If the cause is pursued, and reason prevails, the outcome is certain.

[60] *"Noise corresponds to random, complex patterns, like wind-blown leaves. Chance can do that. Neat noise is like crystals, highly specified, simple, repeating patterns. Natural law "algorithms" can do that. But a picture has specified complexity; the parts are placed in a complex manner with specificity. Can nature alone do that?"*

Specified complexity—the hallmark of information, as discussed above—is at the heart of any theory of intelligent design because it cannot be explained by any known natural causes. Beyond teleological arguments that imply a purpose in natural things, at bottom intelligent design simply posits that "nature points beyond itself."[561] That is, *who* designed, or *why*, is a secondary consideration in intelligent design theory, which simply recognizes that the "big issues," consciousness, the origin of life, and the fine-tuning of universal constants, are not amenable to purely naturalistic explanations.

Probably the most effective proponent of this "branch" of intelligent design theory is William Dembski, a mathematician, philosopher, and author of the book *Intelligent Design, the Bridge Between Science and Religion*. In particular, Dembski has proposed a method of scientifically determining the presence of design in nature. He calls his method the "explanatory filter." The "filter" uses three stages of inquiry to make a judgment as to whether an observed phenomena can be attributed to design. First, one looks for "regularity," or natural law at work. This is the "neat noise" referred to in the quote above. Crystals, pooled water, or electron motion can all be explained by repetitive and predictable natural laws such as chemical attraction, gravitational attraction, and electro-magnetic attraction, respectively. This is not necessarily evidence of design.

Second, one looks for "chance." Probabilistic methods are employed to predict outcomes and phenomena. This is the "noise" referred to in the quote above. Hurricanes, rock slides, earthquakes, and other phenomena are examples of how nature can cause unpredictable, chance occurrences.

Third, one looks for complex realities that defy explanation by either regularity or chance. Dembski refers to such realities as "complex specified information," or CSI. According to Dembski, natural law and chance are incapable of generating CSI, only an intelligent agent can do so. Much of Dembski's work has been aimed at "proving" CSI, a frustrating task shared by all intelligent design

---

561 Witham, *By Design*, p. 142.

theorists due to the double standard of proof that permits Darwinism to be assumed.

Complex specified information in living organisms cannot be explained by chance and law alone. Science must have a third option: "design." Although naturalistic Darwinists resist intelligent design theory as a covert operation that is ardently Christian and politically conservative, some see the possibilities of true scientific progress. Paul Davies, for example, does not appear to believe in a personal God but does make a crucial point regarding Dembski's work:

> Dembski's attempt to quantify design, or provide mathematical criteria for design, is extremely useful. I'm concerned that the suspicion of a hidden agenda is going to prevent that sort of work from receiving the recognition it deserves. Strictly speaking, you see, science should be judged purely on the science and not the scientist.[562]

Therefore, once today's ruling elite, as exemplified by the National Academy of Sciences, for example, can get past the insecurity of a hidden agenda, intelligent design theory may actually get the recognition it deserves.

[61] *"I think a debate can be arranged. And just whom would you like to debate?"*

Debates between Darwinists and anti-Darwinists consistently generate high interest. In some revealing commentary on Darwinists and the strength of their positions, note the following advice (quoting Eugenie Scott)[563] which was found at www.talkorigins.org under the heading "Debating Creationists: Some Pointers":

> Eugenie Scott, director of the National Center for Science Education, argues that public debates with creationists do little to further the cause of quality science education. Among her reasons are that (1)

---

[562] Ibid., p. 149.

[563] Eugenie C. Scott is director of the National Center for Science Education, an organization intent on countering strategies aimed at removing evolution from our schools. Interestingly, according to the NCSE, "the best guarantee of good education is public understanding of the issues" (www. pbs.com). Ms. Scott is also one of the 13 named people in the "Working Group on Teaching Evolution" that authored The National Academy of Sciences booklet entitled *Teaching About Evolution and the Nature of Science*.

> creationists usually fill the audience with their supporters, who are unlikely to have the educational grounding necessary to understand the arguments presented in favor of evolution; (2) evolution is a complex set of ideas that are not amenable to simple explanation in a short debate format, and (3) creationists make many simplistic but convincing-sounding arguments against evolution that take a significant amount of time to refute.[564]

So, there it is. Creationists don't have the "educational grounding" to understand the necessary "complex set of ideas." Then, to top things off, those pesky creationists make "convincing-sounding" arguments against evolution that take too long to refute. Such risible reasoning speaks volumes.

Note that a debate tactic such as "use scientific reasoning to refute creationist ideas" is not one of Ms. Scott's recommendations. If it takes a "significant amount of time" to express what evolutionist Richard Dawkins refers to as "a remarkably simple theory,"[565] then maybe it is not creationists who lack the educational grounding. And if evolution is not amenable to simple explanation, maybe Darwinists should stop insisting that third graders learn the "fact" of evolution.

In a similar manner, and on the same page at www.talkorigins.org, evolutionist Richard Trott[566] gives some advice to people who are considering debating "the Institute for Creation Research's most prominent debator [*sic,* debater], Dr. Duane Gish." Among Mr. Trott's recommendations:

> [T]hat you know your audience, know Gish's standard presentation, be prepared for his charisma and standard evasions, and avoid being a dull or arrogant lecturer.[567]

Unfortunately, it seems that arrogance has become inbred into the higher species of Darwinists, possibly as an allopatric adaptation

---

564 *See,* http://www.talkorigins.org/faqs/debating-creationists.html (April 2004).
565 Dawkins, *Blind Watchmaker,* p. xv.
566 Richard Trott was at one time associated with Rutgers University, and based on internet searches was apparently active in various efforts of skeptics to debunk junk science. At the time of this writing Mr. Trott's whereabouts are unknown.
567 Ibid., www.talk.origins.org in The Talk.Origins Archive, 2003.

that permits their tribe to thrive in their isolated environment.[568] Dullness, though less prevalent in most Darwinist populations, may also be a naturally selected trait found in those capable of understanding such a complex set of ideas as evolution.

[62] *"The question, though, in general is 'How did any of the things we see around us get here?'"*

Why is there something instead of nothing?

This question, which may seem silly at first, is one of the weightiest questions one can consider. All theories aside, a serious mental journey into the distant past, imagining one's creation process of choice, leaves one confounded as to how any "thing" got here in the first place. It is a logical necessity that if there was ever nothing, nothing would now exist. There simply is no physical law known that can produce something from nothing. Any "thing" had to come from somewhere, sometime.

Richard Dawkins rightly notes that such a thought process is not so different even when entertaining the idea of a deity. In *The Blind Watchmaker* he writes:

> To explain the origin of the DNA/protein machine by invoking a supernatural designer is to explain precisely nothing, for it leaves unexplained the origin of the designer. You have to say something like "God was always there," and if you allow yourself that kind of lazy way out, you might as well just say 'DNA was always there', or 'Life was always there', and be done with it.[569]

Dawkins has a point. Rather than cast the inquiry in terms of laziness, however, the question can be reduced to "What is more *reasonable* to believe?" That is, we can't know first hand the truth of the matter, so taking into account all we can know, what is more reasonable to believe? (Which is very different from asking, which is more *comforting* for a particular world view?)

Darwinists obviously find it more reasonable to believe that "something" other than a designer always existed, be it DNA or otherwise. This "something" for Darwinists must be natural, for there

---

[568] Allopatric speciation is a theory according to which a new species may evolve when a population acquires isolating mechanisms while isolated from its parent population. Mayr, *What Evolution Is*, p. 175.

[569] Dawkins, *Blind Watchmaker*, p. 141.

can be no resort to *super*-natural.[570] But for something natural, i.e., molecules or atoms, to have always existed, is problematic. For the atoms of this world, in any configuration, to "be" as we know them and not ever to "have not been" is highly improbable. It is, most likely, impossible—certainly it is impossible by any known or knowable physical or mental processes. The impossibility of "infinite matter" led such philosophers as Aristotle to postulate a "First Cause" as a logical necessity. Again, Darwinists can deny a First Cause, a designer, a God, or any other natural or supernatural intelligence as an explanation for existence, but at some point the position becomes unreasonable as a matter of simple logic.

[63] *"To the lowest available energy state, within the constraints of the system."*

"Available energy" is required for any process to do "work"—a term having a defined meaning in physics of "force multiplied by distance."[571] *Directed* work is required to drive a process of any kind in a direction of increasing order—such as is necessary for Darwinian evolution. Available energy is energy having the potential to exert a force through a distance. Thus a water wheel (as on old mill houses) is driven by the force exerted by moving water as it falls and applies a force through a distance. Before the water fell it had a certain "potential" energy available to do the "work" of turning the wheel. After falling, it no longer had potential energy with respect to the water-water wheel system. Thus, within the constraints of the system (and assuming the water fell to a flat, level surface), the water is at its lowest available energy state, constrained by the ground so that it can do no more work. If the water had anywhere else to flow or fall, it could lose additional "potential," which could likewise be directed as work if harnessed.

What do work and energy concepts have to do with evolution? The concept of available energy, especially "potential energy," has been well developed in science, and is expressed by the laws of

---

[570] At least not in terms of a deity or other God-figure. Space aliens and other extraterrestrial beings are postulated freely as if such beings were self-evident.

[571] The term "work" is central to the ideas that follow, but, unfortunately, it has a technical meaning that is unlike the intuitive notion most people have. For a very readable discussion explaining the concept of "work," as well as "energy," "entropy," and other concepts central to the topic of the Second Law of Thermodynamics, *see*, Isaac Asimov, *Understanding Physics*, 3 Volumes in 1 (New York: Barnes & Noble Books, 1993), pp. 90-92, 220-40. Readers may associate Asimov's name with science fiction, but Asimov also wrote non-fiction works, including the *Understanding Physics* trilogy. The books are written for the lay reader with a minimum of jargon and mathematical equations. Also, although Asimov was an admitted atheist and a full-fledged evolutionist, the book is remarkably free of "up is really down" convoluted explanations in the crucial area of the Second Law of Thermodynamics.

thermodynamics. There are two important laws of thermodynamics, aptly named the First and Second Laws of Thermodynamics. The Second Law of Thermodynamics (the "Second Law") in particular bears on evolution in a way that must be understood, given that on its face it flatly contradicts the ideas of evolution. Darwinists are predictable in their disregard and dismissal of the serious problems presented by the Second Law. Understanding the Second Law (and the evolutionist response) is necessary for an understanding of how it is a powerful argument against natural, unguided evolution. For this reason, the reader is urged to consider carefully the following discussion, where every effort has been made to avoid overly tedious detail.

Water wheels rely on a difference in water level to create the energy "potential" necessary to do work. But falling water is not the only form of potential energy available for work. Differences in temperature create a potential for heat engines, such as the automobile engine. Electrical potential differences in batteries represent available energy, as do chemical potential differences in certain chemical formulations. In short, in whatever form energy exists, work can be obtained only if the energy is present in a state of greater intensity in one portion of the system and lesser intensity in another portion.[572]

In a real sense, the idea of an energy "potential" is the idea of energy being able to go "down hill" to be useful. As long as energy can still go down hill, more work can "potentially" be extracted. Thus, more water wheels can be placed downstream from the first, each one driving more mill stones. Heat can be extracted and used to run a heat engine, batteries can run radios, and chemical reactions can drive to final products. But at some point, the river can no longer run down hill, the temperature differences disappear, the electrical charges equalize, and the chemical reactants are spent. At that point there is no more "available energy" within the "constraints of the system."

Keeping in mind the example of the water wheel, it is easy to see that you can never expect the water wheel to turn faster than the water is falling. Nor can you get the wheel to turn if the forces of friction are too great for the force of the water to overcome. In general, it is clear that one cannot extract from a system more energy than the total energy present in the first place. This is, in fact, is the essence of the First Law of Thermodynamics (the First Law).[573]

In addition to the First Law, the Second Law states that it's impossible to extract more work (i.e., useful, directed energy) from a system than the quantity of available energy will permit. The idea here is that you "can't get something for nothing" or, "you can't get more

---

[572] Ibid., p. 230.
[573] Ibid., p. 228.

than you started with."[574] The practical implications of the First and Second Laws are that no matter how the energy is supplied, for any system *the amount of unavailable energy always increases with time*. We might say that the unavailable energy in any system can remain unchanged under ideal conditions, but *always increases* with time under actual conditions.[575]

The term "entropy" was invented to serve as a measure of the unavailability of energy.[576] Therefore, as Asimov states:

> We can say then that the entropy of a system can remain unchanged under ideal conditions, but always increases with time under actual conditions. And this, too, is an expression of the second law of thermodynamics.[577]

The Second Law and entropy are each expressed in various ways depending on the particular discipline of science to which they are applied. But in any context, entropy is in essence a measure of disorder, or the "evenness" of energy distribution. For example, when an electrical battery discharges, its electrical energy is more and more evenly distributed over its substance and over the material involved in the electrical flow of current.[578] The increasing evenness with which energy is spread out is one manifestation of increasing disorder.[579] This "evenness" is an invariable result of the Second Law and its effects are universal to all aspects of nature. As Asimov states:

> For this reason, when we shuffle a neatly stacked deck of cards into random order, we can speak of an increase in entropy. And, in general, all spontaneous processes do indeed seem (in line with the second law of thermodynamics) to bring about an increase of disorder. Unless a special effort is made to reverse the order of things (increasing our own entropy), neat rooms will tend to become messed up, shining objects

---

[574] Asimov notes another way students have learned to remember the basic idea of the First and Second Laws: "It has been said that the first law of thermodynamics states: 'You can't win,' and that the second law of thermodynamics adds, 'And you can't break even, either'" (Ibid., p. 229).

[575] Ibid., p. 231, emphasis added.

[576] Ibid.

[577] Ibid.

[578] Ibid., p. 238.

[579] Ibid., p. 239.

> will tend to become dirty, things remembered will tend to become forgotten, and so on.[580]

It's the "special effort" referred to by Asimov that is the problem for Darwinists. As mentioned previously, the universality of the Second Law is unquestioned; it is sometimes referred to as the basis for our perception of time itself. The Second Law says that in the absence of directed energy (the "special effort" Asimov referred to), natural laws and physical constraints in the system will always result in a "winding down"; systems always go from a relatively ordered to a relatively more disordered state; from high energy to low energy; from hot to cold; from organized to disorganized.[581]

On its face, therefore, the Second Law stands diametrically opposed to any theory, including biological evolution, which requires matter to go spontaneously from a simple, random form to a more complex, specified form. The stock reply from virtually all Darwinists to such an objection is that the Second Law applies only to closed systems. In our case, Darwinists say, the Second Law's tendency to prevent the incredible ordering necessary for evolution is circumvented by including the sun in our local earth system. Usually the Second Law is dismissed flippantly and casually, as if its inapplicability to evolution is hardly worth elaboration.[582] For example,

---

[580] Ibid.

[581] It is true that natural processes can produce simple ordered structures; therefore, the lowest energy state can be ordered by physical constraints such as molecular attraction or barriers to further potential energy loss. Thus, natural laws and physical constraints on a system can produce well-known "anti-entropic" simple order, such as the order in a crystal structure. The geometrical arrangement of a crystal lattice is determined predominantly by the interatomic forces operating. Ibid., p. 251. Likewise, marbles dropped on a flat, level surface come to rest in an "ordered" manner because all come to rest in the same plane. However, such "ordering" is simply a result of natural laws and physical constraints and is not to be confused with the origin of specified complexity necessary to drive evolution.

[582] *See, e.g.*, the National Association of Biology Teachers stated in their "Statement on Teaching Evolution," as one of their "tenets of science": "Evolution does not violate the second law of thermodynamics: producing order from disorder is possible with the addition of energy, such as from the sun" (NAS, *Teaching About Evolution*, p. 127). Likewise, the subject is dealt with at www.talkorigins.org on a page that starts, "Creationists believe that the second law of thermodynamics does not permit order to arise from disorder …" In addition to being a straw man argument that starts from a false premise (at least for intelligent design theorists, who have no doubt that "order" can arise from disorder in full compliance with the Second Law), the lengthy text-book exposition on entropy does little to address the issue. The entry closes by noting: "Considering the earth as a system, any change that is accompanied by an entropy decrease (and hence going back from higher probability to lower probability) is possible as long as sufficient energy is available. The ultimate source of most of that energy, is of course, the sun" (http://www.talkorigins.org/faqs/thermo/probability.html, October 28, 2002). The assertions of this statement can easily be refuted, but for our purposes two

in evolutionist Ernst Mayr's 318-page book *What Evolution Is*, fully one paragraph of seven lines is employed on page 8 to assure us:

> [T]here is no conflict, because the law of entropy is valid only for closed systems, whereas the evolution of a species of organisms takes place in an open system in which organisms can reduce entropy at the expense of the environment and the sun supplies a continuing input of energy.[583]

The idea referred to by Mayr, which is the standard refutation invariably flashed like a fake ID to get past all but those who actually care, is that the Second Law applies only to systems closed to external sources of energy input, and our earth system has unlimited energy input from the sun. Therefore, because the earth is not a closed system, the sun can supply energy for local entropy reduction on earth at the expense of an increase in entropy elsewhere. Other Darwinists agree. For example, with respect to the Second Law and closed systems Richard Dawkins refers with typical condescension to:

> [A]n irony in the claim, made by lay opponents of evolution, that the theory of evolution violates the Second Law of Thermodynamics … within any closed system.[584]

---

fallacies need only be identified. First, creation of "order" is not at issue—creation of "specified complexity" or "irreducible complexity" needs to be explained by Darwinists. Second it is not true that "entropy decreases" are "going from higher probability to lower probability." In anti-entropic systems such as crystal growth, the decreased entropy states *are* the higher probability states due to restraining forces such as molecular forces.

583 Mayr, *What Evolution Is*, p. 8. The Second Law, or thermodynamics in general, is not even mentioned in the index to Mayr's book. The only other mention of "entropy" is in the glossary in which it is evident that Mayr apparently does not even understand the concept. Entropy is defined by Mayr as "The degradation of matter and energy in the universe to an ultimate state of inert uniformity. Entropy can be reached only in a closed system" (Ibid., p. 285). The first sentence is not a definition of entropy, but it is at least an acceptable description of the effects of the Second Law of Thermodynamics, and can pass for a description of "entropy" for the lay reader. However, the second sentence is simply wrong. Entropy is a *measure* of something, like other measures such as temperature, weight, or distance. The second sentence is analogous to saying "Temperature can be reached only in a closed system." Mistakes are excusable, but such cursory treatment of a topic otherwise given short shrift by someone of Mayr's stature is difficult to understand.

584 Dawkins, *a devil's chaplain*, p. 84. Elsewhere Dawkins refers to arguments that evolution violates the Second Law of Thermodynamics as "ignorant nonsense" (Dawkins, *Blind Watchmaker*, p. 94).

In typical "up is down" thinking, Dawkins explains:

> If anything appeared to violate the law (nothing really does), it would be the *facts*, not any particular explanation of those facts! The Darwinian explanation is the only viable explanation we have for those facts that shows us how they could have come into being *without* violating the laws of physics. [585]

It's difficult to grasp Dawkins' point here, at least rationally. Without question the *facts* appear to violate the Second Law—that is the reason for the inquiry in the first place. We are seeking to understand how any natural process could make water go uphill, so to speak, in the way Darwinism requires. To simply presume as a fact that Darwinism is the *only viable explanation* for these facts is a bald, question-begging assertion—putting forth as a conclusion what is actually the question under consideration.

Other Darwinists inadvertently prove themselves wrong while pretending to give a scientific answer to the question. For example, Dr. Tim M. Berra addressed "Some Creationist Claims" including the claim that "evolution violates the second law of thermodynamics" in his book *Evolution and the Myth of Creationism*.[586] Clumsily brandishing his fake ID, he answers:

> These statements conveniently ignore the fact that you *can* get order out of disorder if you add energy. For example, an unassembled bicycle that arrives at your house in a shopping carton is in a state of disorder. You supply the energy of your muscles (which you get from food that came ultimately from sunlight) to assemble the bike. You have got order from disorder by supplying energy. The Sun is the source of energy input to the Earth's living systems and allows them to evolve.[587]

---

[585] Ibid., emphasis in original. Dawkins here includes a footnote that the facts referred to are "About life's functional complexity or high 'information content.'"

[586] Tim M. Berra, *Evolution and the Myth of Creationism* (Stanford, CA: Stanford University Press, 1990), p. 126.

[587] Ibid., emphasis in original.

Berra's blunder is easy to see;[588] it doesn't take a PhD. to see that Berra's science is doctored with philosophy. Obviously, it is not only energy that is supplied to the bicycle parts; a large dose of intelligent direction is also necessarily present. If Berra wants to use the bicycle analogy, he must explain how raw, undirected energy might combine with *chance* processes to do anything but cause the bicycle parts to decay, rust, or otherwise deteriorate. As he has set up his illustration, Berra has succeeded in proving that *an intelligent being* is necessary to direct energy (even if all the parts are in existence, and the energy ultimately comes from the sun). Like Berra's bicycle, living organisms also need a "maker" otherwise the component parts would simply bask in the sun until they break down into even greater and greater disorder.

For a serious evolutionist perspective on the difficulties of reconciling evolution with the Second Law, the reader should consider Paul Davies' treatment in *The Fifth Miracle*.[589] Subtitled *The Search for the Origin and Meaning of Life*, Davies' book treats this critical topic with the attention it deserves. In Chapter 2, entitled "Against the Tide" Davies notes that, rather than simply being ignorant nonsense among the "lay opponents of evolution," serious evolutionist thinkers see problems with evolution in light of the Second Law:

> Some eminent scientists have been deeply mystified by this contradiction [i.e., natural examples of an increase in order]. The German physicist Hermann von Helmholtz, himself one of the founders of the science of thermodynamics, was one of the first to suggest that life somehow circumvents the second law. … Eddington likewise perceived a clash between Darwinian evolution and thermodynamics, and suggested either that the former be abandoned or that an "anti-evolution principle" be set alongside it. … Even Schrödinger had his doubts. In his book *What Is Life* he examined the relationship between order and disorder in conventional thermodynamics and contrasted it with life's hereditary principle of more order from order.[590]

---

[588] Not *the* blunder referred to as "Berra's Blunder" by Phillip E. Johnson (Berra showed four Corvette automobiles from different years to illustrate descent with modification. Thinking he was defending Darwinian evolution against creationist explanations, Berra unwittingly showed the opposite—creation by design). In the present blunder (perhaps called Berra's Blunder II), Berra successfully argues for the necessity of intelligence in utilizing energy to get "order out of disorder."

[589] Davies, *Fifth Miracle*.

[590] Ibid., p. 52.

To be clear, Davies is an unequivocal evolutionist, and to the question "Is there a problem with the second law of thermodynamics when it comes to biological evolution?" he answers forthrightly, "No, there isn't."[591] Unfortunately, Davies' justification for this conclusion falls short of being persuasive. Discarding the standard-issue fake ID, Davies has a custom-made ID with a more realistic background that requires a little closer inspection. He begins by stating the standard "the second law applies only to closed systems," argument and gives the example of a refrigerator, which decreases entropy inside at the expense of an increase in entropy outside.[592] This, of course, is a correct characterization of the Second Law, and when applied, as Davies does, to *existing* organisms it explains how they can exhibit entropy-decreasing characteristics because *they are designed to do so*.[593] For example, Davies' explanation accounts for how things designed to direct energy, like refrigerators and rhododendrons, can process raw energy into increased order in entropy-decreasing processes such as refrigeration or photosynthesis. Unfortunately, this line of thought misses the point of explaining how the photosynthesis process, much less the rhododendron itself, got there in the first place—and that's what Darwinism must account for.

Nevertheless, Davies does make an honest attempt to apply the entropic principle to biological evolution, rather than deny its applicability:

> The appearance of new species marks an increase in order, but Darwin's theory identifies the price that is paid to achieve this. To evolve a new species requires many mutations, the vast majority of which are harmful and get eliminated by the sieve of natural selection. ... The carnage of natural selection amounts to a huge increase in entropy, which more than compensates for the gain represented by the successful mutant."[594]

This argument is almost persuasive. However, it fails by falling short of explaining *how* the ordering of specified complexity came to be in the lucky mutant. The "carnage of natural selection" is nothing more than the death of already-existing (if not unlucky) complex living things. But the death of the unlucky does not explain the creation of the lucky—the information-rich, complex, entropy-decreasing order necessary for

---

[591] Ibid.

[592] Ibid., p. 53.

[593] Like the person assembling Berra's bicycle, *directed* energy is the question, not the answer.

[594] Ibid., pp. 53-54.

Darwinian survival. Davies' line of reasoning is a specific instance of the general argument that the Second Law permits a local decrease in entropy as long as there is an offsetting increase in entropy somewhere else in the system. This is accurate, but equally accurate and required by the Second Law is that *a decrease (local or otherwise) in entropy comes about only in the presence of some ordering force or principle*. That is, just as photosynthesis locally decreases entropy due to the programmed instructions in the plant, an entropy-decreased organism must have some ordering principle to account for local decreases in entropy. Like trying to claim water can *locally* run uphill if there are enough offsetting downhill runs, such a claim requires an explanation of *how* the uphill run happens.[595] Including the sun in the system does not suffice.

Nevertheless, Davies' in-depth attempt is appreciated as a good-faith effort to reconcile the Second Law with evolution, rather than all but ignore the inquiry as if it was just a misunderstanding among "lay opponents of evolution." Davies clearly comprehends the problems of explaining evolution in the face of the Second Law, and his various theories on the subject, particularly with respect to the origin of life, are worth considering. And while his attempts at "reconciliation" strike an anti-evolutionist as contrived and somewhat far-fetched, at least he's making an honest attempt instead of denying the conflict.[596]

One of Davies most important contributions to the inquiry of the Second Law is his acknowledgment of the role *information* plays in living organisms. Davies is particularly interested in finding an explanation for the source of information as it relates to the origin of life in the first place, but his comments and insights are equally applicable to the Darwinian process of evolution, that must (if it is to be science-based) explain increasing, directional, specified complexity.[597] Specifically, the contribution Davies makes to the inquiry is recognizing that for origin of life, we need to "explain, not the origin of material stuff, but the origin of information."[598] Specifically, he realizes:

---

[595] Or, like claiming that a bicycle could put itself together from parts if there are enough other bicycle parts being destroyed while trying.

[596] For example, in the face of such overwhelming, impossible odds for life forming in the first place, Davies entertains the idea that life has *always* existed. That is, there was no origin of life, because life is eternal, "spread around the universe ... without having originated anywhere in particular" (Ibid., pp. 247-49).

[597] With respect to problems presented regarding the chance origin of life, Davies stated, "If you have found the foregoing argument persuasive, you could be forgiven for concluding that a genome really is a miraculous object. However, most of the problems I have outlined above apply with equal force to the evolution of the genome over time" (Ibid., p. 120).

[598] Ibid., p. 62.

> We seem to be faced with a disturbing contradiction. The second law forbids the total information content of the universe from going up as it evolves, yet, from what we can tell about the early universe, it contained very little information. So where has the information present in the universe today come from? Another way of expressing the problem is in terms of entropy. If the universe started out close to thermodynamic equilibrium, or maximum entropy, how has it reached its present state of disequilibrium, given that the second law forbids the total entropy to go down?[599]

The fact that Davies recognizes the problem in terms of the universe (i.e., the ultimate "system" of interest) is noteworthy. Equally notable is the fact that he also recognizes that simply attributing the energy of the sun to our "earth system" as an explanation for an information-building process falls short of being an explanation—scientific or otherwise. He clearly admits, for example, that protein formation from amino acids requires an "uphill process" that "heads in the wrong direction, thermodynamically speaking."[600] And it is not simply for lack of energy input into the system that such a process is in the "wrong direction." As Davies notes:

> To be sure, there would have been no lack of available energy sources on the early Earth to provide the work needed to forge the peptide bonds, but just throwing energy at the problem is no solution. The same energy sources that generate organic molecules also serve to destroy them. To work constructively, the energy has to be targeted at the specific reaction required. Uncontrolled energy input, such as simple heating, is far more likely to prove destructive than constructive.[601]

Again, not to misrepresent Davies, he is discussing origin of life in the first place, not Darwinian descent with modification. However, the same considerations come into play with the Darwinian simple-to-specified-complex origin of species. The Darwinian process requires massive amounts of entropy-reducing gains in specified complex information to be generated naturally (albeit over long periods of time, but the Second Law is not any less applicable over long

599 Ibid., p. 63.
600 Ibid., p. 89.
601 Ibid., p. 91.

periods of time). Simply attributing such entropy-defying requirements to "energy from the sun," as Dawkins and Berra do, suffers from the same defects as Davies' origin of life explanations would if he also took such a lazy, circumspect route.

To reiterate, simply "throwing energy" at the problem of the Second Law's applicability to evolution is not a solution—natural laws based on chance and necessity are incapable of converting raw energy into information-rich, complex, specified structures.[602] Simply invoking the sun into the earth's system as a cure-all for evolutionary complexity ignores the fact that raw, undirected energy is not known to be capable in itself, i.e., in the absence of a directing law or process (or *person*, such as Berra's bicycle maker), of providing order out of disorder, much less *specified* order, and further much less *specified complexity*. As mentioned previously, and confirmed by Davies, the undirected energy of the sun has the opposite effect—in the absence of an imposed ordering principle (like photosynthesis), the sun's radiation tends to break down matter into less ordered states.

The British astronomer Sir Arthur Eddington regarded the Second Law as occupying the supreme position among the laws of nature. He once wrote:

> [I]f your theory is found to be against the second law of thermodynamics I can give you no hope; there is nothing for it but to collapse in deepest humiliation.[603]

To this sentiment Darwinists have yet to answer. That is, Darwinists have yet to propose any natural law, process, or mechanism that can "get something for nothing" as evolution requires. Specifically, what Darwinists have yet to explain is the origin of the information-rich, specified complexity exhibited by living organisms. Appealing to the sun to explain Second-Law-defying phenomena is like assuring us that somewhere a water wheel is pushing water uphill on its own simply because the sun is shining brightly.

With rare exceptions humility is not a selected trait among leading Darwinists. Nevertheless, Eddington is correct; and, like water running downhill against an earthen dam, the unyielding reality of the Second Law will continue to build up against the leaky dike of Darwinism. Therefore, it is only a matter of time before Darwinism is washed away by a flood of inviolable truth.

---

[602] Of course, things designed to do so can convert energy into complex structures.
[603] Ibid., p. 51, quoting A. S. Eddington, *The Nature of the Physical World* (Cambridge: Cambridge University Press, 1928), p. 74.

[64] *"So, how* do *we know what we know?"*

Although rarely considered by those convinced of the "fact of evolution," the question of the nature of knowledge is central to determining the validity of any theory, scientific or otherwise. The history of philosophy from the earliest Greek philosophers to the present day is the history of men and women attempting to come to terms with what we can know and how we can know it.

A sharp distinction between science and philosophy was unknown to the early Greek thinkers.[604] Both disciplines were born out of a quest for "ultimate reality" and the early thinkers were not above searching for a metaphysical answer to the physical world. Such thinking gave rise to the discipline of *epistemology*, which is the theory of the origin, nature, and boundaries of knowledge.[605] As David Noebel explains with respect to the world around us:

> Epistemology answers the question, 'How much can we know about the world and how does one obtain this knowledge?' Naturalism answers that everything in the physical world (which means *everything*) is knowable, and science is the proper means of knowing it.[606]

If naturalism is the correct philosophy, correct in the sense that it best informs us of ultimate reality, then the epistemological basis for the "fact of evolution" is secure. Contrary facts and evidence can, indeed, be dismissed as either anomalous or only seemingly contradictory based on the incomplete nature of the proposed evidence.

However, if naturalism is not the correct philosophy, then the epistemological basis for biological evolution is lacking, and further study of evolution will unlikely yield important knowledge consistent with ultimate reality. Therefore, those wanting to believe evolution as a theory to explain life on earth should consider as a starting point the epistemological underpinnings of the theory, and determine for themselves if such a basis is *reasonable*, given all the available data from which to draw.

[65] *"Can we say absolutely that there is no absolute truth?"*

---

604 Sproul, *Consequences of Ideas*, p. 13.
605 Noebel, *Understanding the Times*, p. 121.
606 Ibid., p. 120.

To deny objective truth is logically inconsistent. The statement "There is no objective truth" is itself a propositional statement that must be either true or false. Slight reflection shows that, as a formal matter, the statement can only be false—to be true it would by necessity be a nonsensical self-contradiction. Therefore, as a matter of logic, there must be objective truth.

Objective truth is truth that stands unchanged regardless of how it is perceived, when it is perceived, how it is represented (or misrepresented), or how it is used for philosophical purposes. Such truth is not changed by winds of theory or floods of dogma. In fact, objective truth is not even budged by puffed up theories and dripping discourses from self-important philosophers who pretend to scientific reasoning. Like the objective fact of the earth's course in the solar system, truth is not changed by what anybody thinks about it, how anybody argues about it, or because one group silences another with respect to it. And like the motion of the earth, truth continues long after those who ignorantly and arrogantly denied its motion are long gone.

[66] *"Do you think it matters which option is true?"*

Besides simply being an interesting academic and scientific issue, the answer to the question of human origins has far-reaching practical implications. Specifically, the concepts of law and justice are impacted by the truth of the matter. If, as leading evolutionist George Gaylord Simpson says, "Man is the result of a purposeless and natural process that did not have him in mind,"[607] then there is no absolutely legitimate basis for regulating human behavior. Even the vague notions of goodness expressed generally as "don't harm others" cannot be supported rationally in a system that is driven by survival of the fittest.

"Group fitness"—transferring the concept of "fittest" to groups as opposed to individuals, as Darwinists are wont to do—is a weak reed upon which to foist such weighty concepts as law and justice.[608] Nevertheless, it is not surprising that a post-modern society that rejects authority and any restrictions on behaviors that "don't harm anyone" would embrace a theory like Darwinian evolution. Likewise, it is not surprising that Darwinism is the theory of choice for humanists, secularists and atheists. Although each of these philosophical positions may differ in starting premises, each necessarily has the same end point with respect to purpose and hope. In one form or another each takes the position of humanist Blanche Sanders:

---

[607] George Gaylord Simpson, *The Meaning of Evolution* (New Haven, CT: Yale University Press, 1949), pp. 132, 345, as quoted by Wells, *Icons of Evolution*, p. 203.

[608] The abortion question, for example, cannot be resolved based on group fitness alone. Abortion, which is an example of un-natural selection by the fittest, seems unlikely to result in increased group fitness and should be rejected.

> A Humanist has cast off the ancient yoke of supernaturalism, with its burden of fear and servitude, and he moves on the earth a free man, a child of nature and not of any man-made gods.[609]

The end result is a logical inevitability. As atheist and humanist theologian Ernest Nagel explains, "A tragic view of life is thus an uneliminable ingredient in atheistic thought."[610]

Intelligent design theory offers no specifics with respect to purpose and hope in life. However, part-and-parcel of some aspects of intelligent design is the awareness of design for a purpose. Such a purpose may be elusive to locate, but the idea of being designed for a purpose can offer hope and encouragement. Hope and encouragement are the opposite of "a tragic view of life" and this thought alone makes a measurable difference in the quality of life.

Therefore, rather than being an abstract thought without consequences, the truth with respect to human origins is more than an academic question, and one's position on the matter has important and far-reaching consequences.

[67] *"Science? That's easy. Science is concerned solely with what can be explained in natural terms – based on the evidence, of course."*

At the heart of the evolution debate is the question, How is science to be defined? The naturalistic definition of "science" expressed above is consistent with that espoused by scientific institutions, such as the National Science Teachers Association (NSTA). In their "Statement on the Teaching of Evolution," under a section entitled "The Nature of Science and Scientific Theories," the NSTA defines science as "a method of explaining the natural world."[611] So far, so good. But further elaboration by the NSTA assures us that, "Because science is limited to explaining the natural world by means of natural processes, it cannot use supernatural causation in its explanations."[612]

Interestingly, the definition of "science" adopted by the NSTA is not novel. It is a concise definition of a philosophical position called naturalism. Naturalism is a philosophical view that acknowledges only

609 Blanche Sanders, *The Humanist*, no. 5, 1954, p. 226, as quoted in Noebel, *Understanding the Times*, p. 61.

610 Ernest Nagel, "Philosophical Concepts of Atheism," *Critiques of God* (Prometheus Books, 1976), p. 17, as quoted in Noebel, *Understanding the Times*, p. 63.

611 NAS, *Teaching About Evolution*, p. 124.

612 Ibid.

natural elements, causes, and forces, denying the existence of the supernatural.[613] It is the philosophy of humanism and secularism.

Making science synonymous with naturalism is probably harmless error for most scientific disciplines. But when extended to origins research, such a definition is an unnecessary narrowing of the field of inquiry. The NSTA's definition would be much more apt to render useful knowledge if the term "true" were substituted for "natural." Why rule out by fiat one whole branch of inquiry?

[68] *"So, science that, by definition, can concern itself only with natural causes has determined that the Lavoux Cavern cave art must be the result of natural causes?"*

Darwinists constantly stress that evolution is "the best scientific explanation" for life, as if "best explanation" equals "correct explanation." For example, the National Academy of Sciences states that, "Biological evolution is the best scientific explanation we have for the enormous range of observations about the living world."[614] That the NAS would deem biological evolution as "the best scientific explanation" for life is not surprising. It is simply a natural result of limiting "scientific" explanations to explanations involving "natural" causes—a result compelled not by science, but by philosophy. Science has no way to know the ultimate truth of the matter of life's origins. Therefore, to simply choose naturalism as a philosophical base and then boast that evolution is the "best explanation" for life is somewhat unsatisfying. In this instance, "best" and "correct" have no necessary correlation.

That those holding to a philosophy of naturalism find Darwinian evolution to be the "best explanation" for life is no more surprising (or meaningful) than Christians saying that the Genesis-account of creation is the "best religious explanation for the enormous range of observations about the living world." Knowing this provides no indication as to the correctness of the conclusion.

What remains to be resolved, then, is which philosophical system is the most reasonable, given all the available evidence. As long as "science" is limited to philosophical naturalism by fiat, this important aspect of origins research will remain ignored. However, *if* the truth is that an unseen intelligent agent was involved in origins of life through supernatural causes, science can, *by definition*, never arrive at knowledge of such truth. In that case science could not possibly lead to knowledge of ultimate reality. Science need not even make a leap from natural to supernatural to find ultimate reality. Simply permitting "scientific" explanations to go beyond explanations involving

---

613 Noebel, *Understanding the Times*, p. 115.
614 NAS, *Teaching About Evolution*, p. 56.

"natural" causes to explanations involving "intelligent" causes would surely serve science well, as it would open up more possibilities for certain knowledge.

[69] *"It's an idea called themeistic progression."*

Theistic evolutionists are theists who generally accept the entire Darwinian theory of evolution but say that God was invisibly and undetectably behind it. Most theistic evolutionists (or, more accurately, theistic Darwinists) accept methodological naturalism as the basis of scientific thinking, and therefore they also accept the neo-Darwinian picture of evolution as governed at all levels by some combination of physical law, chance and natural selection.[615]

The basic idea behind theistic evolution is difficult to comprehend in a rationally satisfying manner. For one, theistic evolutionists believe that there is design in the universe, but, oddly, believe that we can detect such design only through the eyes of faith. Strangely, theistic evolutionists believe that, for all we can tell, our appearance on planet earth is an accident.[616] That is, design cannot be empirically detected; it's as if God's purpose was to conceal his purpose in creation.[617]

According to William Dembski:

> [I]ntelligent design is incompatible with what typically is meant by "theistic evolution" (or what is also called "creative evolution," "teleological evolution," "evolutionary creation" or most recently "fully gifted creation").[618]

In particular, according to Dembski:

> Intelligent design and theistic evolution ... differ fundamentally about whether the design of the universe is accessible to our native intellect. Design theorists say yes; theistic evolutionists say no.[619]

---

[615] Johnson, *Wedge of Truth*, p. 89. According to Johnson, theistic evolutionists advocate a modernist worldview within the mainstream religious community. The question is, of course, why?

[616] Dembski, *Intelligent Design*, p. 110.

[617] Ibid.

[618] Ibid.

[619] Ibid., p. 111

The larger problem with the position of theistic evolution, however, is its unsatisfying, and unproductive straddling of the origins fence. Why would a theist in the Jewish or Christian tradition not believe in the creative power of a God who is understood to be omnipotent (and truthful)?[620] What consistency of conviction is there in "believing" in God but denying who he said he is and denying what he said he did? Likewise, why would a Darwinist have any interest in a superfluous being, particularly one that is only a small step removed, if removed at all, from an imaginary being? What possible value could such a God add to an otherwise sufficient theory, if whatever hand he had in guiding evolutionary development is invisible?

[70] *"Themeistic progression avoids conflict with science by recognizing that progression occurred, but that the Warrior King may have played a roll in getting the process started."*

Surprisingly (or not, upon reflection), theistic evolution, like a "secular church" or a "Christian humanist," is an oxymoronic concept viewed by both sides of the evolution/design debate as doing little more than providing cover for a few "religious" folks. Proponents of both Darwinism and intelligent design simply find theistic evolution loaded down with unnecessary baggage inconsistent with one view or the other. As William Dembski elaborates:

> For the Darwinian establishment the "theism" in theistic evolution is superfluous. For the hard-core naturalists, theistic evolution at best includes God as an unnecessary rider in an otherwise purely naturalistic account of life.[621]

Dembski continues:

> Unlike full-blooded Darwinists, however, the design theorist's objection to theistic evolution rests not with what the term *theistic* is doing in the phrase "theistic evolution" but rather with what the term *evolution* is doing there.[622]

---

[620] Theistic Darwinists tend to be of Jewish or Christian faiths. However, the reasoning here can be applied to any theistic origins belief.

[621] Ibid., p. 111.

[622] Ibid., p. 112, emphasis in original.

Rather than avoiding conflict with mainstream science, therefore, theistic evolution simply plays into the hands of Darwinists by implicitly admitting the "fact of evolution." To make matters worse, theistic evolutionists are not even respected for their service to Darwinism. Referring to the philosophical tool known as "Occam's razor,"[623] Dembski concludes:

> It's for failing to take Occam's razor seriously that the Darwinian establishment despises theistic evolution. Not to put too fine a point on it, the Darwinian establishment views theistic evolution as a weak-kneed sycophant that desperately wants the respectability that comes with being a full-blooded Darwinist but refuses to follow the logic of Darwinism through to the end. It takes courage to give up the comforting belief that life on earth has no purpose. It takes courage to live without the consolation of an afterlife. Theistic evolutionists lack the stomach to face the ultimate meaninglessness of life, and it is this failure of courage that makes them contemptible in the eyes of full-blooded Darwinists. (Richard Dawkins is a case in point.)[624]

Thus, while theists may wish to make their peace with true Darwinists, the harmony of this morganatic marriage is lost on the royalty of Darwinism. The effort is neither desired nor welcomed by "full-blooded" Darwinists, who do well to humor the comity, errors and all.

[71] *"It's amazing how differently the same world can look based on the simple decision to go away from or toward the sun."*

What people believe about any subject is based more upon what they think they know about the subject than on any direct observation. Apparently contrary "facts" can be easily dismissed for any of a number of reasons. Using a common example, whether or not one believes the sun orbits the earth, or vice versa, depends on what that person "knows" to be true, rather than upon first-hand

---

[623] Occam's razor is a scientific and philosophic rule that entities should not be multiplied unnecessarily. The rule is applied practically by requiring that the simplest of competing theories be preferred to the more complex or that explanations of unknown phenomena be sought first in terms of known quantities. *See, Merriam-Webster's Collegiate Dictionary, Tenth Edition: 1996* (Springfield, MA: Merriam-Webster. Inc., 1996).
[624] Ibid.

observation.[625] Today, it may seem ridiculous that anyone could ever have thought that the earth was the center of the solar system, but that idea sounds silly only because we have been taught that it is not so. To the earth-bound observer, however, the sun traveling across the sky looks the same either way. Granted, the current heliocentric theory is supported by confirmed and repeatable observations and measurements, but without such confirmable data, what one thinks one "knows" about the relationship of the sun and the earth is simply and purely determined by one's assumption-based starting point—geocentric, heliocentric, or other.

The same "knowledge-dictating-assumptions" phenomena can be experienced in any of the sciences. Once one has made some progress in a certain direction, it is very, very difficult to make a change, even when based upon confirmable evidence—all the evidence will simply be interpreted in light of what the person "knows" to be true based upon the starting assumptions. In a classic look at how science works, Thomas S. Kuhn's book *The Structure of Scientific Revolutions* documents the process of scientific discovery, and analyzes how facts are incorporated into theories.[626] Kuhn explores the role of what he termed "paradigms" in the workings of "normal science," i.e., science as it's actually practiced.[627]

Kuhn's concept of a "paradigm" can be roughly equated with an "accepted theory" or a "way of thinking and knowing about a subject."[628] Paradigms, like theories, play a valuable role by focusing the inquiries of scientists working in the various paradigmatic disciplines. As Kuhn says:

> Men whose research is based on shared paradigms are committed to the same rules and standards for scientific practice. That commitment and the apparent

---

[625] No one has ever stepped outside our solar system and observed the orbits of the planets about the sun. All the pictures we see of this arrangement are arrived at indirectly, based upon measurements and calculations.

[626] Thomas S. Kuhn, *The Structure of Scientific Revolutions* (Chicago: The University of Chicago Press, 1996).

[627] The terms "paradigm" and "normal science" are Kuhn's. The former refers to "universally recognized achievements that for a time provide model problems and solutions to a community of practitioners," and the latter refers to "research firmly based upon one or more past scientific achievements [*e.g.*, paradigms] that some particular scientific community acknowledges for a time as supplying the foundation for its further practice" (Ibid., pp. x and 10).

[628] For example, Kuhn speaks of "the eighteenth century paradigm for [the field of optics] provided by Newton's *Opticks*, which taught that light was material corpuscles." This paradigm is contrasted with the later quantum-mechanical characteristics of photons. Kuhn elaborates: "These transformations of the paradigms of physical optics are scientific revolutions, and the successive transition from one paradigm to another via revolution is the usual developmental pattern of mature science" (Ibid., p. 12).

> consensus it produces are prerequisites for normal science, i.e., for the genesis and continuation of a particular research tradition.[629]

However, paradigms are more than theories; paradigms are accepted models or patterns, usually models received by choosing among competing theories. According to Kuhn:

> To be accepted as a paradigm, a theory must seem better than its competitors, but it need not, and in fact never does, explain all the facts with which it can be confronted.[630]

Agreed-upon paradigms, then, become the model on which all further research is based. A new paradigm implies a new, but more rigid, definition of the field of study.[631] Thus, a paradigm provides the beneficial service of establishing the "rules" of normal science, rules that thereafter define the group of scientists associated with the paradigm, as well as their research. Those unwilling or unable to accommodate their work to the new paradigm must proceed in isolation or attach themselves to some other group.[632]

One of Kuhn's primary contributions was recognizing and documenting the necessary components for paradigm change—the required series of events that leads ultimately to what he calls scientific revolutions. Understanding Kuhn's analysis helps to illuminate the current posture of the "normal science" of evolutionary theory. While not doing justice to Kuhn's thorough work, the short summary below is intended to help those convinced of the validity of the intelligent design paradigm that there is hope—scientific revolutions do happen.

Briefly, Kuhn identifies three important steps leading to the overturning of one paradigm for another, i.e., a scientific revolution. The first is the recognition of an anomaly (or, more generally, anomalies) in the first paradigm, i.e., "that nature has somehow violated the paradigm-induced expectations that govern normal science."[633] Second, when the anomaly (which at first can be ignored) continues to resist explanation and violates deeply entrenched expectations, the anomaly precipitates a crisis threatening large-scale paradigm destruction and major shifts in the problems and techniques

---

[629] Ibid., p. 11.
[630] Ibid., pp. 17-18.
[631] Ibid., p. 19.
[632] Ibid.
[633] Ibid., pp. 52-53.

of normal science.[634] Third, when, *and only when*, another paradigm having (or appearing to have) fewer associated anomalies is available to replace the first paradigm, a major shift in "normal science" occurs.[635] Kuhn provides many examples in this anomaly-crisis-replace model; the following discussion will examine his observations in light of the current paradigm of origins of living things: Darwinism.

Darwinism (and naturalistic evolution in general) fits Kuhn's description of a paradigm. To its adherents it "seems better than its competitors" and "it need not, and in fact never does, explain all the facts" of nature with which it is confronted. And those "unwilling or unable" to accommodate the work of Darwinists must "attach themselves to some other group." The normal science of Darwinism proceeds, therefore, "increasing the extent of match between [observed] facts and the paradigm's predictions, and by further articulation of the paradigm itself."[636]

That there are "anomalies" that Darwinism is incapable of explaining cannot be denied. Anomalies include the lack of linking forms in the fossil record, the lack of examples of speciation by natural selection in living organisms, the presence of altruistic instinct in living creatures, the presence of specified complexity (*e.g.*, DNA) and the presence of irreducible complexity. Each of these observable phenomena present Darwinism with anomalies that resist explanation by the current paradigm.

The beauty of a paradigm is its ability to absorb and, to a certain extent, disregard anomalies as research is focused on immediate questions of interest—what Kuhn calls "puzzle-solving." The Darwinian scientific community chooses the problems to solve according to the Darwinian paradigm and, as Kuhn remarks:

> To a great extent these are the only problems that the community will admit as scientific or encourage its members to undertake. Other problems, including many that had previously been standard, are rejected as metaphysical, as the concern of another discipline, or sometimes as just too problematic to be worth the time.[637]

Kuhn's insight is remarkable. Without any thought, it appears, to current Darwinism, his characterization of the state of Darwinian science is 100% accurate. Intelligent design is deemed "unscientific"

---

[634] Ibid., p. 67.
[635] Ibid., pp. 77-81.
[636] Ibid., p. 24.
[637] Ibid., p. 37.

and rejected as metaphysical or the concern of another discipline such as "religion." In reality, intelligent design theory simply presents issues too problematic for Darwinists to deal with, and is therefore not "worth the time" to refute. Further, as Kuhn observes with uncanny accuracy:

> A paradigm can, for that matter, even insulate the community from those socially important problems that are not reducible to the puzzle form, because they cannot be stated in terms of the conceptual and instrumental tools the paradigm supplies.[638]

This "insulation" of the community from those who may be asking other questions can be misunderstood by lay persons who mistakenly believe that scientists have unspoken, but nevertheless legitimate, answers for the questions. Thus, as Kuhn states:

> That scientists do not usually ask or debate what makes a particular problem or solution legitimate tempts us to suppose that, at least intuitively, they know the answer. But it may only indicate that neither the question nor the answer is felt to be relevant to their research.[639]

While perhaps not having reached the stage of crises yet, the current state of the Darwinian paradigm, it appears, matches the situation described above, in which numerous anomalies have been identified, but the paradigm is strong enough at present to resist change. However, Darwinism is approaching (and no doubt will reach) crisis stage as the anomalies build and it becomes clear that Darwinists do not have answers—no solutions to the puzzles. Certainly the signs of crisis are showing. According to Kuhn:

> [T]he emergence of new theories is generally preceded by a period of pronounced professional insecurity. As one might expect, that insecurity is generated by the persistent failure of the puzzles of normal science to come out as they should. Failure of existing rules is the prelude to a search for new ones.[640]

---

[638] Ibid.
[639] Ibid., p. 47.
[640] Ibid., pp. 67-68.

In line with the paradigmatic structure of normal science, most Darwinists would likely deny professional insecurity. But such insecurity is manifested in the shrillness of the response to opposing theories such as intelligent design, the dogmatic manner in which the opposition is silenced, and the arrogant way in which leading Darwinists comport themselves. It should only be a matter of time before full crisis mode yields to a search for a replacement.

Unlike other scientific revolutions, however, Darwinism may be an exception to Kuhn's pattern of anomaly-crisis-replace. As Kuhn suggests, scientists rarely, if ever, abandon one paradigm without another to take its place. In the case of Darwinism, there is another paradigm that presents fewer anomalies: that of intelligent design. There is one very important reason, however, why intelligent design is (and is to be) avoided at any cost by the humanistic Darwinists of current normal science. Unlike other paradigm changes, such as those associated with heliocentricism, the nature of light, the discovery of oxygen, quantum theory, and the like (as discussed in Kuhn's book), the intelligent design paradigm carries with it profound philosophical component with far-reaching implications. Therefore, it is debatable whether Darwinism will ever yield to intelligent design as the in-vogue paradigm, not because of any scientific objection, but simply because of the philosophical repercussions of admitting to a designer. But truth will find a way, and it may just be that, consistent with other outdated theories, it will take a new generation of scientists to make the change.

The history of scientific endeavor makes clear that much of what we believe we "know" in science is more a product of what we think we know to be true about the natural world, than what we actually have data to support. So one who thinks "geocentrism" represents truth looks up at the sun and is *sure* he sees the sun moving around the earth. Likewise, one who thinks "heliocentricism" corresponds to reality is equally sure that the earth is moving around the sun. Such discrepancies of "knowledge" are even more evident with respect to the origin and diversity of life and species—an idea that necessarily invokes philosophical assumptions when interpreting current observations to explain historical events. It should not be surprising that the same physical world can look very, very different to two observers of the same phenomena, based solely on the philosophical starting point of the observers. The materialist will observe clear design but "see" nothing but "evolution" where others not so constrained will observe and see clear design.

To summarize, consider Kuhn's observation:

> [Normal science is an enterprise that] seems an attempt to force nature into the preformed and relatively

> inflexible box that the paradigm supplies. No part of the aim of normal science is to call forth new sorts of phenomena; indeed those that will not fit the box are not seen at all. Nor to scientists normally aim to invent new theories, and they are often intolerant of those invented by others.[641]

Those engaged in the paradigmatic science of Darwinism would do well to recognize that theirs is not a unique discipline, immune from the vagaries of human nature and historical precedent. Ignoring phenomena that don't fit the box and intolerance of new theories pretty much sums up the situation for contemporary Darwinism. Intelligent design theorists should be encouraged. Although scientific revolutions can take decades to come about, as a new generation of scientists put their philosophical biases aside, time will prove once again that truth will win out over falsehood—a falsehood that in hindsight will look as inane as believing the sun orbits the earth.

[72] *"After careful reflection one finds that there is grandeur in Neihil's view of nature, with its several powers, having been originally breathed into a few forms or into one; and that, while this planet has gone cycling on according to the fixed law of gravity, from so simple a beginning endless forms most beautiful and most wonderful have been, and are being, progressed."*

With these words Darwin closed *On the Origin of Species* (with *this* for *Neihil's* and *evolved* for *progressed*). Interestingly, Darwin uses the term "breathed," presumably as a metaphor for the mode of imparting life-giving powers. Darwin did not attempt to provide an explanation for the origin of life. To be fair, this was not his intent. With replicating life forms in place, his theory is intended to explain the "endless forms" of species. But with the term "breathed" Darwin alludes to what more and more scientists are realizing: life is more than the right molecules in the right places. At the moment of death the deceased has the same molecules in the same places as he did moments before, while alive. But something is gone. "Life" is gone and we call such a person "dead."

What is that "something" that exits to render a thing alive one instant and dead the next? Surprisingly, Darwin's metaphor, possibly intentionally, parallels the Genesis account of creation in which God "breathed" life into the first man; Adam was more than atoms. Literal breath or not, God or not, it is becoming increasingly obvious that "life" is more than chemical, material or compositional in nature.

---

641 Ibid., p. 24.

One evolutionist who has publicly acknowledged the immaterial nature of "life" is Paul Davies. In his book *The Fifth Miracle: The Search for the Origin of Life*, Davies states that "we are missing something fundamental about the whole business [of the origin of life by law and chance alone]."[642] Although Davies is convinced that once life got started chance and law could do the rest, he understands the seemingly insurmountable problems Darwinists face in explaining the first step: the origin of life. He realizes correctly that the jump from chemistry to biology requires something in addition to chance and law. Recognizing *information* as key to origin of life, he states:

> A law of nature of the sort we know and love will not create biological information, or indeed any information at all. Ordinary [physical] laws just transform input data into output data. They can shuffle information about but they can't create it.[643]

Thus, Davies clearly understands the limitations of physical laws in creating information. He concludes, with respect to "molecular Darwinism," and theories that seek a natural law that leads to the origin of information:

> When I set out to write the present book, I did not think that such laws were necessary to explain biogenesis. Instead, I assumed it was a case of "Darwinism all the way down." ... I am now much more skeptical. It seems to me very unlikely that all that is necessary is for the right chemical reaction or the right molecule to turn up. Real progress with the mystery of biogenesis will be made, I believe, not through exotic chemistry, but from something conceptually new.[644]

Something "conceptually new?" Davies is close to implying an intelligent agent, and indeed, was personally pressed on this issue by intelligent design theorist Phillip E. Johnson. The episode is recounted in Johnson's book *The Wedge of Truth* where Johnson relates an experience of attending and speaking at a 1998 conference at which Davies also spoke along the lines above.[645] Johnson drew out the

---

642 Davies, *Fifth Miracle*, p. 17.
643 Ibid., p. 254.
644 Ibid., p. 260.
645 Johnson, *Wedge of Truth*, p. 55-56.

implications of Davies' thinking and asked the question that never fails to prick Darwinists: "whether the first commitment of science was to impartial investigation of the evidence, or to upholding materialist preferences regardless of the evidence."[646] Indeed Johnson was sternly reproved for "attacking science," and was lectured on the theme that "evolution has occurred."[647]

It is only a matter of time before even methodological naturalists will be forced to concede that natural selection is not an information creator, much less an information enhancer. Being committed to philosophical first principles will keep most from admitting an intelligent designer, pushing some into more and more exotic theories with names like "chaos theory" or speculations on things like "negentropy." But at least with the inquiry placed correctly, we all, like Davies, can focus on the real issue—the essence of life itself. This, then, will result in a view of life that is truly "grandeur."

[73] *"... climbing a seemingly insurmountable mountain peak by way of a long, gradual slope ..."*

Richard Dawkins' book *Climbing Mount Improbable* uses the metaphor of a mountain to illustrate the workings of evolution. In Dawkins' metaphor "Mount Improbable rears up from the plain, lofting its peaks dizzily to the rarefied sky."[648] Dawkins describes ambitious but frustrated mountaineers that "shake their tiny baffled heads and declare the brooding summit forever unscalable."[649] Of course, this is only because our thwarted mountaineers fail to look "round the other side of the mountain" to find "gently inclined grassy meadows, graded steadily and easily towards the distant uplands."[650]

So Dawkins describes the ease with which flight, sight, and might could inch their respective ways up "the kindly, grassy slopes on the other side of the mountain."[651] Skeptics, according to Dawkins, make the mistake of focusing on the vertical precipice and its dramatic height. It was Darwin's great achievement to discover the gentle gradients winding up the other side. [652]

So compelling are the "kindly" slopes of Mount Improbable that, according to Dawkins' implications, the slopes *completely remove chance* from the formula for Darwinian evolution. Dawkins makes the surprisingly bold statement:

---

646 Ibid., pp. 57-58.
647 Ibid., p. 58.
648 Dawkins, *Climbing Mount Improbable*, p. 73
649 Ibid.
650 Ibid.
651 Ibid., p. 79.
652 Ibid., p. 79.

> It is grindingly, creakingly, crashingly obvious that, if Darwinism were really a theory of chance, it wouldn't work. You don't need to be a mathematician or physicist to calculate that an eye or a haemoglobin molecule would take from here to infinity to self-assemble by sheer higgledy-piggledy luck.[653]

According to Dawkins, Darwinism is not really a theory of chance, but nevertheless uniquely solves the "astronomic improbability" of evolution:

> [By] ... breaking the improbability into small, manageable parts, smearing out the luck needed, going round the back of Mount Improbable and crawling up the gentle slopes, inch by million-year inch.[654]

For those who might confuse the "luck" that must be "smeared out" with chance, Dawkins has little patience. He says of two scientists who claimed complex gene sequences could not come about by chance:

> What [physical scientist Sir Fred] Hoyle and [astrophysicist Chandra] Wickramasinghe miss is that Darwinism is *not* a theory of random chance. It is theory of random mutation plus *non-random* cumulative selection. Why, I wonder, is it so hard for even sophisticated scientists to grasp this simple point?[655]

Darwinism is not a theory of random chance, but it is a theory of random mutation ...? Is not random mutation inherently chance mutation, i.e., random chance? Of course it is, but giving Dawkins the benefit of the doubt, how does he dispense with chance? He doesn't. Dawkins is as disappointing in his follow-through as he is dramatic in his build-up. After assuring us that Darwinism is not a theory of random chance, he continues:

---

[653] Ibid., p. 77. Referring to Sir John Herchel's characterization of Darwinism as, "The Law of Higgledy-Piggledy" (but note that Ernst Mayr attributes the rule of "higgledy-piggledy" to Samuel Butler (in introduction to facsimile of *Origin*, p. xvii).
[654] Ibid.
[655] Ibid., p. 75, emphasis in original.

> [T]o invoke chance, *on its own,* as an explanation, is equivalent to vaulting from the bottom to the top of Mount Improbable's steepest cliff in one bound. And what corresponds to inching up the kindly, grassy slopes on the other side of the mountain? It is the slow, cumulative, one-step-at-a-time, non-random survival of *random* variants that Darwin called natural selection.[656]

Random variants. What are random variants but chance variants? And why would *millions* of cumulative, random variants be more probable than *one* random variant? Dawkins nowhere explains this in his lengthy book that gives inordinately detailed and beautiful descriptions of some of the most intricately designed living systems known. He simply refers to the misunderstandings of Darwinism that have provoked "this canard."[657]

Dawkins himself apparently cannot resist indulging the "canard." In the same book he refers to "accidental improvements" found in each generation of living things. [658] In explaining why living things are so good at doing what they do, he says:

> They are good because of the accumulated wisdom of their ancestors. But it is not wisdom that they have learned or acquired. It is wisdom that they *chanced upon by lucky random mutations,* wisdom that was then selectively, non-randomly, recorded in the genetic database of the species.[659]

Without even stopping to think how one could have wisdom without "acquiring" it, one must be baffled at Dawkins' brazen inconsistency. Now we have "lucky random mutations." What is luck but chance? Dawkins goes on to explain how Mount Improbable "smears out" the "luck" necessary for evolution:

> [By taking] ... what looks like an immense amount of luck—the luck needed to make an eye, say—and explaining it by splitting it up into lots of little pieces

---

[656] Ibid., p. 79, emphasis added.
[657] Ibid., p. 80.
[658] Ibid., p. 28.
[659] Ibid., p. 90, emphasis added.

> of luck, each one added cumulatively to what has gone before.[660]

Like a patient gambler who thinks he can beat the odds by betting one dollar at a time rather than a million (or billion) at once, Dawkins' irrational attempt at making sense of Darwin's theory of chance certainly accounts for the vast amounts of time postulated (by necessity) by Darwinists — but it is a loser nonetheless.[661]

Darwinism is, by any reasonable standard, a theory of chance (or luck, to use Dawkins' terminology). For Dawkins to deny this and be taken seriously by professional peers is truly astounding. For one, it is simply illogical to suggest that a process devoid of intelligent guidance is anything other than a process of chance and luck. If Darwinism is not a chance process, even in part, then it must be guided (at least in part), which requires an intelligent agent. Additionally, if Dawkins prefers to rely on "splitting [one big lucky event] into lots of little pieces of luck," that is his prerogative. But to freely foist such foolishness on the public as a scientific explanation of how to reach a highly improbable event is exactly the kind of shenanigans that draws the ire of thinking people.

[74] *"... is safe to say that if you meet somebody who claims not to believe in progression, that person is either ignorant, stupid, insane or wicked ..."*

Richard Dawkins made this declaration in 1989 and has subsequently defended it. The exact quote from Dawkins is:

> It is absolutely safe to say that if you meet somebody who claims not to believe in evolution, that person is ignorant, stupid, or insane (or wicked, but I'd rather not consider that).[662]

---

[660] Ibid., p. 91.

[661] For example, Richard Dawkins stated that "*Core Darwinism* ... is the minimal theory that evolution is guided in adaptively nonrandom directions by the nonrandom survival of small random heredity changes" (Dawkins, *a devil's chaplain*, p. 81). However, note that this statement is logically equivalent to the following: "*Wealth building* is the minimal theory that wealth is guided in adaptively nonrandom directions by the nonrandom payment of cash for small random rolls of the dice." Ironically, the statements are most probably *true* for both evolution and economics. That is, the "nonrandom directions" will be to death and poverty, respectively.

[662] Richard Dawkins, review of *Blueprints* by Donald C. Johanson and Maitland A. Edey, *New York Times*, April 9, 1989, sec. 7, p. 34.

Such statements are hard to deal with rationally, except to note the psychological tendency to see one's failings more acutely in others. But Dawkins is not ignorant or stupid. It is unlikely that he is insane. He could be considered wicked, depending on one's metaphysical leanings (but we'd rather not consider that).[663] Most likely, however, such a statement is the result of frustration at being unable to understand how anyone would not agree with an entirely reasonable, if not *proven*, proposition. For example, Galileo was probably called ignorant, if not insane. He was certainly called wicked (without Dawkins' reservations, for sure). But, Galileo's steadfast refusal to believe the earth did not move, and his resolute refusal to be intimidated by the forces of tyranny, earned him an honored place in history.

In today's origins controversy, the earth-focused Ptolemaists of the world are firmly in control. But there are "Galileoans" in their midst—mostly under house arrest—but more and more every day, each one resolutely sure of the truth. In the end, after whatever torture and punishment can be inflicted by those that hold to the geo-centric, earth-born ideas of human evolution, the facts, stubborn things that they are, will win the day for truth.

Not all Darwinists agree with irrational bouts of speech like the one quoted above, and most are simply "doing what they know to do" to fulfill their roll as scientists, educators, and administrators. In true evolutionary fashion, it is simply a matter of time until, one by one, each like a modern-day Galileo, they chance upon the data of a modern-day Copernicus of intelligent design, and naturally select a more modern view of the evidence in which to "believe."

[75] *"... is not just some sort of ruse! No, progression is a fact, fact, FACT!"*

In a valiant attempt at defending evolution, evolutionist Michael Ruse instead simply informed his readers that "Evolution is a fact, *fact*, FACT!"[664] As a rhetorical device, such statements can rally the true believers. But for those not so quickly rallied, the validity of this propositional statement can be determined only when one knows how Ruse is using his terms. If by "evolution" he is referring to changing

---

[663] That a Darwinist like Dawkins would acknowledge the idea of "wickedness" is something of an interesting item in itself. "Wickedness" is a moral concept that makes sense only in a framework of absolutes. In a Darwinian scheme, whatever works to reproductive advantage is favored with no thought to such anthropomorphic constructs like "wickedness" or "virtuosity." If "wicked" creationists (to use the lowest form of *Homo sapiens* a Darwinist can fathom) are more successful at reproduction such that their populations increase, then as an evolutionary matter, "wicked" is a good thing. Rather than being shy to consider such a thing, Dawkins should be smugly satisfied at their growing numbers as another example of the "fact of evolution."

[664] Michael Ruse, *Darwinism Defended* (Reading, MA: Addison-Wesley, 1982), p. 58.

finch beaks, peppered moths, insecticide-resistant insects, and the like, then he is absolutely, 100% correct. We can all rally. However, of course, if by "evolution" he means one species changing into another, or the creation of new, complex information-bearing organisms from simpler organisms, or even the creation of life in the first place—then he is simply expressing a wish projection, believing that such bluster might muster a fact.

Ruse is not alone. The rhetoric of the ultra-Darwinists becomes almost ear piercing on this point. In answering the question "Is evolution a fact?" Ernst Mayr answers:

> Evolution is not merely an idea, a theory, or a concept, but is the name of a process in nature, the occurrence of which can be documented by mountains of evidence that nobody has been able to refute. ... It is now actually misleading to refer to evolution as a theory. ... Evolution is no longer a theory, it is simply a fact.[665]

Not one to let others surpass him in hyperbole, Richard Dawkins has also assured us in no uncertain terms of the fact of evolution. Whereas such esteemed organizations as the National Academy of Sciences must muddle through a vague answer when questioned on the "fact" of evolution, Dawkins is not at all reticent. In a review of a book stating that "nearly half the people in the United States don't believe in evolution," he states:

> [W]e are here talking about the fact of evolution itself [as opposed to a particular theory such as Darwinism], a fact that is proved utterly beyond reasonable doubt.[666]

Why then are there so many doubters? According to Dawkins, it is "ignorance" which is "no crime in a country with strong local traditions of interference in the freedom of biology educators to teach the central theorem of their subject."[667] If people would only read more books on evolution, Dawkins thinks, ignorance would go away. Of the "100 million U.S. citizens" who believe in creation (presumably he is

---

[665] Mayr, *What Evolution Is*, p. 275.

[666] Richard Dawkins, review of *Blueprints* by Donald C. Johanson and Maitland A. Edey, *New York Times*, April 9, 1989, sec. 7, p. 34, (bracketed text added).

[667] Ibid. "Ignorance" is just one possibility. Dawkins also stated that one who doesn't believe in evolution may be "stupid or insane or wicked." He adds, "If that gives you offense, I'm sorry" (Ibid.).

referring to creationists who subscribe to the literal Genesis account, which is the only kind of creationist most Darwinists acknowledge), he says:

> They are contradicting—influentially and powerfully—vast fields of learning in which their own knowledge and reading is indistinguishable from zero.[668]

To remedy this sad state of affairs we should all "mug up" on Darwinism. As Dawkins says:

> This is because the more you read, quietly and soberly, the evidence for evolution, the more powerful will you discover that evidence to be.[669]

The problem with this approach is that the same could be said for virtually any subject: the more you read, quietly and soberly, about UFOs, the more powerful you will discover the evidence to be; the more you read about out-of-body experiences, the more powerful you will discover the evidence to be. The list could go on and on. The point is that reading only one side of any issue will, likely, make one a believer—*even if the subject matter is pure conjecture*. What a true scientist would suggest is that one should read evidence *for and against* evolution, and weigh the evidence in the balance of reason.

Of course, the true believers and other keepers of Darwin's flame cannot find any evidence against evolution, so the issue is moot. The theory is true by definition (it's practically a *law*), so to fathom evidence against evolution is like imagining evidence against gravity. Only fools (to humbly add to Dawkins' list of "insane," "stupid," and "wicked") would even entertain the thought of such "evidence."

[76] *Knowing that science practiced objectively cannot rule out an inference of human design, The Rule was nevertheless ruthlessly enforced to limit possible explanations of the facts to exclude any that might infer human design.*

The NAS defines science in one place as follows:

---

[668] Ibid.

[669] Ibid.

> Science is a particular way of knowing about the world. In science, explanations are restricted to those that can be inferred from confirmable data — the results obtained through observations and experiments that can be substantiated by other scientists.[670]

This is an objective definition of science, and because the study of intelligent design meets this definition of science it should be, according to the NAS, "part of science." Explanations of design can be inferred from confirmable data — data that is seen by all, including evolutionists, as confirming at least the appearance of design. Results from the data can be obtained through observations and experiments that can be substantiated by other scientists.

Note also that intelligent design satisfies the NAS's stated position with respect to historical sciences, namely that "Things in science can be studied even if they cannot be directly observed or experimented on."[671] In the context of origins theories, intelligent design is simply the scientific activity of proposing and testing the "design hypothesis" based on an observable group of living organisms, and then investigating whether, upon observation of data regarding that group of living organisms, the design hypothesis seems true. Such a hypothesis should not seem strange to scientists; by all accounts living organisms appear to be designed.

Without question, therefore, intelligent design is a scientific hypothesis based on inferences drawn from observable, empirical data. Not only do evolutionists admit that nature exhibits design, thereby validating the intuitive notion supporting the inference, but the design inference is used in many other disciplines of science. The Search for Extraterrestrial Intelligence (SETI) project, for example, is premised on the fact that design can be recognized and it is the hallmark of intelligence.[672] Archeologists use the design inference to test archeological finds — is it a stone or a tool? Arson investigators use the design inference in fire investigations — was it an accident or arson? Forensic scientists use the design inference to piece together the data in the form of clues — was it accidental or premeditated? In each case,

---

[670] NAS, *Teaching about Evolution*, p. 27.

[671] NAS, *Teaching about Evolution*, p. 55.

[672] *See*, www.seti.org. For example, the scientists at the SETI institute state, "Signals sent by a civilization for its own purposes may be impossible for us to unravel. But one thing we would know irrespective of content is that another *intelligent* civilization is out there." Ibid., emphasis added. Evolutionists hold that hypothesizing an alien intelligence for radio signal information is legitimate science while hypothesizing intelligent design for DNA information is not. This is because an inference of design in life supports the notion of a God, while an inference of design in a radio wave does not. But this is a philosophic or religious reason, not a scientific one. *See*, Note 78 below.

evidence of intelligence in the data is sought, thereby supporting a hypothesis of intelligent causation. In every case, each hypothesis is tested against competing hypothesis for falsification.[673] And in each case there is no question that the pursuit is scientific, and the individual disciplines are "part of science."[674]

Why then do proponents of evolution, particularly educators who insist that only evolution be taught, oppose intelligent design? It is not due to a lack of evidence; it is because of their insistence on a hidden assumption, an assumption that effectively re-defines "science" and serves to block all but unintelligent natural forces as potential causes for life and its diversity. The assumption operates as a rule—the "unwritten rule" of science, as described in a book by popular science writer Robert Wright.[675] In his book, Wright explains that certain assumptions in science result in "unwritten rules of scientific conduct" that require adherents "to scrupulously avoid even the faintest teleological [design] overtones.[676]

The Rule effectively re-defines science in a way that excludes the design hypothesis, not based upon the evidence, but by definition. In a public display of The Rule, consider this quote from the president of the NAS, Mr. Bruce Alberts:

> In evolution, as in all areas of science, our knowledge is incomplete. But the entire success of the scientific enterprise has depended on an insistence that these gaps be filled by natural explanations, logically derived from confirmable evidence. Because 'intelligent design' theories are based on supernatural explanations, they can have nothing to do with science.[677]

How does Mr. Alberts know that there is a natural explanation for life and its diversity? No one, including Mr. Alberts, knows for

---

[673] Falsification is one option. As Carol Cleland elaborates in her paper entitled *Historical Science, Experimental Science and the Scientific Method*, in actual practice historical researchers tend to search for "positive evidence—a smoking gun." According to Cleland, "A smoking gun is a trace that picks out one of the competing hypotheses as providing a better causal explanation for the currently available traces than the other." Carol Cleland, *Historical Science, Experimental Science and the Scientific Method* (Geology, November 2001, Vol. 29, No. 11) p. 988.

[674] NAS, *Teaching About Evolution*, p. 55.

[675] Robert Wright, *Three Scientists and Their Gods* (New York: Times Books, 1988), 70-71.

[676] Ibid.

[677] Bruce Alberts, New York Times editorial, February 12, 2005 (from letter dated February 9, 2005), emphasis added.

sure. What if there is, in fact, no natural explanation? Mr. Alberts' insistence belies a naturalistic assumption not required by the official definition of science, including his organization's definition cited in Note 5 above. That is, Mr. Alberts' statement reflects a usurper definition of "science," making an assumed premise—only natural explanations are valid—into an essential premise of science.

The assumption that only natural, unintelligent causes exist is the essential premise of the philosophy of naturalism. Naturalism is a philosophy precisely because its essential premises cannot be shown to be objectively true. If the underlying essential premise of naturalism—only natural causes exist—could be tested by the scientific method, naturalism would not be a philosophy, it would be science. Likewise, the opposite is true: if an essential premise of science cannot be tested, science has ceased being objective, instead being governed by a philosophical assumption.

The usurper definition of science treats the naturalistic assumption as an essential premise and leads to a presumption: "Science must be restricted to explaining natural phenomena using natural causes." This definitional restriction unnecessarily narrows the range of possible alternative explanations for origins. Because science has no knowledge of the actual, objective truth of the matter, a restriction to natural explanations is just a premise, not an essential premise. A scientific premise can be an essential premise only if science can rule out every competing premise, either by logic or by reason. Short of such a process of elimination, adopting an essential premise can only be based upon an assumption. When assumptions become established opinions put forth as authoritative, they become dogma by definition.[678]

With respect to the dangers of science being beholden to dogma, the NAS stated in their booklet entitled, Science and Creationism: A View from the National Academy of Sciences:

> No body of beliefs that has its origin in doctrinal [i.e., dogmatic] material rather than scientific observation, interpretation, and experimentation should be admissible as science in any science curriculum. Incorporating the teaching of such doctrines into a

---

678 Dogma is something held as an established opinion, a definite authoritative tenet or a code of such tenets, or a point of view or tenet put forth as authoritative without adequate grounds. (*See*, www.m-w.com). Examples of dogma include such tenets as, "supernatural causes created life and its diversity," and "natural causes alone created life and its diversity." Competing groups hold each statement as an established opinion and insist on its authority dogmatically. However, each statement is held as authoritative without adequate factual grounds—neither can be proven scientifically.

> science curriculum compromises the objectives of public education.[679]

Intelligent design proponents agree. However, the NAS is apparently blind to the fact that the belief (part of a body of beliefs) that all natural events have only natural causes is not based on scientific observation, interpretation, or experimentation, but rather on an assumption, and is, therefore, doctrinal. And according to the NAS, incorporating such beliefs into a science curriculum compromises the objectives of public education.

When a philosophic belief having its origin in doctrinal material of the philosophy of naturalism controls, presupposition replaces interpretation, so that observation is limited to confirmation of only the doctrinally acceptable views. Therefore, those who adopt and insist on an essential philosophical premise as an essential scientific premise, destroy objectivity with respect to origins science, effectively removing certain inquiries from the realm of theory, placing them squarely in the realm of dogma.

One method is available to ensure that subjective bias in the form of hidden assumptions is rooted out and eliminated, keeping science objective. The NAS, ironically recognizing that Darwin himself made use of this method, explains:

> By publishing his ideas, Darwin subjected his hypothesis to the tests of others. This process of *public scrutiny* is an *essential part* of science. It works to eliminate *individual bias and subjectivity*, because others must also be able to determine whether a proposed explanation is consistent with the available evidence.[680]

Thus, according to the NAS, an essential part of science to eliminate *bias* is *public scrutiny* to see if a proposed explanation is consistent with the available evidence. This is exactly what intelligent design proposes. This is exactly what Darwinists (excluding Darwin, were he alive), oppose. Evolutionists attribute bias to the challenger, public scrutiny is rejected, and the available evidence is ignored. This is

---

[679] National Academy of Sciences, *Science and Creationism: A View from the National Academy of Sciences* (Washington DC: National Academy Press, 1999), p. 25, emphasis added.

[680] NAS, *Teaching About Evolution*, p. 39, emphasis added.

the result of a true "essential part" of science—subjecting hypotheses to test—being exchanged for a false "essential part" of science—a naturalistic assumption that rules out challengers by definition, rather than on the evidence.

Institutions of science must be open to competing hypotheses to prevent evidence from becoming subject to the dictates of dogma rather than the precepts of principle. When proponents of the hypothesis of no-design in nature, i.e., proponents of Darwinian evolution, fail to heed the published instruction of the NAS, and instead refuse to permit challenges of a contrary hypothesis, i.e., a design hypothesis, such proponents have ceased being scientific. Scientific principle, indeed the scientific method, ceases to operate in principle, having been displaced by the dictates of dogma.

[77] *"Although not fully realizing why Dr. Seeting would wish to suppress such evidence, Jenny came to realize that The Rule imposed an artificial requirement on science – a surreptitiously adopted essential premise of "no-design" that was enforced as dogma."*

Importing an assumption as an essential premise in a modified definition of science results in the propositional statement: "Science is restricted to explaining natural phenomena using only natural causes." However, this statement begs the question. That is, the assertion assumes the truth of the premise that there are only natural causes, and enforcing the assumption excludes the design hypothesis by definition, rather than by evidence. Such a position is inherently unscientific due to its failure to objectively consider all the evidence before ruling out a competing hypothesis. Objective science ceases in the area of origins research when this position becomes the entrenched methodology of scientific investigation (e.g., objective knowledge-seeking) because the "answer" has been effectively provided in advance by the definitional limitation itself.

A key attribute of science, an attribute that serves chiefly to distinguish science from religion, is that science is and must remain theoretical. However, if the claim of the design hypothesis is removed from science simply because it runs afoul of a naturalistic assumption, then the claim of no-design ceases to be theoretical and, ironically, becomes nothing more than a dogma—an unchallengeable assertion, true by definition. For evolution to remain scientific, and, indeed for science to maintain its legitimacy as an objective, knowledge-seeking discipline, it must remain opposed to dogma, objectively permitting the challenge of the design hypothesis. To have any legitimacy, science must be indifferent to the implications of the evidence.

Therefore, to the extent science adopts an essential premise that cannot be tested, this premise is either, (1) *not* actually essential, or (2) science has removed certain inquiries from science, not based on knowledge, but upon a philosophical naturalistic assumption. In both cases, the premise serves only to render science dogmatic in its defense of an assumption—the assumption of naturalism.

[78] *"Just as forensic methods are used in other fields, good historical researchers focus on formulating multiple competing (versus single) hypotheses to explain unrepeatable happenings."*

Integral to the practice of science is the formulation and testing of hypotheses. One area of science is particularly dependent upon the objective formulation and testing of hypotheses. Historical events that cannot be recreated or tested by experimentation cannot be falsified by classical experimental methods, and must be tested against competing hypotheses.

Testing one hypothesis against competing hypotheses is the method used in forensics sciences, for example, as well as in the scientific disciplines of cosmology, archeology, biology (biogenesis), among others. Where a crime scene leaves mere clues, so-called circumstantial evidence, for example, forensic scientists must work from inferences based on the evidence to formulate hypotheses. Under these circumstances, the only remaining test is one that seeks to rule out competing hypotheses on the basis of the evidence. Hypotheses are challenged, reformulated, and possibly falsified by other competing hypotheses. [681]

The NAS recognizes the testing of competing hypotheses as a legitimate and possibly the only method for testing explanations of historical events such as the origin of life and its diversity. Specifically in this context the NAS answers the question: "How can evolution be scientific when no one was there to see it happen?"

> This question reflects a narrow view of how science works. Things in science can be studied even if they cannot be directly observed or experimented on. Archaeologists study past cultures by examining the artifacts those cultures left behind. Geologists can describe past changes in sea level by studying the

---

[681] *See*, Carol Cleland, *Historical Science, Experimental Science and the Scientific Method*, pp. 987-990. Ms. Cleland explores the necessary role of hypothesis testing for scientific disciplines not amenable to classical experimental testing. Hypothesis testing by formulating multiple competing hypotheses is essential in the context of historical sciences, such as cosmology, geology, and biogenesis of life.

> marks ocean waves left on rocks. Paleontologists study the fossilized remains of organisms that lived long ago.
>
> Something that happened in the past is thus not "off limits" for scientific study. Hypotheses can be made about such phenomena, and these hypotheses can be tested and can lead to solid conclusions.[682]

Intelligent design theorists agree, and refuse to take a narrow view of science. Like the above-identified scientific endeavors, intelligent design theory simply posits intelligent causation for living things. Like evolution, intelligent design is a science "that can be studied even if [it] cannot be directly observed or experimented on."[683] Like evolution, intelligent design hypothesizes about "something that happened in the past" and is, according to the NAS, not "'off limits' for scientific study." Like evolution or any study of origins, intelligent design is subject to the scientific activity of testing hypotheses, rather than being strictly held to laboratory experiments. As the NAS confirms, such study "can lead to solid conclusions."[684]

Ernst Mayr, probably the world's greatest modern Darwinist, agrees and has also acknowledged the evidentiary problems inherent in the historical nature of origins research. The nature of historical sciences is such that in many cases historical hypotheses and theories are not falsifiable.[685] According to Mayr, contrasting evolution with such theories as heliocentricity:

> Evolution is a historical process that cannot be proven by the same arguments and methods by which purely physical or functional phenomena can be documented. Evolution as a whole, and the explanations of particular evolutionary event, must be inferred from observations.[686]

---

682 NAS, *Teaching About Evolution*, p. 55.

683 Ibid.

684 Ibid.

685 "Many biologists and philosophers deny the existence of universal laws in biology and suggest that all regularities be stated in probabilistic terms, as nearly all so-called biological laws have exceptions. Philosopher of science Karl Popper's famous test of falsification therefore cannot be applied in these cases." Ernst Mayr, "*Darwin's influence on Modern Thought*," p. 80-82, (July 2000, Scientific American).

686 Mayr, *What Evolution Is.*, p. 13.

Therefore, to be scientific the no-design hypothesis of evolution, the design hypothesis of intelligent design, or any origins-related scientific inquiry must be practiced as defined—by making inferences from observations. Because these inquiries inherently involve historical events, the testing component must include at least the testing of competing hypotheses. Otherwise, a lone hypothesis is not a hypothesis, but an implicitly accepted, non-objective conclusion.[687]

[79] *"Silva supposes what he has to suppose, and Hawkins finds it easy to believe what he wants to believe, but supposing and believing are not enough to make a scientific explanation."*

The above quote is adapted from Phillip E. Johnson's book *Darwin on Trial*.[688] The statement points out the pervasive technique of using "just so" stories to "explain" evolution.[689] When examined closely, almost any account of evolutionary development, such as that of the wing or the eye, involves mostly "supposing" to get from point A to point B. Supposing is fine for imagining; but supposing falls short of proving.

The fact that Darwinists are quick to presume elaborate imaginations to be akin to factual accounts for the origin of something as complex as a working eye is both amusing and troubling. On the one hand, like ancient *andabatae*,[690] Darwinists make great spectacle as they thrash about in their imaginations (imagination being their only guide). Formidable in appearance only, alternating probes and thrusts in impressive form, Darwinists seem oblivious to the futility imposed by their armored helmets of philosophy. On the other hand, unlike the ill-fated *andabatae* (all of whom no doubt would risk removing the helmet

---

[687] Imagine a murder case where a child is found dead in the basement of a home occupied only the mother and father. Three reasonable competing hypotheses are that the death was caused by (a) the mother, (b) the father, or (c) and intruder. Until the prosecutor finds evidence that rules out any two she cannot prosecute the third. Imagine further that the prosecutor wishes to deal only with a lone hypothesis, possibly for reasons of expediency. She may ask the judge to instruct the jury that, not on the basis of the evidence, but on the basis of expediency, it must *assume* that the father or an intruder did not commit the crime. Although the jury would necessarily convict if it followed the judge's instructions, the result would not rest on any valid fatal foundation and should be reversed on appeal.

[688] Johnson, *Darwin on Trial*, p. 42.

[689] The reference is to Rudyard Kipling's book, *Just So Stories*, originally published in 1902. The book is a collection of fanciful tales with titles such as "How the Whale Got His Throat," "How the Camel Got His Hump," and the like.

[690] *Andabatae* were Roman-era gladiators that were heavily armored, but their helmets had no eye holes; they fought without the benefit of their eyesight.

for the ability to see, if permitted), the groping blindness of today's Darwinists is self-imposed, evoking a certain pity. What purpose can be served by steadfastly insisting on a view of reality that does not admit certain lines of *scientific* inquiry, regardless of how reasonable?

Unfortunately, the answer to that question is that science has evolved to the precarious predicament of being guardian of a worldview. Changes in scientific understanding often trigger a change in worldviews. In his book *The Structure of Scientific Revolutions,* Thomas Kuhn includes a chapter entitled "Revolutions as Changes in World View" which explores how worldviews change with scientific developments.[691] Simple examples such as how Aristotle and Galileo understood the motion of pendulums can illustrate how fundamental shifts in scientific understanding affect how one views all of nature, i.e., one's worldview.

But the change in worldview brought on by understanding the difference between one view or another with respect to pendulum dynamics hardly rises to the level of change that theories of origins require. Discussing theories of origins may make for wearisome academic debates, but the truth of the matter (i.e., the actual, historic happening of the origin of living beings) has profound implications with respect to all areas of life—legal, political, and ethical—to name a few. At bottom, the fact of our existence requires a theory of origins that is either matter-focused—matter is all that exists, or intelligence-focused—there is intelligence behind the matter.

Today, mainstream science has chosen to stake its flag squarely and immutably in the worldview associated with naturalism—matter is all that exists, or, at least, matter is all that matters. Such a stance is understandable when the object of study is applied science, particularly in light of historical developments. But with respect to origin of life theories, science is championing the cause of naturalism unnecessarily. Contrary to the oft-repeated rhetoric, Darwinism is not necessary to study and understand any area of science any more than is intelligent design, or even special creation. For example, photosynthesis, planetary motion, life cycles, even genetics, can each be studied and understood, as well as applied to solve practical problems, without recourse to any theory of origins, including the study of Darwinism.

Why, then, do mainstream scientists guard Darwinism so fiercely? First, it seems that, by definition, one can't *be* a mainstream scientist without paying due homage to Darwin. Second, many people are convinced that Darwinism represents the truth with respect to origins. Not ever having been exposed to contrary evidence, such scientists are under the impression that there *is* no contrary evidence.

---

691 Kuhn, "Revolutions as Changes in World View," *The Structure of Scientific Revolutions,* p. 111.

Even more so, Darwinism has a symbolic value as the defining discipline dividing science and religion, and any attack on Darwinism is an attack on science itself. The attacks are all the more threatening because they always seem to come from "religious" people. Finally, like most religious people and their beliefs, many people who believe Darwinism have never studied their chosen dogma; they believe it because that's what they have been taught and to think otherwise makes them look like charter members of the flat-earth society.

But pure scientific considerations are insufficient to explain the religious ferocity with which the evolutionist elites protect their Darwinian domain. There is more in play here than a simple scientific controversy, such as whether light is a particle or a wave, to give another example of a question that divided the scientific world for a time. No, the question of origins brings into play the question of ultimate worldview—which philosophy is correct for understanding reality? If naturalism is not the correct philosophy, i.e., matter is *not* all that exists, then weighty questions arise as to just what else may exist. The "what else" implies an intelligent "who else" that leads to a "why else" which causes one to see the world very, very differently from what naturalism would require.

Like *andabatae* who may survive by removing their unduly restrictive helmets, scientists willing to consider alternatives to naturalism can remove undue restrictions on the mind and can see the world in an entirely unobstructed way. Such a paradigm shift has its professional risks, but those who are willing to honestly consider the alternatives to naturalism will no doubt find that the new way of seeing permits new ways of solving problems—solutions with fewer anomalies than naturalism. Theories of origins can, then, join other scientific paradigm shifts. As Kuhn says:

> Scientists then often speak of the "scales falling from the eyes" or of the "lightening flash" that "inundates" a previously obscure puzzle, enabling its components to be seen in a new way that for the first time permits its solution.[692]

But, as long as science is held hostage by the philosophy of naturalism the scales will stay, the puzzles will remain obscure, imaginations will continue, and "supposing" will have to suffice for explanation. Sophisticated "just so" stories will have to carry the load, feebly substituting for scientific reasoning. That such "just so" stories go unchallenged from the mind's eye of Darwinists to the reading eyes of an unwary public is unfortunate. But that science would condone

[692] Ibid., p. 122.

such behavior, risking its reputation for the honor of a philosophy, is tragic.

Let him who has eyes to see, see.

Index

E

NOTES

FURTHER REFERENCE

Online Resources:

www.intelligentdesignnetwork.org -- The Intelligent Design network, inc. main site provides additional resources, including links to books and video materials that can be purchased online. In particular it is highly recommended that one obtain and read ***Intelligent Design: The Scientific Alternative to Evolution,*** by William S. Harris, Ph.D. and John H. Calvert, J.D. (National Catholic Bioethics Quarterly, Autumn 2003).

www.arn.org -- For information on just about every aspect of Intelligent Design, as well as links to excellent book and video resources, a must-visit site is that of the Access Research Network.

www.discover.org/csc/ -- The Discovery Institute's Center for Science and Culture is a wealth of information on all aspects of the topic of evolution, free speech and academic freedom.

www.idthefuture.com -- For the latest from the heavy-hitters of Intelligent Design, go to ID the Future.

www.thewhitepath.com -- For an international flavor with a unique non-US perspective don't miss The White Path

Printed in the United States
89155LV00003B/26/A

9 781931 796279